GEOLOGICAL SURVEY OF CANADA

PAPER 87-2

RADIOGENIC AGE AND ISOTOPIC STUDIES: REPORT 1

1987

Energy, Mines and Resources Canada
Énergie, Mines et Ressources Canada

Available in Canada through

authorized bookstore agents and other bookstores

or by mail from

Canadian Government Publishing Centre
Supply and Services Canada
Ottawa, Canada K1A 0S9

and from

Geological Survey of Canada offices:

601 Booth Street
Ottawa, Canada K1A 0E8

3303-33rd Street N.W.,
Calgary, Alberta T2L 2A7

A deposit copy of this publication is also available for reference
in public libraries across Canada

Cat. No. M44-87/2E	Canada: $12.00
ISBN 0-660-12674-5	Other countries: $14.40

Price subject to change without notice

Cover

Electron microscope photomicrographs showing polished and etched sections of zircons from the Protoerozoic Thelon Tectonic Zone. Both zircons are about 100 μm across.

Top:

Synneusis twin from a charnokitic rock.

Bottom:

A bizarre crystal from a mylonitic granulite gneiss, featuring zonal irregularities attributed to growth during dynamic conditions.

Original manuscript received: 1987 - 06
Final version approved for publication: 1987 - 08

CONTENTS

INTRODUCTION

"Radiogenic Age and Isotopic Studies" is an annual collection of reports presenting data from the Geochronology Section of the Lithosphere and Canadian Shield Division. The main purpose of this collection is to make available promptly geochronological and other radiogenic isotope data at a time when these are increasingly critical to the interpretation and integration of many geological maps and compilations. Reports make full presentations of the data, relate these to field settings and make comparatively short interpretations. Readers are cautioned that many of the data reported here are part of work in progress and that minor enhancements of results and further interpretation will in some cases be published at a later date. A list of references provides a further record of all other data produced in the laboratory but published in outside journals or separate GSC publications. The present reference list covers the gap between this collection and the "Rubidium-Strontium and Uranium-Lead Isotopic Age Studies" Report 5 in Current Research, Paper 82-1C and "K-Ar Isotopic Ages" Report 16 in GSC Paper 82-2.

The list of reports is headed by those covering advances and/or reviews of analytical techniques. The new equipment and advances in the Geochronology Section would not have been possible without the encouragement and support of J.C. McGlynn, Director of the Lithosphere and Canadian Shield Division, and the senior management of the GSC. The most significant progress at the GSC of the last few years has been in the U-Pb isotopic analysis of zircons and other accessory minerals. Fundamental to this advance has been the construction of a positive clean air laboratory, the purchase of two state-of-the-art solid source mass spectrometers, the refinement of operational software and the production of an extremely pure ^{205}Pb tracer (spike). These facilities have also enabled the expansion of radiogenic tracer studies, including common Pb and Rb-Sr isotopic studies on various minerals and whole-rocks. Recent advances in high performance liquid chromatograpy techniques have been applied to the extraction and purification of Sm and Nd and this capability is nearing completion.

La collection intitulée « Radiogenic Age and Isotopic Studies » (âge radiométrique et analyses isotopiques) est une collection annuelle de rapports dans lesquels sont présentées des données fournies par la Section de la géochronologie de la Division de la lithosphère et du Bouclier canadien. Cette collection a pour but essentiel de fournir rapidement des données géochronologiques et d'autres données radiométriques, au moment où ces dernières deviennent d'une importance critique pour l'interprétation et l'intégration de nombreuses cartes et compilations géologiques. Les rapports présentent les données de façon complète, établissent une corrélation entre celles-ci et les conditions réelles *in situ*, et donnent des interprétations relativement brèves. Il est à noter qu'un grand nombre des données figurant dans le rapport font partie de travaux en cours, et que parfois, sont ultérieurement publiées quelques améliorations apportées aux résultats, et des interprétations plus poussées. Une liste de références constitue un autre inventaire de toutes les autres données obtenues en laboratoire, mais publiées dans des revues externes ou dans divers articles de la CGC. La présente liste bibliographique comble la lacune qui existait entre cette collection et le rapport 5 de Recherches en cours, Étude 82-1C, intitulé « Rubidium-Strontium and Uranium-Lead isotopic age studies » (études de datation radiométrique par les méthodes au rubidium-strontium et à l'uranium-plomb), ainsi que le rapport de l'Étude 82-2 de la CGC, intitulé « K-Ar isotopic ages » (études de datation radiométrique par la méthode au K-Ar).

En tête de la liste des rapports, figurent les rapports traitant des progrès accomplis dans les techniques d'analyse ou les révisions de ces techniques, ou les deux à la fois. La Section de la géochronologie doit l'acquisition de son nouveau matériel et ses récents progrès à l'encouragement et à l'appui que lui ont témoignés M. J.C. McGlynn, directeur de la Division de la lithosphère et du Bouclier canadien, et les cadres de la CGC. Au cours des dernières années, l'une des réalisations les plus importantes de la CGC a été l'analyse radiométrique des zircons et autres minéraux accessoires par la méthode à l'U-Pb. Ces réalisations ont été rendues possibles par la construction d'un laboratoire propre sous pression, par l'achat de deux spectromètres de masse avec source en état solide du type le plus moderne, par le perfectionnement du logiciel d'exploitation, et par la production d'un traceur extrêmement pur, le ^{205}Pb (isotope de dopage). Ces installations ont permis d'élargir le champ des analyses par traceurs radioactifs, y compris les analyses isotopiques courantes par les méthodes à l'U-Pb et au Rb-Sr, effectuées sur divers minéraux et sur la roche entière. On a appliqué les progrès récents obtenus dans le domaine des méthodes de chromatographie en phase liquide à rendement élevé, à l'extraction et à la purification des éléments Sm et Nd; cette méthode se trouve actuellement à l'étape de la mise au point finale.

The $^{40}Ar/^{39}Ar$ technique has also been introduced; in-house design and construction of the Ar extraction system and mass spectrometer is in the final stages. While systematic K-Ar and $^{40}Ar/^{39}Ar$ studies are reported in individual papers, a continuing demand for limited K-Ar and $^{40}Ar/^{39}Ar$ ages in pilot studies is anticipated. These limited data sets are compiled and reported annually in a manner analogous to the old series GSC Paper 82-2.

On a aussi introduit la méthode à l'$^{40}Ar/^{39}Ar$; la conception et la construction, dans les laboratoires de recherche de la CGC, du système d'extraction de l'Ar et du spectromètre de masse en sont aux étapes finales. Tandis que dans des articles individuels, sont décrites des analyses systématiques par les méthodes au K-Ar et à l'$^{40}Ar/^{39}Ar$, on prévoit une demande continue de datations limitées effectuées selon les méthodes au K-Ar et à l'$^{40}Ar/^{39}Ar$ aux fins des études-pilotes. Ces groupes de données limitées, une fois compilés, paraissent chaque année dans une publication semblable à l'ancienne Étude 82-2 de la CGC.

O. van Breemen

Uranium-lead analytical techniques at the geochronology laboratory, Geological Survey of Canada

R.R. Parrish, J.C. Roddick, W.D. Loveridge, and R.W. Sullivan[1]

Parrish, R.R., Roddick, J.C., Loveridge, W.D., and Sullivan, R.W., Uranium-lead analytical techniques at the geochronology laboratory, Geological Survey of Canada; in Radiogenic Age and Isotopic Studies: Report 1, Geological Survey of Canada, Paper 87-2, p. 3-7, 1987.

Abstract

Analytical procedures for U-Pb isotope dilution analyses at the Geological Survey of Canada geochronology laboratory are described in detail. The procedures apply to zircon, monazite, sphene, rutile, baddeleyite, thorite, apatite, allanite, and garnet. Zircon in particular is carefully selected and generally air abraded, and is characterized using polishing and etching techniques. Reagent preparation and chemical processing are done in a clean air laboratory. Samples are spiked with a mixed ^{205}Pb-^{233}U-^{235}U tracer, dissolved using steel jacketed Teflon™ pressure vessels or Teflon™ screw cap containers. U and Pb are separated using anion exchange resin columns. Blanks for typical sample amounts of 0.01 to 1.0mg are less than 10pg U and 6-40pg Pb. Rapid 10 minute mass analysis of the U and Pb are carried out using a Finnigan MAT 261 variable multicollector mass spectrometer solely in the static mode. Data reduction incorporates consideration of all sources of error in the analysis, and discordant U-Pb results are assessed on concordia plots using published regression techniques, taking into consideration the particular mineral systematics.

Résumé

La présente étude décrit en détail les méthodes d'analyse par dilution isotopique d'U-Pb au laboratoire de géochronologie de la Commission géologique du Canada. On applique ces méthodes au zircon, à la monazite, au sphène, au rutile, à la baddeleyite, à la thorite, à l'apatite, à l'allanite, et au grenat. Le zircon en particulier est choisi très soigneusement, poncé généralement au jet d'abrasif et caractérisé à l'aide de techniques de polissage et d'attaque micrographique. La préparation du réactif et le traitement chimique sont effectués en salle blanche. Les échantillons sont dopés à l'aide d'un traceur constitué d'un mélange de ^{205}Pb-^{233}U-^{235}U, dissous dans des récipients sous pression en Teflon^MC recouverts d'acier ou dans des contenants en Teflon^MC à couvercle fileté. On sépare l'U et le Pb dans des colonnes à résine échangeuse d'anions. Des blancs destinés à des quantités typiques d'échantillon de 0.01 mg à 1.0 mg sont plus petits que 10 pg de U et 6-40 pg de Pb. On effectue une analyse rapide (10 minutes) de l'U et du Pb dans un spectromètre de masse Finnigan MAT 261 à multicollecteur variable, uniquement dans le mode statique. La réduction des données tient compte de toutes les sources d'erreurs dans l'analyse et les résultats discordants de l'analyse par U-Pb sont évalués sur des courbes de concordance à l'aide des techniques publiées de régression, en tenant compte de la classification particulière des minéraux.

INTRODUCTION

The recent design and construction of a clean chemistry laboratory, the acquisition of two state of the art mass spectrometers with custom-modified operating software, and the synthesis of a highly purified ^{205}Pb isotopic tracer have all contributed significantly to the technical advances at the geochronology laboratory of the Geological Survey of Canada. Current U-Pb geochronological techniques employed at the geochronology laboratory are specified in detail in this paper. These procedures include mineral preparation and selection, addition of tracer (spiking), chemical dissolution, chemical separation and purification, mass spectrometry, error analysis, and linear regressions. This paper will be useful to other laboratories in both comparing procedures and assessing U-Pb results produced by this laboratory.

MINERAL SEPARATION AND SELECTION

Minerals dated by U-Pb methods include zircon, titanite (sphene), monazite, baddeleyite, allanite, garnet, apatite, thorite and rutile. These minerals are separated from specimens weighing from 1 to 50 kg by standard crushing, grinding, Wilfley table (utilizing a custom fabricated plexiglass top), heavy liquid, and magnetic techniques (employing a Frantz™ LB-1 separator), followed by sieving in ethyl

[1] Geological Survey of Canada, 601 Booth Street, Ottawa, Ontario K1A 0E8

alcohol into a number of size fractions. All minerals are selected for analysis by handpicking in ethyl alcohol. The criteria used in mineral selection, particularly zircon, optimize the clarity and diagnostic morphology of the crystals (*see* van Breemen et al., 1986, 1987), the lack of cracks and inclusions, and the general absence of cored grains unless these are specifically of interest. A portion of many zircon mineral separates is mounted in FEP Teflon™ [1] or epoxy grain mounts, polished to an optical finish, and etched in HF vapour to reveal details of the internal structure. Optical microscopy, photomicrography, and scanning electron photomicrography are utilized to observe and document the morphological aspects which assist in the interpretation of U-Pb zircon results (as for example in van Breemen et al., 1986, 1987).

The majority of zircon fractions greater than 100 Ma old are air abraded following the technique of Krogh (1982) using pyrite or less commonly pyrrhotite as a medium in abrasion cells modified after the Krogh (1982) design. Abraded zircons are generally leached 15 minutes in luke warm 3N HNO_3 to remove abrasion residues prior to the final mineral selection stage.

CHEMISTRY FACILITIES, REAGENTS

The clean chemistry laboratory consists of three separate temperature and humidity conditioned, high efficiency particulate air filtered (HEPA), positive air pressure laboratories for (1) U-Pb, (2) common Pb and (3) Rb-Sr, K, and Sm-Nd chemistry. Critical chemical work and weighing are done in specially designed, mainly polypropylene, hoods with their own separate source of HEPA air. Air quality in these hoods exceeds class 100 conditions, and is monitored with a particle counter. Laboratory air is not recirculated, and the laboratory has a minimum of exposed metal.

High purity water is produced on demand by a commercially available integrated water purification system (Millipore Corporation). The system employs (1) coarse filtration, (2) reverse osmosis, (3) organics absorption, (4) mixed-bed deionization, and (5) 0.22 micron membrane filtration. Systems of this type with a dedicated reverse osmosis section supplying pretreated water to a ''polishing'' section are considered an ideal solution for producing ultrapure water (Sampson and Hawkins, 1987). The lead and uranium concentrations in water from our system are consistently less than 0.5 pg/ml and 0.2pg/ml, respectively.

High purity acids are produced by single sub-boiling distillation in Teflon™ bottles after Mattinson (1972). The source reagents for this final distillation are previously double distilled (sub-boiling) in quartz, except for HF, which uses reagent grade HF. Pb blanks in final reagents range from about 1pg/ml (6.2N HCl) to 2.5pg/ml (48 % HF). Reagents from Seastar™ Chemicals (Sidney, British Columbia) have also been used, and the Pb blank levels are similar to those quoted above. Silica gel was made by precipitating silica gel with HNO_3 from a solution of sodium metasilicate, followed by repeated boiling and suction filtering in pure water, and shaking and settling of coarse particles leaving a supernatant suspension for use in mass spectrometric analysis of Pb.

Minerals are weighed using aluminum boats in a microbalance which has a quoted precision of $\pm$ 0.2 micrograms. In practice when weighing samples, experience indicates the uncertainty to be $\pm$ 0.5-1.0 micrograms. Tracer solution is dispensed into Teflon™ dissolution crucibles and weighed by difference using a balance which has a quoted precision of $\pm$ 0.02 mg. Evaporative re-equilibration within the drop dispensing bottle causes an uncertainty in dispensed tracer of approximately 0.1mg for 50mg aliquots.

Following chemical separation and purification, Pb and U solutions are collected and evaporated in 30ml PMP (polymethylpentene) beakers. The collection beakers are precleaned by boiling in reagent grade 6N HCl followed by refluxing for at least 1-2 hours on a hot plate using 6-8N HCl which was double distilled in a quartz sub-boiling still.

U-Pb SPIKING, DISSOLUTION AND COLUMN CHEMISTRY

Typical weights of minerals analyzed range from 0.1-1.0mg for sphene to 0.01 to 0.2mg for zircon. Minerals (except for apatite) are first acid washed in warm high purity 3N HNO_3 for about 20 minutes, then rinsed in high purity H_2O and acetone. After drying, they are weighed in aluminum foil boats using the micro-balance and loaded into pre-cleaned TFE Teflon™ crucibles. The crucibles are either of 3ml capacity after a design by J. Mattinson (similar in style to those of Krogh (1973) but smaller) which are fitted into a 6-sample steel holder or the 0.4ml capacity micro-capsules of Parrish (1987) which utilize a steel jacketed, 125 ml, TFE Teflon™ dissolution vessel made by the Parr Instrument Company which can accommodate up to 8 micro-capsules. Mineral fractions are then spiked with a small amount of mixed ^{205}Pb-^{233}U-^{235}U tracer (^{233}U = ^{235}U, $^{235}U/^{205}Pb$ = 122), approximately 30 to 200pg of ^{205}Pb being added. The ^{205}Pb tracer was made by the Geological Survey of Canada (Parrish and Krogh, 1987) with the assistance of Atomic Energy of Canada Limited.

MINERAL DISSOLUTION

Zircon, baddeleyite, rutile and monazite dissolution and chemistry follow methods modified from Krogh (1973) employing for the most part TFE Teflon™ vessels and dissolution conditions described by Parrish (1987). For zircon, baddeleyite, rutile, and thorite, 0.1 to 0.5 ml of high purity 48 % HF and 0.01ml of high purity 70 % HNO_3 are added; capsules are then sealed and heated in an oven at 230-240°C for 30-60 hours. Monazite is dissolved in 12N HCl at 150-180°C for 30-60 hours in TFE Teflon™ containers (same as for zircon) within steel jacketed pressure vessels.

Dissolution conditions for sphene, allanite, garnet and apatite are less severe than for zircons, and in general these minerals are dissolved in 3ml Savillex™ screw cap vessels at lower temperatures. For sphene and allanite, the loosely capped Savillex™ vessels are either placed inside the large 125 ml Teflon™ dissolution container and heated in an oven about 30 hours at 210°C or heated about 24 hours on a hot plate at about 150°C. For garnet and apatite, the capped vessels

[1] The use of specific commercial products does not imply endorsement by the Geological Survey of Canada.

are heated overnight on a hot plate with surface temperature of 150°C. Sphene and allanite are dissolved in a 0.5 ml mixture of 1 part 48% HF to 6 parts 6.2N HCl, apatite is dissolved in 0.5 ml of 6.2N HCl, and garnet is dissolved in 0.5 ml of 48% HF and 0.01ml 70% HNO_3. Fluoride salts are usually present following the dissolution stage in the HF-HCl mixture, but they dissolve readily during the HCl or HBr steps prior to column chemistry, as outlined below.

COLUMN CHEMISTRY

Following dissolution of zircon, baddeleyite, rutile, monazite and thorite and evaporation to fluoride salts (or chloride salts in the case of monazite), approx. 1ml 3.1N HCl is added and the salts are dissolved by heating in an oven at 210°C overnight. Subsequent element separation employs 3.1N HCl anion exchange chemistry outlined by Roddick et al. (1987), modified from procedures of Krogh (1973), as follows:

Column; resin: 8cm FEP Teflon™ column, 1.25ml capacity; 0.25ml BioRad AG1x8, 200-400 mesh, Cl^- form, anion exchange resin;

Resin wash: 4x1ml H_2O, 2x1ml 6.2N HCl, 1ml H_2O, 1ml 3.1 N HCl, followed by loading of 3.1N HCl sample solution;

Sample wash: 3x0.5ml 3.1N HCl;

Elute Pb: 2x0.5ml 6.2N HCl, add 0.01ml 0.3N H_3PO_4, evaporate;

Elute U: 3x0.5ml H_2O, add 0.01ml 70%HNO_3, evaporate.

Following dissolution of garnet, sphene, allanite, and apatite, the dissolution solutions (which may contain fluoride salt residues) are evaporated and redissolved in 6.2N HCl (hot plate at 150°C surface temperature for garnet and apatite; overnight in 210°C oven for sphene and allanite), subsequently evaporated to dryness, and redissolved in 0.5ml of 1N HBr (hot plate for all four minerals). A clear solution should be present in 1N HBr.

Pb is separated on anion exchange columns (identical to those above) in 1N HBr and 6.2N HCl, and U is purified using another anion exchange column in 8N HNO_3, and eluted with H_2O as follows:

Pb column resin wash: 4x1ml H_2O, 2x1ml 6.2N HCl, 2x1ml H_2O, 1ml 1N HBr, followed by loading of 1N HBr sample solution;

Sample wash: 4x0.5ml 1N HBr, collect wash for U column, evaporate just to dryness, redissolve in 0.5ml 8N HNO_3 for uranium column;

Elute Pb: 1x0.5ml H_20, 2x0.25ml 6.2N HCl, 2x0.5ml 6.2HCl, add 0.01ml 0.3N H_3PO_4, evaporate;

U column resin wash: 4x1ml H_2O, 2x1ml 6.2N HCl, 1ml H_2O, 1ml 8N HNO_3, followed by loading of 8N HNO_3 wash from above (the Pb column can also be used after eluting Pb; if done this way the resin wash consists only of 2x1ml H_2O, 1ml 8N HNO_3, followed by loading of 8N HNO_3 wash from above);

U sample wash: 0.5ml 8N HNO_3, 0.75ml 8N HNO_3;

Elute U: 3x0.5ml H_2O, evaporate. (U column procedure should be repeated for allanite to further purify the U).

Total procedural Pb blanks have varied considerably over the period 1983 to 1987 and dropped from values of about 100-400pg (1983) to present levels of about 6-40 pg, the lowest values being for analyses which employ the 0.4ml capacity micro-capsules described by Parrish (1987). Uranium blanks are less than 10pg, most being less than 4pg. Mass spectrometric Pb loading blanks are about 2pg (Fall 1987).

MASS SPECTROMETRY, ERROR PROPAGATION, AND DATA REDUCTION

Isotopic compositions of Pb and U are measured on a Finnigan-MAT 261 variable multicollector mass spectrometer as described by Roddick et al. (1987).

The lead fraction is loaded in clean (HEPA filtered) air as a phosphate on a single degassed rhenium filament using the silica gel-phosphoric acid method. The mass spectrometric technique employs both Faraday cups and a secondary electron multiplier (SEM). The Faraday cups are positioned to simultaneously collect the ion beams of the major Pb isotopes (206, 207, 208 and 205 tracer) while the small ^{204}Pb beam is collected in the SEM. The static collection of all Pb isotopes effectively minimizes the effect of any beam instabilities and allows rapid collection of Pb data. Calibration of the SEM for bias characteristics is described by Loveridge (1986) while variations in SEM gain are measured as follows. The ^{205}Pb beam is magnetically switched into the SEM before and after each data block and simultaneous 4 second measurements are taken of the ^{205}Pb in the SEM and ^{206}Pb in the adjacent Faraday cup. The SEM gain for the data block is determined by comparing the average ratio obtained in this manner (corrected for SEM biases) with the same ratio measured wholly in the Faraday cups. This gain is then applied to the $^{205}Pb/^{204}Pb$ ratio. The total time required for collection of 6 blocks of data is about 10 minutes.

Uranium is loaded in clean (HEPA filtered) air as the nitrate on the evaporation filament of a degassed double rhenium filament pair and is ionized as the metal ion. Uranium is also analyzed in the static mode with three uranium isotopes (233, 235, 238) simultaneously collected in Faraday cups. Measurement of three isotopes permits mass fractionation corrections and thus better precision on U concentrations.

In calculating the age of a mineral, corrections are made for the small amount of common Pb in the sample and for blank Pb and U. The common Pb correction uses a model Pb isotopic composition of age equal to the interpreted age of the sample. For ages less than 2400 Ma, model Pb isotopic compositions are derived from the growth curve of Stacey and Kramers (1975), and for ages older than 2400 Ma, that of Cummings and Richards (1975) is employed. In our experience the Stacey and Kramers (1975) model is a good approximation of crustal Pb evolution in the Proterozoic and Phanerozoic whereas Cumming and Richards (1975) better approximates Pb evolution for rocks of Archean age. In some cases (i.e. sphene, allanite) the common Pb correction is determined from feldspars from the same rock sample. U and Pb blanks are determined from blank analyses along with each group of 12 to 14 mineral analyses, and are generally assumed to apply equally to all mineral fractions.

The overall uncertainty quoted for isotopic ratios and resultant ages should ideally incorporate uncertainties in all the components involved in an analysis. Roddick (1987) has detailed a numerical procedure for propagating the uncertainties in all the relevant components in the calculation of U-Pb isotopic ratios and ages. Table 1 presents an error analysis using the uncertainties derived from current U and Pb analyses on a typical zircon fraction, in this case being 1049 Ma old. Estimates of uncertainties in other parameters

Table 1. Error analysis of U/Pb data

(No.)	NAME	VALUE	% ERROR	206*/238	207*/235	207*/206*	COR. COEF.†
			PARAMETER:	206*/238	207*/235	207*/206*	COR. COEF.†
			VALUE	.17608	1.8032	.07427	.5872
			% ERROR	.112	.216	.176	
(No.)	VARIABLE NAME	VALUE	% ERROR	%^{2} VARIANCE CONTRIBUTIONS			
1	207/206m	.09367	.052	.00291	Z	.00291	Z
2	207/205m	.27446	.006	.00004	.00006	.00000	.00005
3	205/204m	237.5	.307	.00006	.00913	.00770	.00074
4	238/235m	.14689	.014	.00021	.00021	Z	.00021
5	Pb Frac %/mu	.090	30	.00082	.00626	.00254	.00227
6	206/205sp	.03792	.03	.00000	Z	.00000	Z
7	233/235sp	1.0194	.05	.00572	.00572	Z	.00572
8	U Blank ng	.004	30	.00051	.00051	Z	.00051
9	Pb Blank ng	.046	30	.00109	.00097	.00000	.00103
10	206/204bl	18.4	1.5	.00044	.00000	.00046	−.00001
11	207/204bl	15.6	.7	.00000	.01303	.01306	−.00002
12	206/204cm	16.98	2.4	.00081	Z	.00081	Z
13	207/204cm	15.50	.71	Z	.01097	.01097	Z
(No's)	COR. COEF.		% ERRORS	%^{2} COVARIANCE			
10,11	.7		1.5, .7	−.00000	−.00006	−.00343	.00168
12,13	.7		2.4, .71	.00000	.00000	−.00417	.00209
	TOTAL VARIANCE:			.01261	.04678	.03087	.01426
			AGE (Ma):	1045.5	1046.6	1048.8	
			2σ (Ma):	2.2	2.8	7.1	

† = Correlation coefficient for first two parameters
Z = Parameter not dependent on this variable
m = measured; sp = spike; bl = blank; cm = common Pb
* = radiogenic Pb; σ = standard error
mu = mass unit; ng = 10^{-9} gram
%^{2} = percentage squared

such as Pb fractionation per mass unit, magnitude of U and Pb blanks, and composition of Pb blank are also indicated. The numerical values for some of the error parameters (such as uncertainty in the common Pb composition) are still under review and subject to change. Errors here and in data tables elsewhere in this volume reflect one standard error of the mean whereas all errors in final ages and in error ellipses on plots are reported at the 2 sigma limit.

Correlations between two pairs of variables are also present: $^{207}Pb/^{204}Pb$ vs. $^{206}Pb/^{204}Pb$ ratios for blank Pb and common Pb. The common Pb and the blank isotopic compositions have been given the same correlation coefficients of 0.7 as suggested by Ludwig (1980).

Uncertainties in tracer U and Pb concentrations are not included in the table since a mixed U-Pb tracer is employed. The estimated (i.e. calibrated) U/Pb ratio for the tracer is the same for each mineral fraction analyzed. Therefore the uncertainty in the tracer U/Pb ratio will not cause any differentials in the calculated U/Pb ratios of samples analyzed with the same tracer. It will, however, impose an absolute uncertainty on all such ratios which must be taken into consideration in comparison with results from other laboratories. In these cases the uncertainty in the U/Pb of the tracer (0.06 %, 1 sigma from calibration with reference solutions from National Bureau of Standards metals SRM 960 (U) and SRM 982 (Pb)) should be included in the final ages. The significance of this uncertainty to a regression line is dependent on how much the lower intercept deviates from the origin (zero age). One possible method of applying this is to perform two additional regressions with positive and then negative deviations of tracer uncertainty applied to all U/Pb ratios, and incorporating the resultant deviations in age into the final age uncertainty.

Decay constants used are from Steiger and Jäger (1977). Linear regressions for data sets with little or no scatter (regression line passing through all two sigma error ellipses) utilize either the regression of Davis (1982) or York (1969). If a Davis (1982) regression is used, the probability of fit is quoted; if this parameter is less than about 10 %, significant scatter outside of the assumed analytical error exists in the data, and regression parameters should be treated with caution. In cases of scatter in the data (MSWD > 1.0), we generally employ a modified York (1969) regression which increases the error in the slope and intercept by multiplying these errors by the square root of MSWD, a procedure similar to using the realistic error of York (1969). An hyperbola is fitted to the error envelope and its intersections with concordia are determined by an algorithm of Ludwig (1980). If there is a data point that is reasonably concordant (less than several per cent) in addition to scattered data, the Davis (1982) model for non-linear data may be employed. When the lower intercept is the age of interest as in cases of inheritance, the York (1969) regression (or its modified version for excess scatter, see above) is employed.

An analysis is not considered concordant unless the plotted error ellipse substantially overlaps concordia at the two-sigma level. Where concordant analyses are present, we generally accept the mean of intersections of the error ellipse with concordia as the age and the deviations of the error ellipse intersections from the mean as the error. In young (less than about 400 Ma) rocks, this would correspond generally to U/Pb error whereas in older concordant analyses, it would be dominated by the $^{207}Pb/^{206}Pb$ error. All plotted error ellipses in figures in this volume reflect two-sigma error limits.

If analyses of thorium-rich minerals such as monazite or thorite plot above the concordia curve, we follow Schärer (1984) in interpreting the reverse discordance as a result of excess or unsupported ^{206}Pb from the decay of initially incorporated ^{230}Th. In most cases the $^{207}Pb/^{235}U$ age would be accepted as a valid age with geologic or thermal significance.

SUMMARY

The recent design and construction of our clean chemistry facilities, the acquisition of state of the art mass spectrometers with custom-modified operating software, and the synthesis of a highly purified Pb isotopic tracer have all contributed significantly to the technical advances at the geochronology laboratory. The procedures for U-Pb geochronology have been detailed for the benefit of other laboratories and those using or evaluating the data produced in this laboratory. A direct result of the technical improvements made in the last several years has been a great increase in the production of isotopic data which has significantly augmented the capacity of the Geological Survey of Canada to utilize geochronology and isotope geology as research tools in a wide variety of earth science projects.

ACKNOWLEDGEMENTS

We thank the entire staff of the geochronology section, and in particular O. van Breemen for helping develop the techniques described in this paper. We also thank T. Krogh of the Royal Ontario Museum for advice on a number of aspects involved in U-Pb mineral preparation and chemistry.

REFERENCES

Cummings, G.L. and Richards, J.R.
1975: Ore lead isotope ratios in a continuously changing earth; Earth and Planetary Science Letters, v. 28, p. 155-171.

Davis, D.
1982: Optimum linear regression and error estimation applied to U-Pb data; Canadian Journal of Earth Sciences, v. 19, p. 2141-2149.

Krogh, T.E.
1973: A low contamination method for hydrothermal decomposition of zircon and extraction of U and Pb for isotopic age determinations; Geochimica et Cosmochimica Acta, v. 37, p. 485-494.
1982: Improved accuracy of U-Pb ages by the creation of more concordant systems using an air abrasion technique; Geochimica et Cosmochimica Acta, v. 46, p. 637-649.

Loveridge, W.D.
1986: Measurement of biases in the electron multiplier ion detection system of a Finnigan-MAT model 261 mass spectrometer; International Journal of Mass Spectrometry and Ion Processes, v. 74, p. 197-206.

Ludwig, K.
1980: Calculation of uncertainties of U-Pb isotope data; Earth and Planetary Science Letters, v. 46, p. 212-220.

Mattinson, J.M.
1972: Preparation of hydrofluoric, hydrochloric, and nitric acids at ultralow lead levels; Analytical Chemistry, v. 44, p. 1715-1716.

Parrish, R.R.
1987: An improved micro-capsule for zircon dissolution in U-Pb geochronology: Isotope Geoscience, v. 66, p. 99-102.

Parrish, R.R., and Krogh, T.E.
1987: Synthesis and purification of ^{205}Pb for U-Pb geochronology; Isotope Geoscience, v. 66, p. 103-110.

Roddick, J.C.
1987: Generalized numerical error analysis with applications to geochronology and thermodynamics. Geochimica et Cosmochimica Acta, v. 51, p. 2129-2135.

Roddick, J.C., Loveridge, W.D. and Parrish, R.R.
1987: Precise U/Pb dating of zircon at the sub-nanogram Pb level; Isotope Geoscience, v. 66, p. 111-121.

Sampson, R.L. and Hawkins, A.L.
1987: Pretreatment of HPLC water by reverse osmosis; American Laboratory, v. 19, no. 2, p. 184-187.

Schärer, U.
1984: The effect of initial ^{230}Th equilibrium on young U-Pb ages: the Makalu case, Himalaya; Earth and Planetary Science Letters, v. 67, p. 191-204.

Stacey, J.S. and Kramers, J.D.
1975: Approximation of terrestrial lead isotope evolution by a two-stage model; Earth and Planetary Science Letters, v. 26, p. 207-221.

Steiger, R.H. and Jäger, E.
1977: Subcommision on geochronology: convention on the use of decay constants in geo- and cosmochronology; Earth and Planetary Science Letters, v. 36, p. 359-362.

van Breemen, O., Davidson, A., Loveridge, W.D., and Sullivan, R.W.
1986: U-Pb zircon geochronology of Grenville tectonites, granulites and igneous precursors, Parry Sound, Ontario, *in* New Perspectives on the Grenville Problem, ed. J.M. Moore, A. Davidson and A.J. Baer; Geological Association of Canada, Special Paper 31, p. 191-207.

van Breemen, O., Henderson, J.B., Loveridge, W.D., and Thompson, P.H.
1987: U-Pb zircon and monazite geochronology and zircon morphology of granulites and granite from the Thelon tectonic zone, Healey Lake and Artillery Lake map areas, N.W.T. *in* Current Research, Part A, Geological Survey of Canada, Paper 87-1*A*, p. 783-801.

York, D.
1969: Least squares fitting of a straight line with correlated errors; Earth and Planetary Science Letters, v. 5, p. 320-324.

Preliminary investigation on spatial distributions of elements in zircon grains by secondary ion mass spectrometry

J.A. Jackman[1], P.A. Hunt[2], O. van Breemen[2], and R.L. Hervig[3]

Jackman, J.A., Hunt, P.A., van Breemen, O. and Hervig, R.L., Preliminary investigations on spatial distributions of elements in zircon grains by secondary ion mass spectrometry; in Radiogenic Age and Isotopic Studies: Report 1, Geological Survey of Canada, Paper 87-2, p. 9-12, 1987.

Abstract

Trace variations of U, Th, Hf, Eu and Y in zoned zircon have been obtained as line scans and direct-ion images using an ion probe. Y appears to be especially sensitive to conditions of growth and together with U/Pb age determinations may help to characterize the environment in which the zircon grew at a particular time.

Résumé

Les variations des éléments en traces de U, de Th, de Hf, de Eu et de Y dans du zircon zoné sont présentées sur des diagrammes d'analyse linéaire et à l'aide d'images prises en direct au moyen d'une sonde ionique. Y semble être particulièrement sensible aux conditions de croissance et, conjointement avec les datations établies à l'aide de la méthode U-Pb, peut aider à caractériser le milieu géologique associé à la croissance du zircon à un moment donné.

INTRODUCTION

Zircons not only play a primary role in U-Pb geochronology but are important hosts for rare earth elements and Yttrium. The concentrations of these elements have previously been measured by isotope dilution mass spectrometry (Nagasawa, 1970), a technique which requires the complete dissolution of one or several zircon grains. In this manner average concentrations can be obtained but no information is provided on the spatial distribution of elements within an individual zircon grain.

Interpretation of U-Pb zircon ages is facilitated by observing polished and etched sections of mounted grains which often reveal an internal structure characteristic of environments of growth, resorption and/or abrasion (van Breemen and Parrish, 1986). Most critically a distinction can be made between zircons of igneous and metamorphic origin. While multiple facets may characterize a more resistant medium of growth (van Breemen et al., 1987) it would be useful to characterize zircon growth by trace element composition.

The capability to image trace elements from a polished zircon cross-section may soon be provided by the installation of a CAMECA IMS 4f secondary ion mass spectrometer (SIMS) in CANMET's metallurgical laboratories. To ascertain whether useful data could be obtained from an instrument of this type, preliminary studies were undertaken using the IMS 3f SIMS spectrometers at Arizona State University (Tempe, Arizona) and Surface Science Western (London, Ontario). The following paragraphs contain a description of the experimental techniques and illustrations of the encouraging results obtained for Y, Hf, Eu, Th, and U variations in two zircon grains from a tonalitic granulite gneiss of the Thelon Tectonic zone of the northwestern Churchill Structural Province.

EXPERIMENTAL

Samples consisting of several dozen grains were prepared in epoxy mounts and polished to reveal internal surfaces. For comparison to the SIMS results, relative U concentrations were inferred from the extent of etching related to the degree of cumulative radiation damage caused by the decay of U^{238} and U^{235}. Strongly etched high-U regions appeared dark in backscattered electron images of the sample (Fig. 1A).

A thin coating of Au-Pd was deposited on the surface to prevent excessive sample charging during SIMS analysis. A 12.5 kV O^- primary beam, focussed to about 10 um (micrometres) and scanned across the crystal, was used to sputter secondary ions which were subsequently mass-separated using a mass resolution (M/▲M) of about 300. Variations in secondary ion intensity across the diameter of the grain were measured by two different methods: line scans and direct-ion images.

[1] Metal Physics Section, Canada Centre for Mineral and Energy Technology, 568 Booth Street, Ottawa, Ontario K1A 0G1
[2] Geological Survey of Canada, 601 Booth Street, Ottawa, Ontario K1A 0E8.
[3] Center for Solid State Science, Arizona State University, Tempe, Arizona, U.S.A.

The line scans (Fig. 2) were obtained at Arizona State University by translating the sample under a stationary primary beam in steps of 10 um, using an electron multiplier as the secondary ion detector. Direct-ion images (Fig. 1C,D) were obtained at Surface Science Western by rastering the primary beam over a 500 um square area enclosing the entire zircon. Mass-filtered secondary ions were detected with a high-gain dual channel plate coupled to a fluorescent screen. This system has a 10^3 – fold improvement in sensitivity over the standard single channel plate. The direct-ion images were recorded with a 35 mm camera.

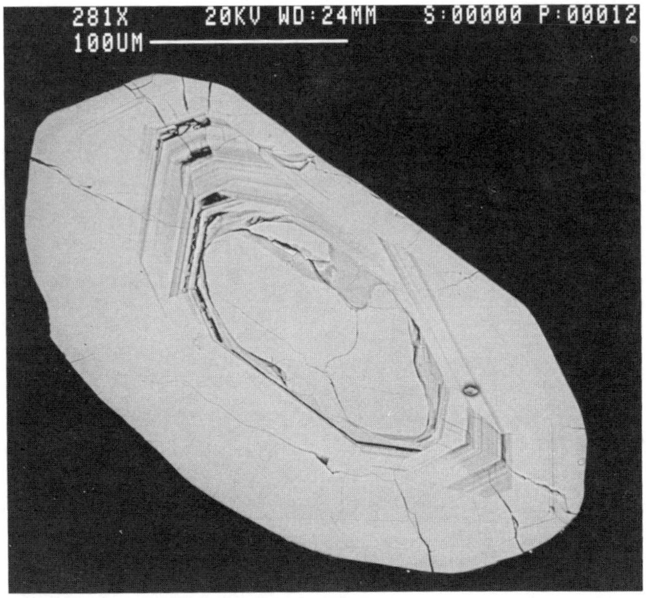

Figure 1A: Scanning electron microscope image (backscattered electron mode) of a polished and etched (HF) zircon section from the tonalitic granulite gneiss (sample 1). Dark regions correspond to more strongly etched, high U, zircon.

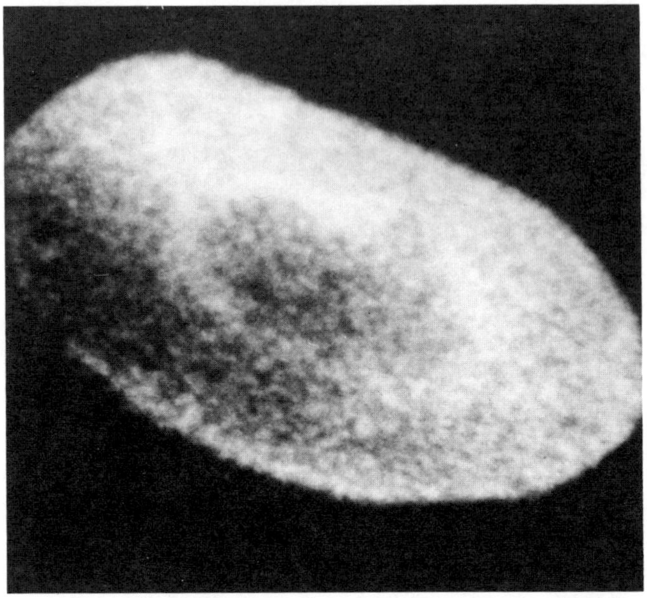

Figure 1C: Direct-ion image of uranium distribution in the zircon of figure 1A from a SIMS ion probe. Bright regions correspond to high uranium concentrations.

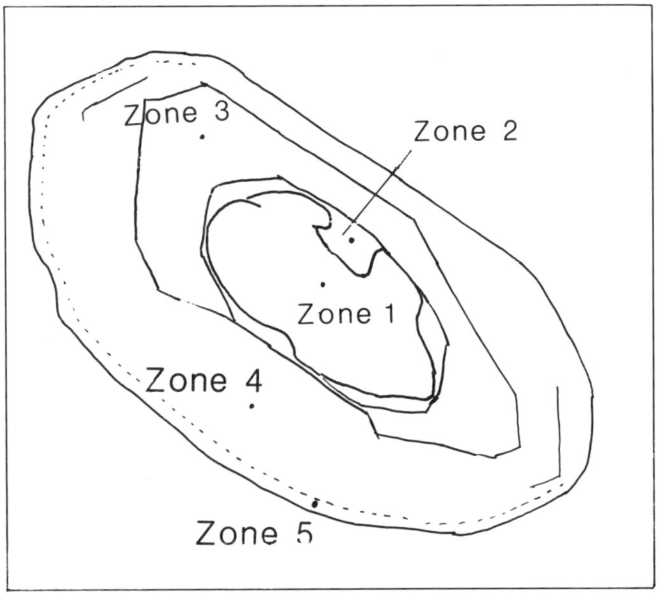

Figure 1B: Map of zircon in Figure 1A showing five zones of zircon growth: 1) core of unkown origin, 2) high U zone with discordant zonation, 3) high U zone with regular igneous zonation 4) mantle with faint zonation, 5) thin outer rim of metamorphic origin.

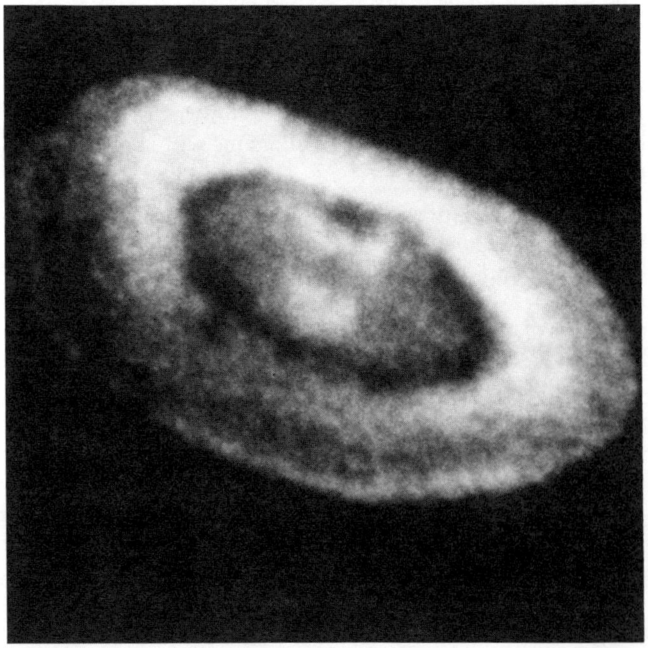

Figure 1D: Direct-ion image of yttrium distribution in the zircon of Figure 1A from a SIMS ion probe. Bright regions correspond to high yttrium concentrations.

RESULTS

Data are presented from two zircon grains from a tonalitic granulite gneiss of the Thelon Tectonic Zone as described in van Breemen et al. 1987 (sample 1). These zircons had a complex history as revealed by both U-Pb isotope systematics and the internal structure visible in polished and etched grain mounts. Most grains from this sample show a core with euhedral prismatic zoning characteristic of igneous zircon (Silver, 1969) followed by a euhedral zone of high U and an outer rim of low U content.

Electron images

Backscattered electron images (Fig.1A) of the specimen used for direct-ion imaging showed the strongly etched high-U regions as dark zones compared to lighter, relatively undamaged and unetched areas. Five zones were identified from these images (Fig.1B). The zircon consisted of a low-U inner core (Fig.1A,1B, zone 1) which had undergone resorption, a high-U outer core (zone 2) which resulted from subsequent minor growth, and showed discordance which may be related to the resorption of the inner core. New, temporally-distinct zircon growth of clearly igneous origin formed the high-U euhedral zone 3. This zone had been previously interperated to have grown either during igneous crystallization or migmatization (van Breemen et al., 1987, Fig. 83.5 a,b). On close examination the euhedral zoning seems to be truncated against the mantle, zone 4. Further fine zoning can be observed close to the edge of the zircon grain in zone 4. A narrow overgrowth can just be detected along the outer rim (zone 5) which has been interpreted in terms of metamorphic growth (van Breemen et al., 1987). The zircon itself has been rounded and has lost its euhedral shape.

SIMS results

The capability to detect variations in U, Th, Y and Hf across the diameter of a single grain was first demonstrated by line-scan measurements, as depicted in Figure 2. U- and Th-rich bands were clearly resolved, encircled by an Y-rich band of larger diameter. The average U concentration in this sample was estimated to be 270 ppm by intergrating and normalizing the ion yield in comparison with the yield from the Sri Lankan standard zircon studied with the ANU SHRIMP Ion Probe (Compston et al., 1984). This concentration is close to the average U concentration of five zircon fractions (135-395 ppm) analyzed by isotope dilution (van Breemen et al., 1987; sample 1, Table 83.1).

The ion-probe images of the whole zircon for U and Y (Fig.1C,D, where bright areas correspond to regions of high concentration) can be directly compared with the micrograph of the polished and etched zircon.

It is clear from both the line scans and the whole-zircon images that the U distribution was consistent with the main divisions observed in the polished and etched section. The zircon inner core (zone 1) was distinctly depleted in U, whereas this element was enriched in zone 2 and to a lesser extent in the outer rim (zone 5).

However, the great advantage of this technique lies in its ability to image many elements in addition to U. The Th distribution (Fig.2) was congruent with U to some extent as was Eu, for which no data is presented. Hf images (not shown) were similar to the line scan (Fig.2) showing no apparent systematic change. The Y distribution was most interesting. Two spots of Y enrichment appeared within zone 1 (Fig.1D). These regions were not distinguishable on the etched sample and they implied the existence of previously unsuspected fine structure within the inner core region. In contrast to U, Y was distinctly depleted in the outer core (zone 2), and strongly enriched in the igneous euhedral zone (zone 3). The main outer mantle (zone 4), was depleted of both U and Y, whereas the outer rim (zone 5) which surrounds zone 4 was high in both U and Y.

SUMMARY

These preliminary ion-probe analyses of complex multi-age zircons from a granulite facies terrane have revealed distinct trace element variations conforming to zones recognizable from backscattered electron images of polished and etched grains. This technique shows great promise for providing

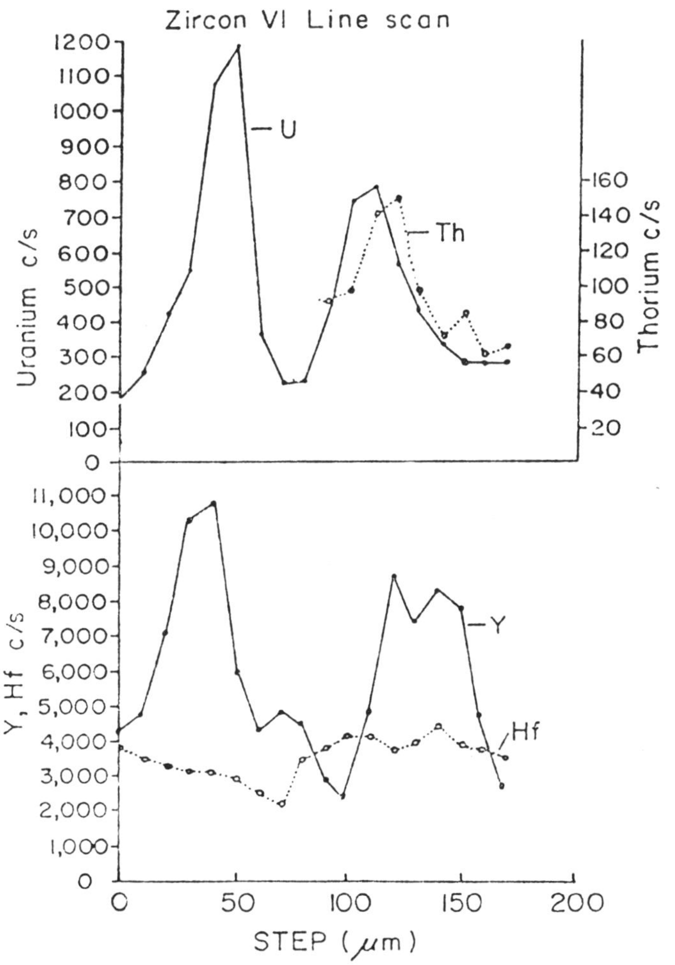

Figure 2: Line-scan measurements across a zircon grain showing variations in U, Th, Y and Hf concentrations.

information on trace element zonation which may in turn throw light on the origin of these or other multiple stage zircons, particularly when correlated with trace element concentrations of the major mineral assemblages.

ACKNOWLEDGMENTS

Assistance in obtaining direct-ion images was received from G. Mount at Surface Science Western. The Sri Lankan standard zircon was provided by W. Compston of the Australian National University. We are also indebted to P. Williams for useful discussions and assistance and to R.R. Parrish for critically reading this manuscript.

REFERENCES

Compston, W., Williams, I.S. and Meyer, C.
1984: U-Pb geochronology of zircon from lunar breccia 73217 using a sensitive high mass-resolution ion microprobe; Journal of Geophysical Research, v. 89, Supplement, p. B525-B534.

Nagasawa, H.
1970: Rare earth concentrations in zircons and apatites and their host dacites and granites; Earth and Planetary Science Letters, v. 9, p. 359-364.

Silver, L.T.
1969: A geochronological investigation of the anorthosite complex, Adirondack Mountains, New York; *in* Origin of Anorthosite and Related Rocks, ed. Y.W. Isachsen; New York State Museum and Science Service, Memior 18, p. 233-251.

van Breemen, O. and Parrish, R.R.
1986: Zircons record ancient geological processes; GEOS, v. 15, p. 18-21.

van Breemen, O., Henderson, J.B., Loveridge, W.D., and Thompson, P.H.
1987: U-Pb zircon and monazite geochronology and zircon morphology of granulites and granite from the Thelon Tectonic Zone, Healey Lake and Artillery Lake map areas, N.W.T.; *in* Current Research, Part A, Geological Survey of Canada, Paper 87-1A, p. 783-801.

A Pb isotopic study of the Valhalla complex and its surrounding rocks, southeastern British Columbia: preliminary interpretations

Mary Lou Bevier[1]

Bevier, M.L., A Pb isotopic study of the Valhalla complex and its surrounding rocks, southeastern British Columbia: preliminary interpretations; in Radiogenic Age and Isotopic Studies: Report 1, Geological Survey of Canada, Paper 87-2, p. 13-20, 1987.

Abstract

The Valhalla complex, southeastern British Columbia, is exposed in the footwall of outward-dipping Eocene normal faults, and contains high grade metasedimentary rocks of uncertain age that are structurally interleaved with granitic orthogneiss sheets that range in age from middle Cretaceous to Paleocene-Eocene. Pb isotopic studies indicate that the orthogneiss bodies were derived from older, radiogenic crust, including some of Precambrian age, although each orthogneiss body is of a distinct age and Pb isotopic composition. Granitoid bodies in the hanging wall above the Valhalla complex, including Middle Jurassic Nelson batholith and other plutons of Jura-Cretaceous age, did not sample the same source rocks as Valhalla orthogneisses, although the granitoids in the hanging wall also display an upper crustal Pb signature. Eocene lamprophyre and basalt dykes and Coryell syenite, which intrude both upper and lower plate rocks, were generated from relatively radiogenic mantle.

Résumé

Le complexe de Valhalla, dans le sud-est de la Colombie-Britanique, affleure au mur des failles normales de l'Éocène à pendage extérieur; il renferme des roches métasédimentaires fortement métamorphisées, d'âge incertain et finement interstratifiées avec des feuillets d'orthogneiss granitique dont l'âge varie du Crétacé moyen au Paléocène et à l'Éocène. Les études de datation radiométrique par analyse des rapports des isotopes du Pb indiquent que les massifs d'orthogneiss ont pour origine une croûte plus ancienne composée d'éléments radiogéniques, et que certains seraient d'âge précambrien, bien que chacun de ces massifs ait un âge différent et une composition isotopique distincte relativement à Pb. Les massifs granitoïdes qui se trouvent au toit, au-dessus du complexe de Valhalla, y compris le batholite jurassique de Nelson et d'autres plutons d'âge jurassique et crétacé, ne proviennent pas des mêmes roches mères que les orthogneiss de Valhalla, bien que les granitoïdes du toit présentent une signature Pb typique de la croûte supérieure. Les dykes lamprophyriques et basaltiques de l'Éocène et la syénite de Coryell, qui pénètrent à la fois les roches de la plaque supérieure et celles de la plaque inférieure, ont pour origine le manteau composé d'éléments relativement radiogéniques.

INTRODUCTION

The Valhalla complex (Reesor, 1965; Parrish et al., 1985a; Carr, 1986; Carr et al., 1987) is an elongate, domal, north-trending metamorphic complex east of Lower Arrow Lake, British Columbia, that is bounded on all sides by outward-dipping Eocene faults (Fig. 1). It contains uniformly high grade metamorphic rocks of uncertain age and several orthogneiss bodies of middle Cretaceous to Paleocene-Eocene age, disposed in arched sheets. Rock units in the hanging wall of the Eocene normal faults (''upper plate rocks'') are generally of low metamorphic grade and in part consist of Middle Jurassic Nelson batholith and other plutons of Jura-Cretaceous age, as well as Paleozoic, Triassic, and Eocene sedimentary and igneous rocks. Valhalla complex has many similarities to ''metamorphic core complexes'' of the western United States (Coney, 1980; Armstrong, 1982).

Many questions regarding the origin, timing, and tectonic significance of core complexes remain unanswered (Parrish et al., 1985b). Specific to this study is the fact that many metamorphic core complexes contain deformed granitic bodies of variable age and composition, and Pb isotopes are useful as tracers of the sources of granitic rocks.

Initial Pb isotopic ratios have been used elsewhere with success in determining the age and type of source material for granitoid rocks (c.f. Doe and Delevaux, 1980; Gariépy and Allègre, 1985; Ayuso, 1986; Mukasa, 1986).

[1] Geological Survey of Canada, 601 Booth Street, Ottawa, Ontario K1A 0E8

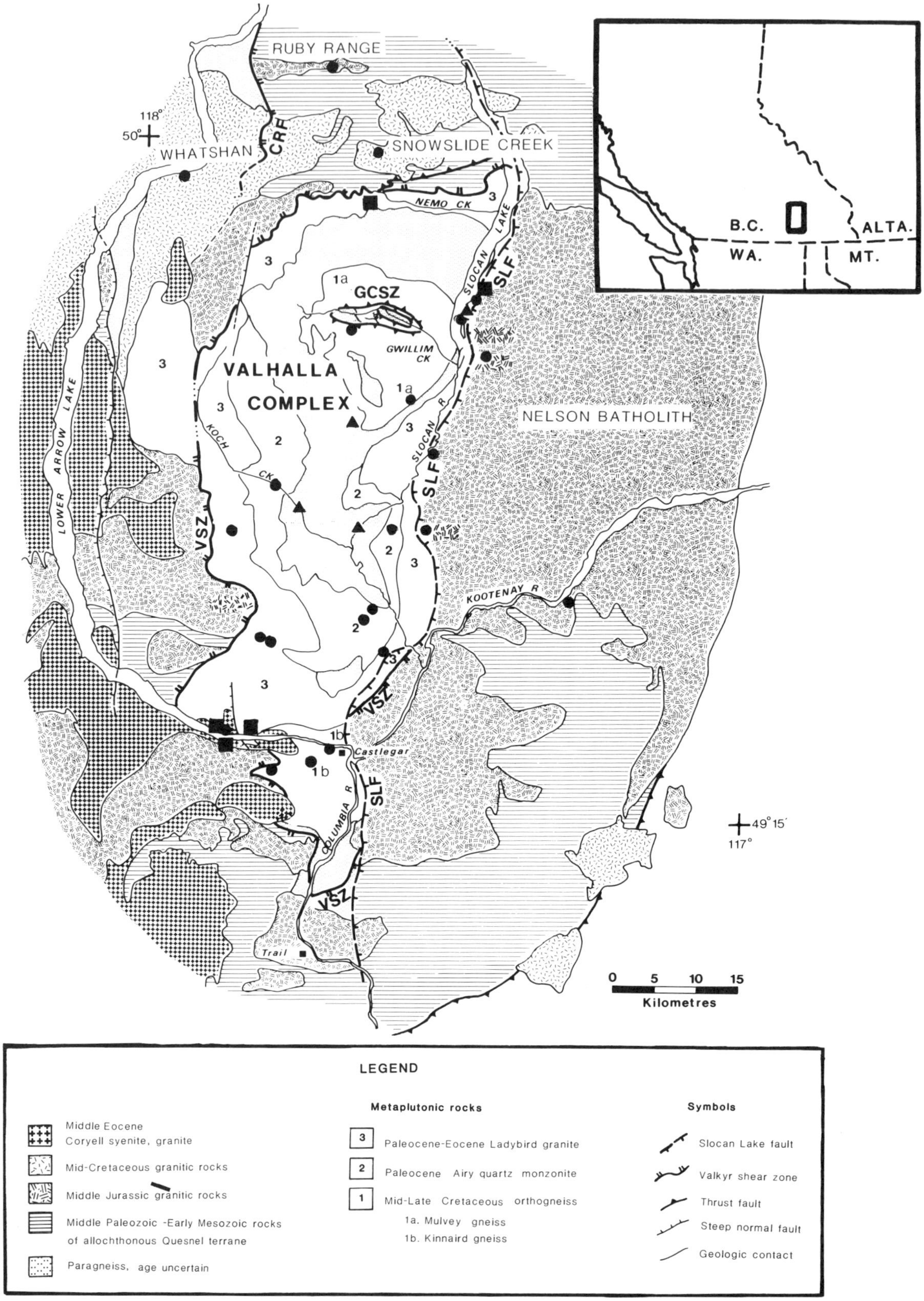

Figure 1. Geological map of Valhalla complex, modified from Carr et al. (1987) and including additional mapping adjacent to Lower Arrow Lake west of Valhalla complex (R.R. Parrish, unpublished data, 1986). Symbols for sample localities are: filled circles- granitoid rocks; filled triangles- leucocratic veins in paragneiss; filled squares-Eocene dykes. VSZ- Valkyr shear zone; SLF- Slocan Lake fault; GCSV-Gwillim Creek shear zones; CRF- Columbia River fault.

The purpose of this paper is to present Pb isotopic data from Valhalla complex and from a number of upper plate rock units, and to use these data to interpret the nature and age of the source(s) for deformed plutons that comprise the orthogneiss sheets of the complex. These data, in particular when combined with information obtained from U-Pb zircon ages of the granitic rocks, permit an assessment of possible genetic relationships (i.e. same source?) for lower and upper plate rocks, and detection of the presence or absence of a Precambrian crustal component in the granitic rocks.

GEOLOGY AND AGE RELATIONSHIPS

The geology of Valhalla complex is discussed in detail by Reesor (1965), Parrish (1984), Parrish et al. (1985a), Carr (1986), and Carr et al. (1987). Summaries of the geology of the surrounding rocks are found in Little (1960), Hyndman (1968), and Fyles (1984). Relevant geochronological data for Valhalla complex and the upper plate units sampled for this study are summarized in Carr et al. (1987) and in Table 1. No Precambrian crystalline basement outcrops in the complex or in the immediate vicinity, although it is assumed to exist in the relatively shallow subsurface on the basis of crustal cross-sections (Parrish and Carr, 1986).

Figure 2 is a schematic east-west cross-section of Valhalla complex drawn to represent relations between rock units in the vicinity of Gwillim Creek and Slocan. Within the main part of the complex, polydeformed sillimanite-orthoclase paragneiss of uncertain age occurs in three sheets up to 2 km thick. Leucocratic quartzofeldspathic veins within paragneiss are predominantly concordant with the foliation, particularly on the east side of the complex. The lower two sheets are structurally overlain and in sheared contact with middle Cretaceous Mulvey granodiorite gneiss, the structurally lowest orthogneiss in the complex (Parrish et al., 1985a; Carr et al., 1987). A third paragneiss sheet structurally overlies Mulvey gneiss. Other minor paragneiss units (Castlegar gneiss, Trail gneiss) outcrop in the southern part of the complex. Detailed descriptions of the Castlegar gneiss, which includes Late Cretaceous Kinnaird orthogneiss and minor quartzofeldspathic paragneiss, amphibolite, and pegmatite, are provided by Simony (1979) and Halwas and Simony (1986).

The uppermost paragneiss in the main part of the complex is structurally overlain by two additional orthogneiss sheets, the Paleocene Airy quartz monzonite and the Paleocene-Eocene Ladybird granite (Fig. 1, 2). These Paleocene and Eocene orthogneiss sheets are variably deformed, and intrusion appears to have both predated and outlasted Eocene deformation and metamorphism (Parrish et al., 1985a). In the southern part of the complex, Ladybird granite interfingers with Late Cretaceous Kinnaird orthogneiss.

The ductile, east-directed Valkyr shear zone (Fig. 1, 2) is about 2 km thick, forms the roof of Valhalla complex (Carr et al., 1987), and juxtaposes strongly foliated Middle Jurassic Nelson batholith, other Late Jurassic to Late Cretaceous plutons, and related metasedimentary rocks of the Mt. Roberts group and Nemo Lakes belt (Parrish, 1981) against and over Ladybird granite. On the east side of the complex the slightly younger and more brittle Slocan Lake fault zone cuts the Valkyr shear zone (Fig. 1, 2). Extensive U-Pb and Rb-Sr dating of granitoids involved in and crosscutting the Valkyr shear zone and Slocan Lake fault zone demonstrates an Eocene age of displacement for both structures (Carr et al., 1987).

The Middle Jurassic Nelson batholith (Fig. 1) contains granodiorite, syenite, and minor granite phases (Little, 1960). Late Jurassic to Late Cretaceous plutons exposed in the upper plate north of Valhalla complex (Whatshan, Ruby Range, and Snowslide Creek plutons; Fig. 1) range in composition from quartz monzonite to granodiorite (Hyndman, 1968). One small stock of Eocene Coryell syenite intrudes the complex on its southern edge and numerous Eocene (?) lamprophyre and basalt dykes intrude both upper and lower plate rocks.

Because a wide variety of ages and compositions of granitic rocks are present in Valhalla complex and its surroundings, it is unlikely that all the granitic rocks came from one source area in the crust. Pb isotopic ratios from feldspar and whole-rock samples and U-Pb systematics in zircons provide a means for distinguishing different source materials.

RESULTS

Analytical techniques.

Feldspars are U-poor and Pb-rich minerals; therefore, their $^{238}U/^{204}Pb$ values are very low, and changes in Pb isotopic composition as a function of age are usually very small. Hence measured Pb isotopic ratios on feldspars can be interpreted as initial ratios.

A clean K feldspar or plagioclase fraction was separated from each granitic and gneissic sample. After initial handpicking, the feldspar fraction was ground with a boron carbide mortar and pestle and sieved to separate a 75-150μ split, which was then subjected to a severe acid-leaching procedure in order to remove grain-boundary contaminants, following Gariépy and Allègre (1985). Fresh chips of fine grained dyke rocks (from which it was not possible to separate a feldspar fraction) were ground to powders using the boron carbide mortar and pestle.

Approximately 15 mg of the powdered sample was digested with a mixture of high-purity 48 % HF and 70 % HNO_3 in clean, covered teflon beakers in a clean air environment. Lead was separated by anion exchange chromatography using HCl and HBr, following procedures described in Barreiro (1982). Purified Pb samples were loaded onto previously outgassed Re filaments using the silica gel-phosphoric acid technique. Analyses were performed on a Finnigan MAT 261 mass spectrometer equipped with fully adjustable multiple collectors. Measurements were made in static multicollection mode with four Faraday collectors positioned to simultaneously collect ^{208}Pb, ^{207}Pb, ^{206}Pb, and ^{204}Pb. Measured Pb isotopic compositions were corrected for mass fractionation of 0.12 ± 0.04 % per a.m.u., determined by the average deviation of NBS 981 from its absolute value (Todt et al., 1984). Within-run precision averages 0.03 % (1σ) for all measured Pb isotopic ratios. Twelve Pb procedural blanks averaged 255 picograms and represent less than 0.17 % of the

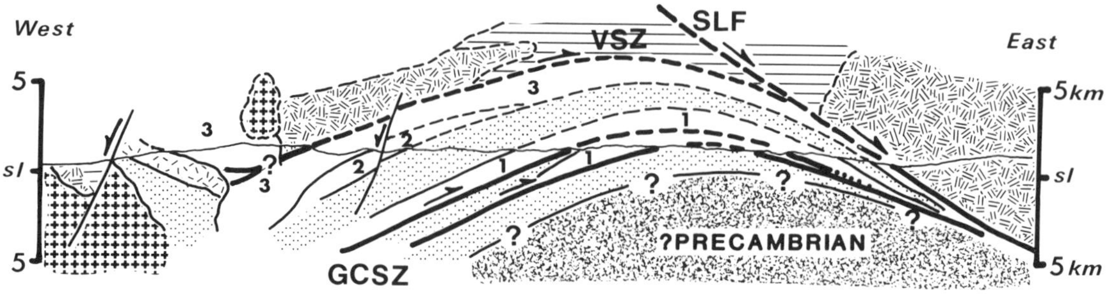

Figure 2. East-west cross-section across Valhalla complex in the vicinity of Gwillim Creek and the south end of Slocan Lake (*see* Fig. 1) (modified from Parrish and Carr, 1986). The topography is approximate and generalized, and faults and contacts above ground are dashed. The depiction of Precambrian basement below the Gwillim Creek shear zones is speculative but permissible given its presence in other nearby metamorphic complexes (Evenchick et al., 1984; Brown et al., 1986). Sl on scale bar refers to sea level.

Table 1. Summary of geochronological data, Valhalla complex.

Unit	U-Pb zircon age	Inheritance
	Upper plate	
Lamprophyre and basalt dykes	Middle Eocene (?)	none
Coryell syenite	51.7±0.5 Ma Middle Eocene	none
Snowslide Creek quartz monzonite	≈75 Ma Late Cretaceous	Precambrian
Whatshan granodiorite	79±2 Ma Late Cretaceous	Precambrian
Ruby Range granodiorite	≈150 Ma Late Jurassic	Precambrian
Nelson batholith granodiorite	169±3 Ma Middle Jurassic	Precambrian
	Lower plate	
Ladybird granite	56.5±1.5 Ma, 59±1 Ma Paleocene-Eocene	Precambrian
Airy quartz monzonite	62±1 Ma Late Paleocene	Precambrian
Kinnaird ortho-gneiss (Castlegar gneiss)	87±8 Ma Late Cretaceous	Paleozoic
Mulvey gneiss	100±5 Ma middle Cretaceous	Precambrian
paragneiss, Gwillim Creek	age uncertain; detrital zircons give Archean to Late Proterozoic ages	—
Castlegar paragneiss	age uncertain	—

Data from Carr et al. (1987) and R.R. Parrish (unpublished data, 1986, 1987). Errors are $\pm 2\sigma$. Time scale used is that of Palmer (1983).

estimated total Pb in the samples; therefore no blank correction was made. Measured Pb isotopic ratios of Eocene whole-rock dykes were corrected for an estimated 60 Ma of decay in order to obtain approximate initial ratios. Because U, Th, and Pb concentration data were not available, analyses of whole-rock lamprophyre dykes were corrected using average U, Th, and Pb concentrations from lamprophyres in Montana (Fraser et al., 1985/1986) and one analysis from a basalt dyke was corrected using average U, Th, and Pb concentrations from Chilcotin Group basalts in south-central British Columbia (Bevier, 1983).

Pb isotopic data.

Results for 33 samples from Valhalla complex and upper plate rocks are shown diagramatically as initial ratios in Figure 3 and listed in Table 2. Samples that are 50-60 Ma old (Airy, Ladybird, Coryell, and Eocene dykes) are directly comparable to each other on Figure 3. Analyses of 87-169 Ma old samples (Mulvey, Kinnaird, Nelson, Late J — Late K upper plate plutons; Fig. 3) should be corrected to include additional radiogenic Pb produced from radioactive decay of U and Th in the interval up to 50-60 Ma, in order to be directly

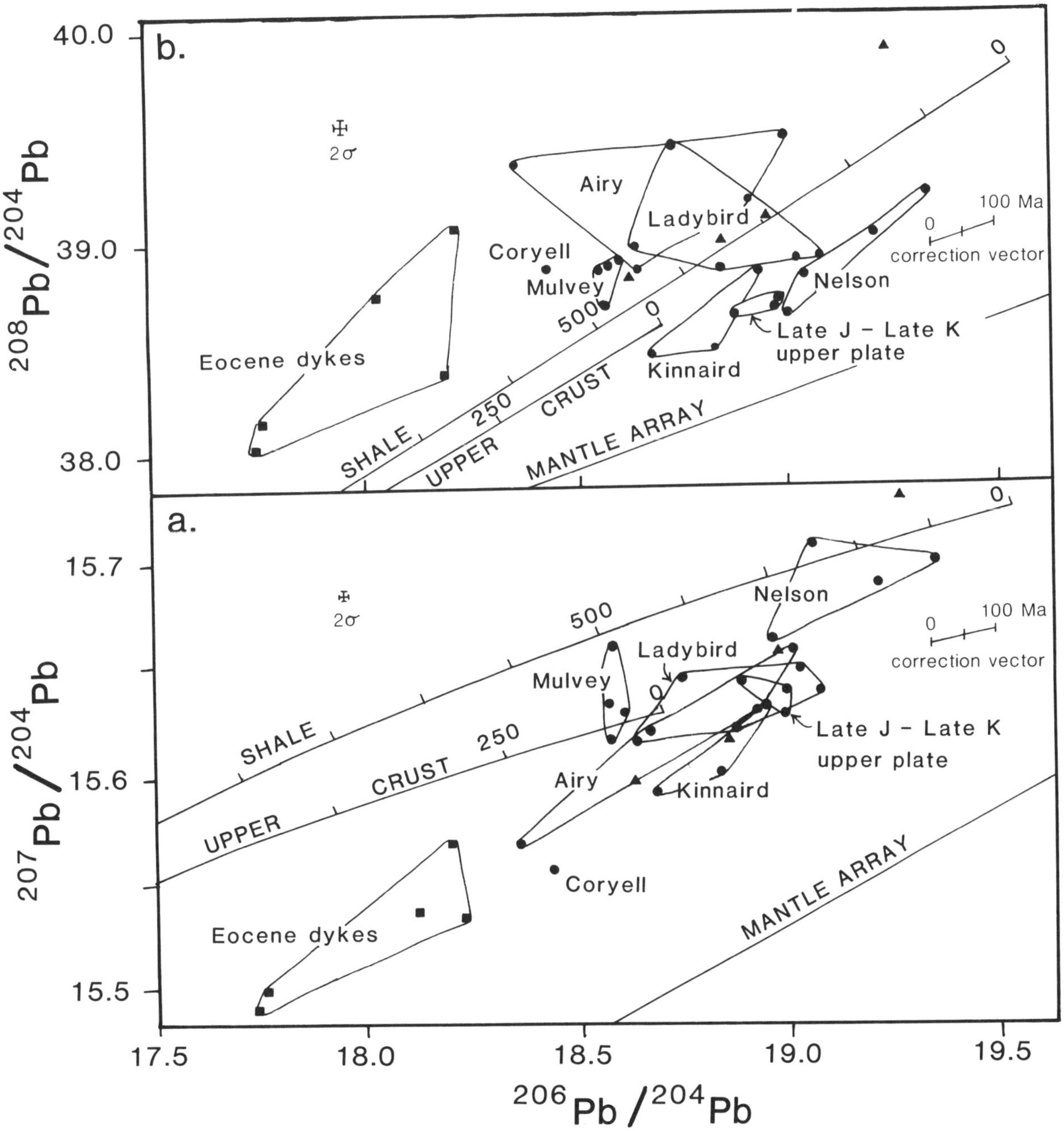

Figure 3. Pb isotope correlation diagrams for Valhalla complex and upper plate rock units. Also shown are Pb isotope evolution curves for the upper crust (two-stage model of Stacey and Kramers, 1975) and North American "shales" (Godwin and Sinclair, 1982), along with the mantle array for the northeast Pacific ocean (Church and Tatsumoto, 1975). Symbols as for Figure 1.

Table 2. Pb isotopic ratios, Valhalla complex and surrounding rocks.

Unit and sample number(s)	$^{206}Pb/^{204}Pb$	$^{207}Pb/^{204}Pb$	$^{208}Pb/^{204}Pb$
Upper plate			
Lamprophyre dykes[1]			
PCA-01-85D	17.758	15.499	38.158
PCA-40-84D	18.219	15.534	39.085
PCA-230-D	18.031	15.534	38.753
L-DYKE	17.737	15.491	38.046
Basalt dyke[1]			
Cape Horn	18.057	15.562	38.263
Coryell			
21	18.426	15.556	38.867
Late J-Late K upper plate			
SCS	18.972	15.629	38.693
WHAT	18.979	15.642	38.716
RRE	18.874	15.646	38.646
Nelson			
80	19.201	15.692	39.022
112	19.042	15.710	38.829
452	18.949	15.665	38.659
NB5	19.336	15.702	39.206
Lower plate			
Ladybird			
38	18.638	15.616	38.971
228	18.868	15.625	38.862
519	18.733	15.647	39.447
MLB-7	19.068	15.642	38.919
MLB-8	19.019	15.651	38.909
Airy			
156	18.996	15.651	39.488
178	18.362	15.576	39.379
410	18.912	15.630	39.181
SL-3e-1	18.645	15.624	38.864
Kinnaird			
58	18.930	15.632	38.853
339	18.829	15.603	38.494
9-85	18.678	15.593	38.464
Mulvey			
121B	18.582	15.652	38.881
121C	18.559	15.634	38.870
529	18.599	15.629	38.903
MLB-2	18.565	15.618	38.713
paragneiss, Gwillim Creek			
133	18.853	15.624	39.009
143	18.633	15.607	38.859
MLB-1a	19.245	15.733	39.908
MLB-3	18.950	15.662	39.121

[1] Initial Pb isotopic ratios for whole-rock lamprophyre and basalt dykes were calculated from measured ratios for an estimated age of 60 Ma (*see* **Analytical techniques** in text).

comparable with data from Paleocene-Eocene units. Approximate correction vectors for granitic rocks (calculated using average crustal values of μ=9.5 (Zartman and Doe, 1981) and Th/U=4 (Clark et al., 1966) are shown in Figure 3.

Analyses from most geologic units in the field area produce discrete and overlapping clusters of points on the Pb-Pb correlation diagrams. In particular, Valhalla orthogneiss sheets are isotopically distinct, although there is significant overlap between the Airy quartz monzonite and the Ladybird granite, and some overlap of the Kinnaird orthogneiss and Ladybird granite. Feldspars from concordant veins in Valhalla paragneiss have a wide range in Pb isotopic composition, and overlap with several of the orthogneiss and granitic bodies. Nelson batholith has higher $^{206}Pb/^{204}Pb$ and $^{207}Pb/^{204}Pb$ than Valhalla orthogneisses. For a given $^{208}Pb/^{204}Pb$, Nelson batholith samples display higher $^{206}Pb/^{204}Pb$ than Valhalla orthogneisses. Analyses from the Whatshan, Ruby Range, and Snowslide Creek plutons and Kinnaird orthogneiss, if corrected to 60 Ma (see correction vectors, Fig. 3), would overlap the field for Ladybird granite. Kinnaird orthogneiss does not overlap any other Valhalla complex orthogneiss. Eocene lamprophyre and basalt dykes, as well as Coryell syenite, have the least radiogenic Pb isotopic compositions of any measured from the area.

DISCUSSION AND SUMMARY

The three main orthogneiss sheets in Valhalla complex (Mulvey gneiss, Airy quartz monzonite, and Ladybird granite) are different in age (Table 1) and their initial Pb isotopic compositions, especially their $^{207}Pb/^{204}Pb$ values, suggest that they represent different magma batches derived from melting of pre-existing upper crustal rocks. U-Pb zircon systematics from these units indicate the presence of xenocrystic zircon cores within magmatic zircon crystals; the presence of these cores and the resultant discordant geochronological data (references in Table 1) indicate that a Precambrian component is present in the orthogneisses. The Pb isotopic compositions of Mulvey gneiss approximately fit a two-stage model for upper crustal Pb evolution (Stacey and Kramers, 1975) for $^{207}Pb/^{204}Pb$ versus $^{206}Pb/^{204}Pb$ but not for $^{208}Pb/^{204}Pb$ versus $^{206}Pb/^{204}Pb$. Airy quartz monzonite and Ladybird granite have higher $^{206}Pb/^{204}Pb$ for a given $^{207}Pb/^{204}Pb$ than the model permits (Fig. 3a). This suggests that the source rocks for Valhalla orthogneisses had a more complicated Pb isotopic history than the two-stage model, and that the source area underwent many relatively recent (Proterozoic to Paleozoic?) differentiation events that affected its Pb isotopic evolution. The partial overlap in Pb isotopic composition between Airy quartz monzonite and Ladybird granite and the fact that these two sheets only differ in age by 3 million years together suggest that they may represent two phases of a differentiated and zoned magma batch that was emplaced in two pulses.

Nelson batholith contains the most radiogenic rocks sampled for this study. If corrected (*see* correction vectors, Fig. 3) to 60 Ma, the gap between Nelson batholith and all other units is even larger than shown. None of the phases in Nelson batholith are isotopically similar to Valhalla orthogneisses which implies that the orthogneisses and the volumetrically-dominant Middle Jurassic granitoid suite in the upper plate did not sample the same source material. Parrish et al. (1985a) and Carr et al. (1987) suggest that Valhalla complex rocks were exhumed from depth by Eocene extension, and upper plate rocks originated to the west of their current positions relative to the complex. This reinforces the conclusion drawn from Pb isotopic data that Nelson intrusives did not sample the same source area as Valhalla orthogneisses.

If one assumes average μ and Th/U for granitic rocks and adds radiogenic Pb from an additional 15-90 million years of decay to the initial Pb isotopic compositions of the Whatshan, Ruby Range, and Snowslide Creek plutons (*see* correction vectors, Fig. 3), their age-corrected Pb isotopic compositions at 60 Ma are sufficiently close to those of Ladybird granite that a similar source is possible.

It is unclear from the field relations whether leucocratic quartzofeldspathic veins in Valhalla paragneiss are *in situ* partial melts of the paragneiss or injections (perhaps from a plutonic precursor to one of the orthogneisses) into the paragneiss. Veins in the paragneiss, including those sampled for this study, are predominantly concordant with the foliation. Based on a limited number of samples from the paragneiss, Pb isotopic compositions do not distinguish between formation of the veins as pegmatitic offshoots of either Airy quartz monzonite or Ladybird granite, or derivation of Airy quartz monzonite or Ladybird granite by partial melting of Valhalla paragneiss.

Eocene lamprophyre and basalt dykes and Coryell syenite may be representative of mantle-derived magmas with little or no crustal component, although they are enriched in $^{207}Pb/^{204}Pb$ and $^{208}Pb/^{204}Pb$ relative to the northeast Pacific mantle array (Fig. 3); (Church and Tatsumoto, 1975). Elsewhere in the world the presence of lamprophyre dyke swarms has been interpreted to result from a mantle metasomatic event (Laughlin et al., 1985/1986) that added significant radiogenic Pb, among other components, to the lamprophyre source region in the mantle.

In summary, all Valhalla orthogneisses and upper plate plutons have a radiogenic, mostly Precambrian, upper crustal component in their source area(s), based on their Pb isotopic compositions and U-Pb zircon systematics. The inferred Precambrian upper crustal source in this area must be heterogeneous as shown by the wide range of Pb isotopic compositions of samples from different units. The isotopic data support the concept of crustal growth and evolution by recycling of Precambrian crust in later partial melting events.

ACKNOWLEDGMENTS

I am grateful to Randy Parrish and Sharon Carr for providing the samples for this study, and to Randy Parrish for supplying several of the figures. The manuscript was improved as a result of a critical review by Randy Parrish and comments by Jim Mortensen.

REFERENCES

Armstrong, R.L.
1982: Cordilleran metamorphic core complexes — from Arizona to southern Canada; Annual Review of Earth and Planetary Sciences, v. 10, p. 129-154.

Ayuso, R.A.
1986: Lead-isotopic evidence for distinct sources of granite and for distinct basements in the northern Appalachians, Maine; Geology, v. 14, p. 322-325.

Barreiro, B.A.
1982: Lead isotope evidence for crust-mantle interaction during magmagenesis in the South Sandwich island arc and in the Andes of South America; unpublished Ph.D. thesis, University of California, Santa Barbara, 173 p.

Bevier, M.L.
1983: Implications of chemical and isotopic composition for petrogenesis of Chilcotin Group basalts, British Columbia; Journal of Petrology, v. 24, p. 207-266.

Brown, R.L., Journeay, J.M., Lane, L.S., Murphy, D.C., and Rees, C.J.
1986: Obduction, backfolding, and piggyback thrusting in the metamorphic hinterland of the Southeastern Canadian Cordillera; Journal of Structural Geology, v. 8, p. 255-268.

Carr, S.D.
1986: The Valkyr shear zone and the Slocan Lake fault zone: Eocene structures that bound the Valhalla gneiss complex, southeastern British Columbia; unpublished M.Sc. thesis, Carleton University, Ottawa, 106 p.

Carr, S.D., Parrish, R.R., and Brown, R.L.
1987: Eocene structural development of the Valhalla complex, southeastern British Columbia; Tectonics, v. 6, p. 175-196.

Church, S.E. and Tatsumoto, M.
1975: Lead isotope relations in oceanic ridge basalts from the Juan de Fuca-Gorda Ridge area, N.E. Pacific Ocean; Contributions to Mineralogy and Petrology, v. 53, p. 253-279.

Clark, S.P., Jr., Peterman, Z.E., and Heier, K.S.
1966: Abundances of uranium, thorium, and potassium; *in* Handbook of Physical Constants — revised edition, Clark, S.P., Jr.; Geological Society of America, Memoir 97, p. 521-541.

Coney, P.
1980: Cordilleran metamorphic core complexes: an overview; *in* Cordilleran Metamorphic Core Complexes, ed. M.D. Crittenden, P.J., Coney, and G.H. Davis, Geological Society of America, Memoir 153, p. 7-31.

Doe, B.R. and Delevaux, M.H.
1980: Lead-isotope investigations in the Minnesota River Valley — Late-tectonic and posttectonic granites; *in* Selected Studies of Archean Gneisses and Lower Proterozoic Rocks, Southern Canadian Shield, ed. G.B. Morey, and G.N. Hansen; Geological Society of America, Special Paper 182, p. 105-112.

Evenchick, C.A., Parrish, R.R., and Gabrielse, H.
1984: Precambrian gneiss and late Proterozoic sedimentation in north-central British Columbia; Geology, v. 12, p. 233-237.

Fraser, K.J., Hawkesworth, C.J., Erlank, A.J., Mitchell, R.H., and Scott-Smith, B.H.
1985/
1986: Sr, Nd, and Pb isotope and minor element geochemistry of lamproites and kimberlites; Earth and Planetary Science Letters, v. 76, p. 57-70.

Fyles, J.T.
1984: Geological setting of the Rossland Mining Camp; British Columbia Ministry of Energy, Mines, and Petroleum Resources, Bulletin 74, 56 p.

Gariépy, C. and Allègre, C.J.
1985: The lead isotope geochemistry and geochronology of late kinematic intrusives from the Abitibi greenstone belt, and the implications for late Archaean crustal evolution; Geochimica et Cosmochimica Acta, v. 49, p. 2371-2383.

Godwin, C.I. and Sinclair, A.J.
1982: Average lead isotope growth curve for shale-hosted lead-zinc deposits, Canadian Cordillera; Economic Geology, v. 77, p. 675-690.

Halwas, D. and Simony, P.S.
1986: The Castlegar gneiss complex, southern British Columbia; *in* Current Research, Part A, Geological Survey of Canada, Paper 86-1A, p. 583-587.

Hyndman, D.W.
1968: Petrology and structure of the Nakusp map area, British Columbia; Geological Survey of Canada, Bulletin 161, 95 p.

Laughlin, A.W., Aldrich, Jr., M.J., Shafiqullah, M., and Husler, J.
1985/
1986: Tectonic implications of the age, composition, and orientation of lamprophyre dikes, Navajo volcanic field, Arizona; Earth and Planetary Science Letters, v. 76, p. 361-374.

Little, H.W.
1960: Nelson map-area, west half, British Columbia; Geological Survey of Canada, Memoir 308, 205 p.

Mukasa, S.B.
1986: Common Pb isotopic compositions of the Lima, Arequipa, and Toquepala segments in the Coastal Batholith, Peru: Implications for magmagenesis; Geochimica et Cosmochimica Acta, v. 50, p. 771-782.

Palmer, A.R.
1983: The Decade of North American Geology 1983 Time Scale; Geology, v. 11, p. 503-504.

Parrish, R.R.
1981: Geology of the Nemo Lakes Belt, northern Valhalla Range, southeast British Columbia; Canadian Journal of Earth Sciences, v. 18, p. 944-958.
1984: Slocan Lake Fault: a low angle fault zone bounding the Valhalla gneiss complex, Nelson map area, southern British Columbia; *in* Current Research, Part A, Geological Survey of Canada, Paper 84-1A, p. 323-330.

Parrish, R.R. and Carr, S.D.
1986: Extensional tectonics of southeastern British Columbia: new data and interpretations; Geological Association of Canada, Program with Abstracts, v. 11, p. 112.

Parrish, R.R., Carr, S.D., and Brown R.L.
1985a: Valhalla gneiss complex, southeast British Columbia: 1984 Fieldwork; *in* Current Research, Part A, Geological Survey of Canada, Paper 85-1A, p. 81-87.

Parrish, R.R., Carr, S.D., and Parkinson, D.
1985b: Metamorphic Complexes and Extensional Tectonics, Southern Shuswap Complex, southeastern British Columbia; *in* Field Guides to Geology and Mineral deposits in the Southern Canadian Cordillera, Geological Society of America, Cordilleran Section Meeting, p. 12.1-12.15.

Reesor, J. E.
1965: Structural evolution and plutonism in Valhalla gneiss complex, British Columbia; Geological Survey of Canada, Bulletin 129, 128 p.

Simony, P.S.
1979: Pre-Carboniferous basement near Trail, British Columbia; Canadian Journal of Earth Sciences, v. 16, p. 1-11.

Stacey, J.S. and Kramers, J.D.
1975: Approximation of terrestrial lead isotope evolution by a two-stage model; Earth and Planetary Science Letters, v. 26, p. 207-221.

Todt, W., Cliff, R.A., Hansen, A., and Hofmann, A.W.
1984: ^{202}Pb + ^{205}Pb double spike for Pb isotopic analyses; Terra Cognita, v. 4, p. 209.

Zartman, R.E. and Doe, B.R.
1981: Plumbotectonics — The Model; Tectonophysics, v. 75, p. 135-162.

Geochronology of Neogene volcanic rocks in the northern Garibaldi Belt, British Columbia

J.C. Roddick[1] and J.G. Souther[2]

Roddick, J.C. and Souther, J.G., Geochronology of Neogene volcanic rocks in the northern Garibaldi Belt, British Columbia; in Radiogenic Age and Isotopic Studies: Report 1, Geological Survey of Canada, Paper 87-2, p. 21-24, 1987.

Abstract

Five samples of volcanic rocks from the Bridge River upland, the northernmost segment of the American high Cascades continental calc-alkaline arc have K-Ar ages ranging from 400 to 800 ka. Though several samples may contain excess argon, the ages are in good agreement with related segments of the arc and support the suggestion that late Pleistocene volcanism in the Cordillera may have been triggered by isostatic adjustment associated with major fluctuations in Pleistocene ice thicknesses.

Résumé

On a établi que l'âge de cinq échantillons de roches volcaniques receuillis dans les hautes terres de la rivière Bridge, soit la région la plus septentrionale de l'arc continental à caractère calco-alcalin de la chaîne élévée des Cascades en Amérique, se situe entre 400 et 800 ka. Bien que quelques échantillons peuvent contenir un excédent d'argon, les âges obtenus correspondent à ceux établis dans le cas de parties connexes de l'arc et viennent appuyer l'hypothèse selon laquelle l'épisode volcanique survenu dans la Cordillère au Pléistocène inférieur aurait été dû à la compensation isostastique liée à d'importantes variations de l'épaisseur de la glace au cours du Pléistocène.

INTRODUCTION

The Garibaldi Volcanic Belt, a northern extension of the American high Cascades, is a continental calc-alkaline arc related to latest Pliocene and Quaternary subduction of the Juan de Fuca Plate (Fig. 1). It comprises three, north-trending, en echelon segments: a southern segment which includes Mount Garibaldi (Green, 1981), a central segment which includes Mount Cayley (Souther, 1980), and a northern segment which includes Meager Mountain (Read, 1979). Dated rocks from the Garibaldi Volcanic Belt (Garibaldi Group) are all younger than 2 Ma. Locally they rest disconformably on deeply dissected Late Miocene plutons, such as the 7.9 Ma Salal Pluton (Stephens, 1972), which are subvolcanic remnants of the older, Pemberton Volcanic Belt (Souther, 1975). Farther east, on the Interior Plateau, relatively alkaline, Late Miocene to Late Pliocene Chilcotin Group basalts are believed to be a back-arc facies of Pemberton and Garibaldi arc volcanism (Bevier, 1983).

Meager Mountain, the principal volcano in the northern segment of the Garibaldi Belt, is a composite pile of andesite, dacite and rhyodacite flows and pyroclastic rocks with a total volume of about 50 km^3. The oldest lavas issued from a vent near the southern edge of the complex about 1.9 Ma ago. Subsequent activity migrated to vents successively farther north and culminated with the eruption of dacitc tephra and ash flows 1340 years ago (^{14}C) from a vent on the north side (Read, 1979). Several tens of smaller volcanoes, necks, tuyas, and flow-remnants occupy a zone extending north from Meager Mountain, across Salal Pluton to the Bridge River upland. Remnants resting on Salal Pluton are predominantly alkaline basalt and hawaiite which have yielded whole-rock K-Ar dates of 0.59 and 0.97 Ma (Lawrence et al., 1984). In 1986 volcanic remnants on the Bridge River upland were mapped and sampled for chemical and isotopic study. This paper reports the whole-rock K-Ar results.

GEOLOGY OF BRIDGE RIVER UPLAND

The Bridge River upland is north of Salal Pluton, at the extreme northern end of the Garibaldi Volcanic Belt. It is underlain by granodiorite and quartz diorite of the Coast Plutonic Complex which, in adjacent areas of the Pemberton area (Woodsworth, 1977), have yielded K-Ar dates ranging from 47 to 76 Ma.

Garibaldi Group volcanic and related subvolcanic intrusions on the Bridge River upland display a variety of landforms which reflect a complex igneous and glacial history

[1] Geological Survey of Canada, 601 Booth Street, Ottawa, Ontario K1A 0E8
[2] Geological Survey of Canada, 100 West Pender Street, Vancouver, B.C. V6B 1R8

(Fig. 1). The oldest rocks occur as dyke swarms and cupolas which bear no relationship to the present topography. A younger group of rocks forms volcanic necks and flow remnants which are deeply dissected and glaciated but retain vestiges of their original form. Still younger, intra-glacial volcanic landforms have been only moderately altered by post-glacial erosion and the youngest eruptive rocks post-date the retreat of ice from the main valleys. Basalt and andesite are the principal lithologies but minor rhyolite is present in the older units.

EXPERIMENTAL PROCEDURES

Samples were prepared using routine rock crushing and sieving techniques. Mineral separates were concentrated by magnetic, heavy liquid and where necessary hand picking procedures. Whole-rocks typically were sieved to $-40+60$ mesh. For the present study of young volcanic rocks procedures previously used by Souther et al. (1984) were used. A $-50+80$ mesh fraction was washed in cold 10% reagent grade HCl for 5 minutes in an ultrasonic bath and then repeatedly washed with water. After a final rinse in filter paper the sample was dried at 60 °C for one hour. This procedure was followed to remove carbonates and reduce atmospheric Ar which may fractionate during sample bakeout and thus result in ages which are too old (Baksi,1974). While this procedure may minimize atmospheric fractionation effects, previous experience suggests that it is not 100% reliable (Souther et al., 1984).

Potassium concentrations were determined by standard atomic absorption spectrometry techniques. Representative sample aliquots of about 100 mg were weighed in duplicate, dissolved in an HF-HNO_3 open beaker digestion, evaporated to dryness and taken up in a 1N HCl solution. Dilution of mineral solutions using pipettors and volumetric flasks, and confirmed by weighing, was done in triplicate for each mineral solution. Dilution was designed to produce K concentrations between 0.5 and 1 ppm K. Atomic absorption measurements of samples were linearly interpolated between standard solutions of 0.5, 1.0 and 1.5 ppm K which maintain linearity in absorption for these concentrations. Small blank corrections were also made. Results were monitored by including one or more of the standards LP-6 biotite, MMHb-1 hornblende, or SY-2 syenite, in duplicate with every batch of samples. Precision of analysis is typically 1.0% (2σ).

Argon concentrations were determined by isotope dilution with an ^{38}Ar tracer dispensed from a gas pipette system. Normally 0.20 to 1.0 g of mineral separate or 1 to 10 g of coarse (0.5 to 1.0 mm) whole-rock was used for argon extraction. Small samples were wrapped in aluminum foil and positioned in a multiple sample tree above a molybdenum crucible mounted on a silica pedestal in the extraction furnace. Aliquots of whole rock samples greater than 1 g were loaded directly into crucibles before evacuation of the extraction line. Prior to extraction, the samples were vacuum baked at 180 to 200 °C for 16 to 40 hours to minimize

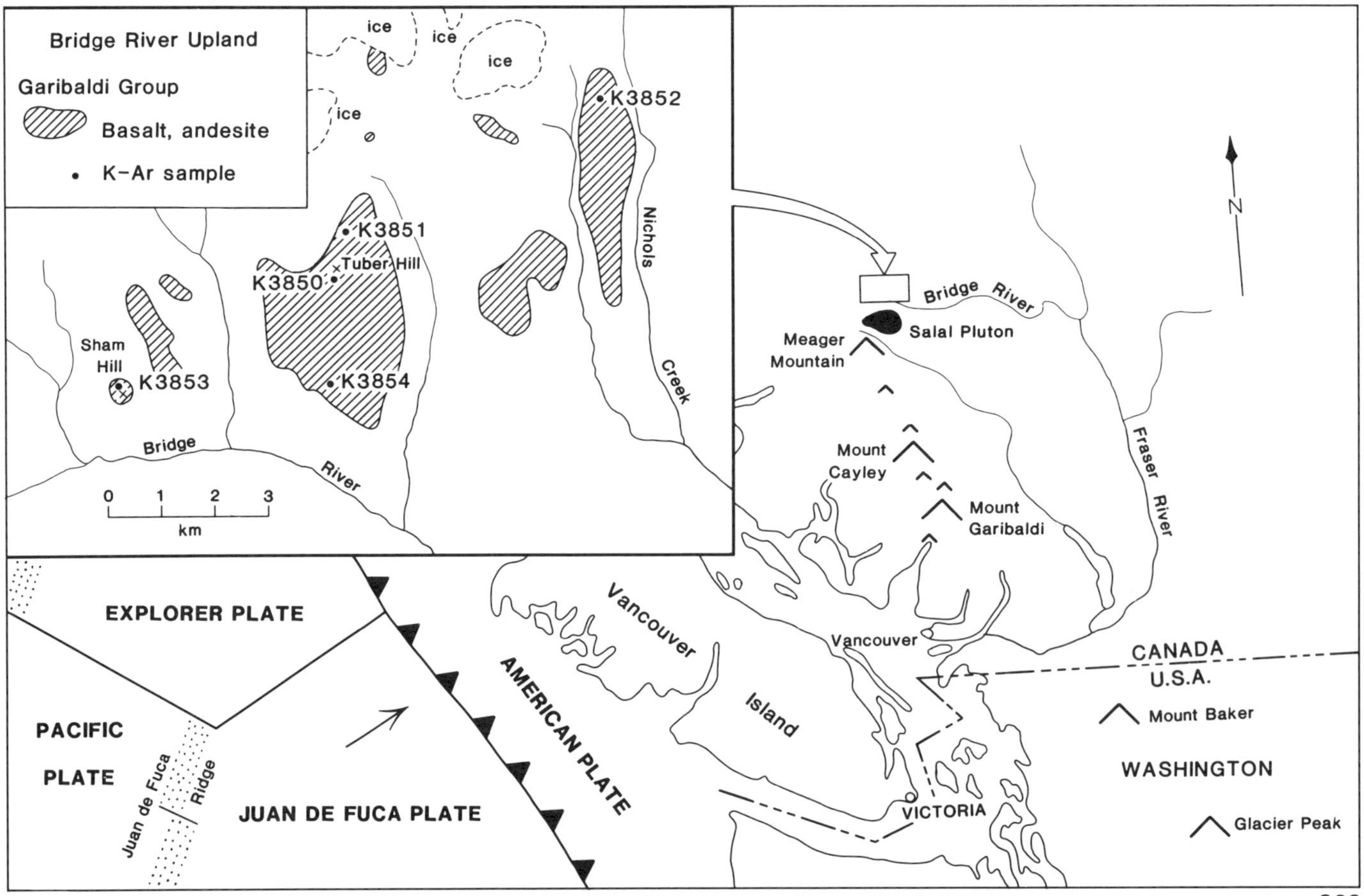

Figure 1. Tectonic setting of the northern Garibaldi Volcanic Belt and locations of dated samples.

atmospheric contamination. In the case of mineral separates the samples were dropped into the previously degassed crucible prior to fusion. ^{38}Ar spike was dispensed prior to sample fusion, as a full aliquot or a 1/6 or 1/38 aliquot using valving, to ensure optimum sample to spike mixtures. Sample fusion was by radio frequency induction heating in a water-cooled pyrex vacuum furnace. Following fusion a two stage purification of the gas was used. The first stage consisted of a 3L titanium sublimation chamber and a separate Ti getter heated to 800 °C. The purified gas was then drawn into a second section on activated charcoal and a second clean-up using a Zr-Al getter was used. For analysis a portion of the gas was equilibrated with the on-line mass spectrometer thus avoiding orifice corrections for isotopic fractionation.

Argon isotopes were measured on a modified AEI MS10 mass spectrometer with a standard 1.8 kG magnet operated in a static mode. A computer controlled high voltage supply was used to sequentially focus the ion beams for individual masses and base lines on to the Faraday cup collector. The ion beams were measured with a Cary 401M electrometer (10^{10} ohm resistor) and Solartron digital volt meter. A peak switching interval of 5 seconds delay and 4 seconds integration was employed. The peak intensities were corrected for: (1) amplifier decay response to varying beam intensities, (2) linear extrapolation to gas inlet time, (3) spectrometer mass discrimination using measurements of air Ar from an on-line pipette system. Ar blanks were typically 3×10^7 cc STP Ar and were combined with the usual correction for atmospheric Ar within the sample. Errors on the Ar concentration and age are calculated by numerical error propagation (Roddick, 1987) of errors in the measured ratios, sample weight, spike aliquot volume and K concentration. Ages were calculated using the decay constants, branching ratio and ^{40}K abundance recommended by the IUGS Subcommission on geochronology (Steiger and Jäger, 1977). The average radiogenic ^{40}Ar content of biotite LP-6 for 12 analyses since April 1984 is 429.1×10^{-7} cc/g STP with a standard deviation of 0.9% (2σ mean of .54%). The concentration, though lower than the recommended value of 431.0×10^{-7} cc/g STP is within the error limits (Odin et al., 1982).

DESCRIPTION OF SAMPLES AND GEOCHRONOLOGY

The five dated specimens with ages ranging from 374 to 760 ka (Table 1) were collected from three separate volcanic piles on the Bridge River upland: Tuber Hill, Sham Hill and the Nichols valley flows. In all cases the samples were petrographically fresh with original unaltered igneous mineralogy.

Tuber Hill is a small composite volcano about 2 km across that still retains much of its original morphology. The lower part of the pile consists of hyaloclastite and pillow lava, formed during an initial subglacial stage of activity. The hyaloclastite is overlain by subaerial columnar flows deposited soon after the edifice was free of glacial meltwater. Moderate erosion has exposed a tabular intrusive body, clearly part of the main feeder system, along the crest of the summit ridge. Three specimens from Tuber Hill were dated. Specimen SE-060186 (598 ka) is basalt from pahoehoe toes near the top of the hyaloclastite succession. Specimen SE-070586 (731 ka) is from a thick flow of columnar basalt near the top of the upper subaerial succession, and specimen SE-070386 (760 ka) is intrusive basalt from the summit ridge.

Sample SE-040486 (752 ka) is from Sham Hill a steep-sided volcanic neck, about 300 m across at the base, which rises more than 60 m above the southern edge of Bridge River upland. Its bare, glaciated surface is strewn with glacial erratics and no vestiges remain of any associated flows or pyroclastic deposits. Based on its state of erosion this neck appears to be older than Tuber Hill.

Specimen SE-210686, the youngest age determined (374 and 405 ka) was collected from a columnar basalt succession forming a median ridge in the upper part of the Nichols Creek valley. The source of the flows has been removed by erosion and the remaining flow succession is flanked deeply by incised canyons.

Table 1. K-Ar data for volcanic rocks of the northern Garibaldi Belt, British Columbia.

Field no.	Lab. no.	Locality	Material	K wt% ± 1σ%	Rad ^{40}Ar cc/g × 10^{-7}	% Rad ^{40}Ar	Age ± 2σ ka
SE-070386	3850	Tuber Hill	Whole rock	1.041 ± .60	0.3075	16.7	760 ± 33
SE-070586	3851	Tuber Hill	Whole rock	0.998 ± .55	0.2836	10.1	731 ± 27
SE-060186	3854	Tuber Hill	Whole rock	0.981 ± .70	0.2279	21.2	598 ± 15
SE-040486	3853	Sham Hill	Whole rock	1.051 ± .65	0.3074	23.9	752 ± 18
SE-210686	3852	Nicols valley	Whole rock	0.419 ± .35	0.0610	6.1	374 ± 23
	3852-2	Nicols valley	Whole rock		0.0660	7.9	405 ± 23

σ = standard error; cc/g = cubic centimetres per gram at standard temperature and pressure.

DISCUSSION

Variations among the five K-Ar dates from the Bridge River upland do not reflect their apparent relative ages based on stratigraphic order or the geomorphology of the terrain. At Tuber Hill the youngest age (598 ka) is from the base of the pile while two samples from higher in the section have the same ages, within error, of 731 and 760 ka. Based on the state of dissection, Tuber Hill appears to be much less modified by erosion than either Sham Hill (752 ka) or the Nichols Creek flows (replicates of 374 and 405 ka).

There are three possible explanations for the apparent discrepancies in the ages. The ebb and flow of major glaciations may occur over a period of 5 to 10 ka, a time span which is much less than both the range of measured ages and their errors. Depending on the severity of the glaciations their erosional effects may be variable and thus may not be a good index of relative age. The other two explanations relate to the analysis of argon within the samples.

K-Ar ages may be too old because of the presence of fractionated atmospheric argon or the incorporation of excess radiogenic argon during crystallization. Baksi (1974) has shown that high bakeout temperatures may enhance the fractionation of loosely held atmospheric argon. To assess the potential of this effect one sample, with the lowest radiogenic argon content (SE-210686), was analyzed in duplicate. The first analysis was baked out in the extraction system at 180 to 200 °C as the other samples, while the second analysis was baked at only 100 to 120 °C. No difference was detected between the ages of the duplicates and in fact the lower temperature bakeout has produced the higher radiogenic Ar content.

The presence of excess argon may account for older ages for Tuber Hill, if in fact it is the youngest volcanic pile. Because of the young age and relatively low K content these samples are sensitive to the incorporation of any excess argon. Previously Souther et al. (1984) showed that some young volcanics ($<$ 1Ma) at Mount Edziza had incorporated sufficient excess argon to more than double their apparent ages. In that case the samples also had high $^{87}Sr/^{86}Sr$ reflecting contamination with older crustal rocks. Further K-Ar analyses will have to be carried out to confirm the apparently older ages for Tuber Hill.

Despite these potential problems with the K-Ar dating it is clear that the ages of the volcanics of the Bridge River upland are within or younger than the 0.59 to 0.97 Ma range of ages previously reported by Lawrence et al. (1984) for volcanics from the Salal Creek area. These data suggest that volcanism in the northern Garibaldi Belt, from north of Meager Mountain, was concentrated during a relatively short interval between 400 and 970 ka. This corresponds to a time of major flucuation in the thickness of Pleistocene ice and associated isostatic adjustment. Mathews (1958) proposed that widespread late Pleistocene volcanism in the Cordillera may have been triggered by such isostatic adjustment. An alternative hypothesis, that the intensity of volcanic activity may reflect changes in the rate of Juan de Fuca Plate subduction, is not borne out by Riddihough (1984) who reported a steady decrease in convergence rates between 6.5 and 0.5 Ma.

ACKNOWLEDGMENTS

We thank D. Bellerive, F.B. Quigg and R.J.G. Seguin for their technical expertise in performing the K-Ar analyses.

REFERENCES

Baksi, A.K.
1974: Isotopic fractionation of a loosely held atmospheric argon component in the Picture Gorge Basalts; Earth and Planetary Science Letters, v. 21, p. 431-438.

Bevier, M.L.
1983: Regional stratigraphy and age of Chilcotin Group basalts, south-central British Columbia; Canadian Journal of Earth Sciences, v. 20, p. 515-524.

Green, N.L.
1981: Geology and petrology of Quaternary volcanic rocks, Garibaldi lake area, southwestern British Columbia; Geological Society of America, Bulletin, v. 92, pt. 1, p. 697-702 and pt. 2, p. 1359-1470.

Lawrence, R.B., Armstrong, R.L. and Berman, R.G.
1984: Garibaldi Group volcanic rocks of the Salal Creek area, southwestern British Columbia: alkaline lavas on the fringe of the predominantly calc-alkaline Garibaldi (Cascade) volcanic arc; Journal of Volcanology and Geothermal Research, v. 21, no. 3/4, p. 255-276.

Mathews, W.H.
1958: Geology of the Mount Garibaldi map-area, southwestern British Columbia, Canada; Geological Society of America, Bulletin, v. 69, p. 179-198.

Odin, G.S. and 35 collaborators
1982: Interlaboratory standards for dating purposes; *in* Numerical Dating in Stratigraphy, ed. G.S. Odin; Wiley, p. 123-148.

Read, P.B.
1979: Geology, Meager Creek geothermal area, British Columbia; Geological Survey of Canada, Open File 603, (map and descriptive notes).

Riddihough, R.
1984: Recent movements of the Juan de Fuca plate system; Journal of Geophysical Research, v. 89, no. B8, p. 6980-6994.

Roddick, J.C.
1987: Generalized numerical error analysis with applications to geochronology and thermodynamics; Geochimica et Cosmochimica Acta, v. 51, p. 2129-2135.

Souther, J.G.
1975: Geothermal potential of western Canada; *in* Proceedings, Second United Nations Symposium on the Development and Use of Geothermal Resources, v. 1, p. 259-267.

Souther, J.G.
1980: Geothermal reconnaissance in the central Garibaldi Belt, British Columbia; *in* Current Research, Part A, Geological Survey of Canada, Paper 80-1A, p. 1-11.

Souther, J.G., Armstrong, R.L. and Harakal, K.
1984: Chronology of the peralkaline, late Cenozoic Mount Edziza Volcanic Complex, northern British Columbia, Canada; Geological Society of America, Bulletin, v. 95, p. 337-349.

Steiger R.H. and Jäger, E.
1977: Subcommission of geochronology: convention on the use of decay constants in geo- and cosmochemistry; Earth and Planetary Science Letters, v. 36, p. 359-369.

Stephens, G.C.
1972: The geology of the Salal Creek pluton, southwestern British Columbia; unpublished Ph.D. thesis, Lehigh University.

Woodsworth, G.J.
1977: Geology, Pemberton (92J) map-area; Geological Survey of Canada, Open File 482.

The ca.162 Ma Galena Bay stock and its relationship to the Columbia River fault zone, southeast British Columbia

R. R. Parrish[1] and R.L. Armstrong[2]

Parrish, R.R., and Armstrong, R.L., The ca.162 Ma Galena Bay stock and its relationship to the Columbia River fault zone, southeast British Columbia; in Radiogenic Age and Isotopic Studies: Report 1, Geological Survey of Canada, Paper 87-2, p. 25-32, 1987.

Abstract

The Galena Bay stock is a biotite-muscovite medium grained quartz monzonite which straddles Upper Arrow Lake (Columbia River), southeast British Columbia. Results of detailed geological mapping show that it does not intrude the high grade Monashee (Shuswap) complex as was previously assumed. Instead, it post-tectonically intrudes rocks of the Kootenay Arc and is faulted in places by, and lies in, the hanging-wall of the Eocene Columbia River fault, an east-dipping extension fault bounding the Monashee complex on the east. U-Pb geochronology on zircon and monazite from the stock demonstrate that its age is 161.6 ± 0.5 Ma, slightly older than a previous, less precise 156.7 ± 2.4 Ma Rb-Sr whole-rock isochron. Age and field relationships clearly show that the stock does not constrain the age of deformation, metamorphism, or development of mylonite within the Monashee complex.

Résumé

Le massif intrusif de Galena Bay est constitué d'une monzonite quartzifère à biotite et muscovite, à grains moyens; il chevauche la zone d'Upper Arrow Lake (fleuve Columbia), dans le sud-est de la Colombie Britannique. Les travaux de cartographie géologique détaillée, établissent qu'il ne pénètre pas dans le complexe fortement métamorphisé de Monashee (Shuswap), comme il a été supposé antérieurement. Au contraire, il a pénétré, au cours d'une phase post-tectonique, les roches de l'arc de Kootenay, et est traversé par endroits par la faille d'âge éocène de Columbia River; il s'agit d'une faille d'extension à pendage est et bordant le complexe de Monashee à l'est. La datation radiométrique par la méthode U-Pb, effectuée sur le zircon et la monazite provenant du massif, montre que l'âge de ce dernier est de 161,6 ± 0,5 Ma, soit un peu plus vieux que l'ancienne date, moins précise, de 156,7 ± 2,4 Ma, déterminée d'après l'isochrone Rb-Sr de la roche entière. Le rapport qui existe entre l'âge établi et les données observées sur le terrain montrent clairement que le présent massif n'a eu aucune influence sur l'époque de la déformation, du métamorphisme, ou de l'évolution de la mylonite au sein du complexe de Monashee.

INTRODUCTION

The Middle Jurassic Galena Bay stock is a biotite-muscovite medium grained quartz monzonite located on the shores of Upper Arrow Lake (Columbia River), 50 km south of Revelstoke, southeast British Columbia (Fig. 1). Although the stock has been widely cited as being intrusive into highly deformed, metamorphosed, and in part mylonitized gneiss of the Monashee complex to the west (Read and Brown, 1981, Brown and Read 1983; Okulitch, 1984, 1985; Brown et al., 1986; Journeay and Brown, 1986; Brown and Journeay, 1987), its actual geological relationship has neither been presented nor described in detail. The widespread occurrence of ca. 50 Ma K-Ar mica dates restricted to the Shuswap complex to the west and the documentation of the importance of Eocene normal faults in the southern Omineca belt has cast a certain amount of suspicion on this inferred relationship (Fig. 1; Price 1981a; Parrish et al., 1985; Parrish and Carr, 1986; Carr et al., 1987). This paper provides a detailed geological map of the Galena Bay area which encompasses both the stock and the Columbia River fault, and presents the often cited Rb-Sr geochronology of the stock with additional and more precise U-Pb zircon and monazite data.

GEOLOGICAL ELEMENTS OF GALENA BAY AREA

The geology in the vicinity of the stock consists of two tectonostratigraphic domains separated by the east-dipping Columbia River fault (Fig. 2).

[1] Geological Survey of Canada, 601 Booth Street, Ottawa, Ontario K1A 0E8
[2] Department of Geological Sciences, University of British Columbia, 6339 Stores Road, Vancouver, British Columbia V6T 2B4

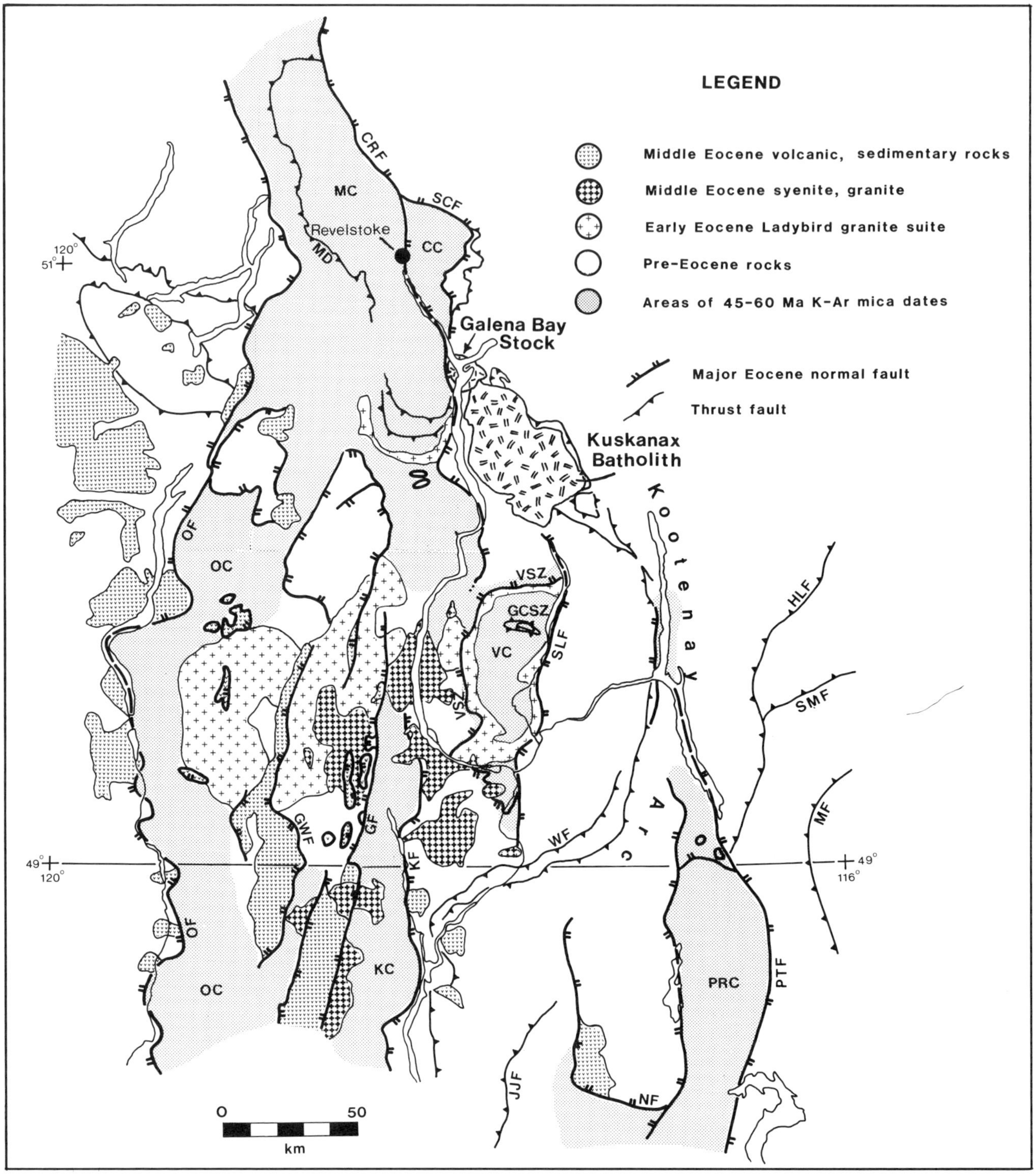

Figure 1. Tectonic map of southern Omineca belt, compiled from Parrish et al. (1985), Parrish and Carr (1986) and Carr et al. (1987), and Journeay and Brown (1986). Features shown with patterns are the Early Eocene Ladybird granite suite, Middle Eocene syenite and granite, Middle Eocene volcanic and sedimentary rocks, and areas of older rocks characterized by K-Ar mica dates of 45-60 Ma. In the hanging walls of the major Eocene normal faults K-Ar dates are generally older than 100 Ma. Major thrust faults within Monashee and Valhalla complexes are the Monashee décollement and Gwillim Creek shear zones, respectively. Metamorphic complexes and faults are designated as follows: MC, Monashee complex; VC, Valhalla complex; CC, Clachnacudainn complex; OC, Okanagan complex; KC, Kettle-Grand Forks complex; PRC, Priest River complex; OF, Okanagan fault; MD, Monashee décollement; CRF, Columbia River fault; SCF, Standfast Creek fault; SLF, Slocan Lake fault; VSZ, Valkyr shear zone; GCSZ, Gwillim Creek shear zones; GWF, Greenwood fault; GF, Granby fault; KF, Kettle fault; NF, Newport fault; PTF, Purcell Trench fault; WF, Waneta fault; HLF, Hall Lake fault; SMF, St. Mary Fault.

Footwall rocks

In the footwall of the fault are high grade metamorphic rocks, pegmatite and orthogneiss of the Monashee complex (Read and Wheeler, 1976; Read and Brown, 1981), exposing both Precambrian basement and metasedimentary paragneiss which in places unconformably overlies it. Adjacent to the Galena Bay area, the rocks consist predominantly of highly deformed, in part mylonitized, paragneiss and pegmatite. Pegmatite preferentially crops out on hillsides, and it probably forms a significant proportion of the local lithology in the footwall domain. Structures preserved in these rocks include a prominent east-plunging mineral and stretching lineation contained within the gently to moderately (15-40°) east-dipping foliation. This foliation is consistent in general orientation within the mapped area (Fig. 2) and is probably subparallel with the moderately east-dipping Columbia River fault (Read and Brown, 1981). Superposed folds on all scales are common in the complex (Duncan, 1984; Read and Brown, 1981; Read and Klepacki, 1981; Okulitch, 1984), but are generally overprinted by mylonitic fabrics in the map area. In close proximity to the Columbia River fault, footwall rocks are mylonites which contain kinematic indicators such as S-C fabrics (Berthé et al., 1979; Lister and Snoke, 1984) indicative of transport of hangingwall and structurally higher rocks to the east. These fabrics adjacent to the fault have been interpreted as of Eocene age and related to normal faulting along the Columbia River fault.

Hangingwall rocks

Rocks which constitute the hangingwall of the Columbia River fault in this region are part of the Kootenay Arc, a major belt of Precambrian to early Mesozoic stratified rocks intruded by mostly Middle Jurassic and middle Cretaceous plutons (Fig. 1). Within the map area the regional metamorphic grade is greenschist facies with higher grade aureoles developed in the immediate vicinity of stocks of the Kuskanax batholith. Locally (Fig. 2) the main rock units (Read and Wheeler, 1976) are the Kuskanax batholith (Parrish and Wheeler, 1983) and its outliers, the Galena Bay stock, and metavolcanic and metasedimentary rocks of probable middle to upper Paleozoic and/or early Mesozoic age.

The outliers of the Kuskanax batholith, not previously reported on the west side of Upper Arrow Lake near the Galena Bay stock (c.f. Read and Wheeler, 1976), are highly foliated and fairly fine grained leucocratic alkalic quartz monzonitic sill-like bodies. They are characterized by very flattened biotite-rich clots which have a certain similarity to a highly strained conglomeratic rock; these clots are actually characteristic of the marginal phases of the Kuskanax batholith in general (Read, 1973; Parrish and Wheeler, 1983), and appear different in the map area only because of the more highly deformed and foliated nature of the local, relatively small intrusions. The higher grade (amphibolite facies) nature of the Kuskanax wall rocks is characteristic of its contact relations in general (Klepacki, 1986; Read, 1973). The batholith and the outlying intrusions in the Galena Bay area appear to have been emplaced syn- or pre-kinematically relative to folding and deformation of Middle Jurassic age within the Kootenay Arc (hangingwall) domain.

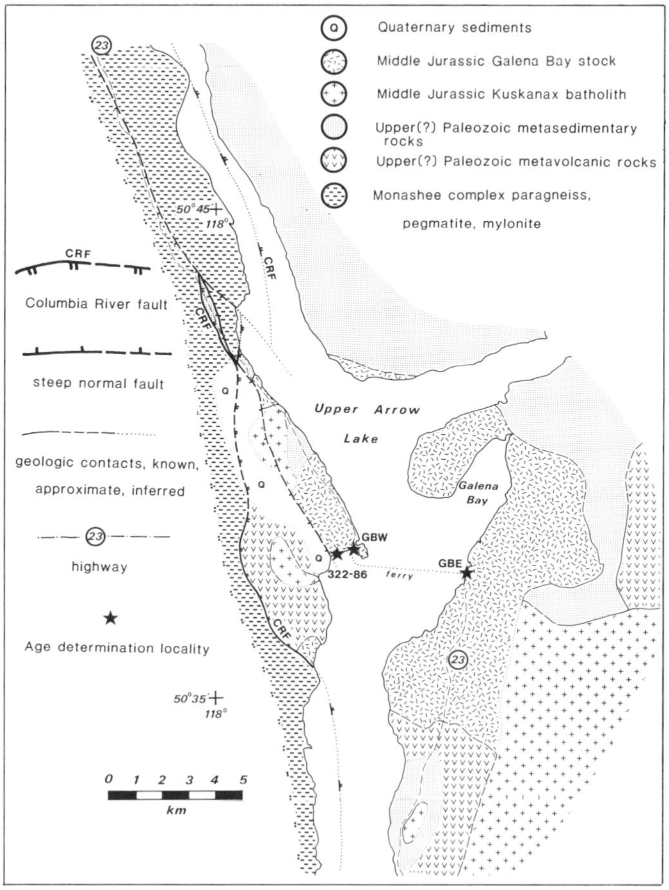

Figure 2. Geological map of the Galena Bay area showing the extent of the Galena Bay stock, Kuskanax batholith, Monashee complex, and Columbia River fault. Most traces of geological contacts on the eastern side of Upper Arrow Lake are from Read and Wheeler (1976). GBE and GBW refer to sample localities for Rb-Sr age determinations.

Metamorphic rocks comprise moderately foliated but relatively uniform actinolitic amphibolite and subordinate metasedimentary rocks. Phyllitic foliation and cleavage, in places refolded and/or cataclastically deformed, are commonly observed in outcrop. Original bedding is rarely visible in metasedimentary rocks. Near and within the Columbia River fault zone cataclastic deformation and chloritic alteration is widespread. Read and Wheeler (1976) mapped metavolcanic rocks as the Permo-Triassic Kaslo formation, but recently Klepacki (1986) and Klepacki and Wheeler (1985) have assigned some metavolcanic rocks previously thought to be Kaslo formation to the upper Paleozoic McHardy assemblage of the Milford Group which also contains abundant metasediments. It remains uncertain, however, which group the local rocks should be assigned to, and this ambiguity is implicit in Figure 2.

The Galena Bay stock is a massive, but locally fractured body of leucocratic biotite-muscovite quartz monzonite. Its relationship to country rocks and Kuskanax intrusions is not exposed on the west shore of Upper Arrow Lake. However, its younger age and massive nature leave little doubt that it crosscuts the foliated Middle Jurassic Kuskanax stocks and deformed country rocks in the hangingwall. The Galena Bay stock is clearly a post-tectonic intrusion relative to Middle Jurassic metamorphism and deformation widely developed in the Kootenay Arc.

Columbia River fault

The Columbia River fault (Read and Wheeler, 1976; Read, 1979; Read and Brown, 1981; Brown and Read, 1983) is an east-dipping normal fault generally confined to the Columbia River valley from north of Revelstoke to south of Nakusp (Fig. 1). Although it generally lies beneath the Columbia River valley, it is exposed near the Revelstoke damsite at Revelstoke, and on the west shore of Upper Arrow Lake west of the Kuskanax batholith, and in the study area.

Its characteristic brittle, normal faulting was considered by Read and Brown (1981) and Lane (1984) to coincide with a shear zone termed the Monashee décollement (Read and Brown, 1981; Brown and Read, 1983) of presumed Jurassic age. The coincidence of mylonitic rocks and superposed brittle faulting was viewed as a structurally controlled but temporally unrelated superposition. Recent data (Parrish and Carr, 1986) suggest that mylonite development subjacent to the fault zone and the superposed brittle faulting are related both in time and process, and are caused by crustal extension of large magnitude.

Within the area studied, the Columbia River fault is a highly faulted and sheared zone with considerable chloritic alteration in both hangingwall and footwall separating pegmatite-rich gneiss (footwall) from metamorphic and plutonic rocks of the hangingwall. Highly sheared fault rocks are present in highway outcrops 10km northwest of the ferry landing and are typical of the fault zone (Fig. 2). The distribution of rocks on the west side of the lake is complicated by later, high angle, probable normal faulting of smaller magnitude which has offset the trace of the Columbia River fault, causing in places footwall rocks to be almost surrounded by rocks of the much lower grade hangingwall domain (Fig. 2, northwest part of map). These later faults have controlled the location of creeks and valleys on the west shore of the lake and account for the prominent ridge of gneiss which separates Highway 23 from the western shore of the lake.

AGE OF THE GALENA BAY STOCK

Early attempts to date the stock were confined to K-Ar dates on mica from exposures on the eastern shore. These dates were cited in Read and Brown (1981), but are generally considered unreliable, and have not been verified. The Middle Jurassic age of the body was established on the basis of Rb-Sr whole rock and muscovite dating some years ago which provided a ca. 157 Ma age (cited in Read and Brown, 1981; Brown and Read, 1983; Okulitch, 1984; Brown et al., 1986; Journeay and Brown, 1986). These Rb-Sr data, which include samples of quartz monzonite, pegmatite and aplite from both sides of Upper Arrow Lake at each ferry landing, are presented in Table 1 and Figure 3, and are based on analytical procedures summarized as a footnote to Table 1. Eight whole-rocks yield an isochron age (MSWD = 1.2) of 156.7 ± 2.4 Ma with $^{87}Sr/^{86}Sr$ initial of 0.7056 ± 0.0001. A muscovite-whole-rock pair from the east shore gives an age of 150 ± 3 Ma.

U-Pb results

A sample of typical biotite-muscovite quartz monzonite, (sample #322-86) was collected in 1986 from a roadcut on the main logging road which follows the west shore of the lake, about 800m west of the western ferry landing (Fig. 2). The sample is moderately fractured with some sericitic and chloritic alteration on fracture surfaces, but it is otherwise fresh. Zircon and monazite were dated by U-Pb methods using analytical procedures summarized by Parrish et al. (1987), and the data are listed in Table 2.

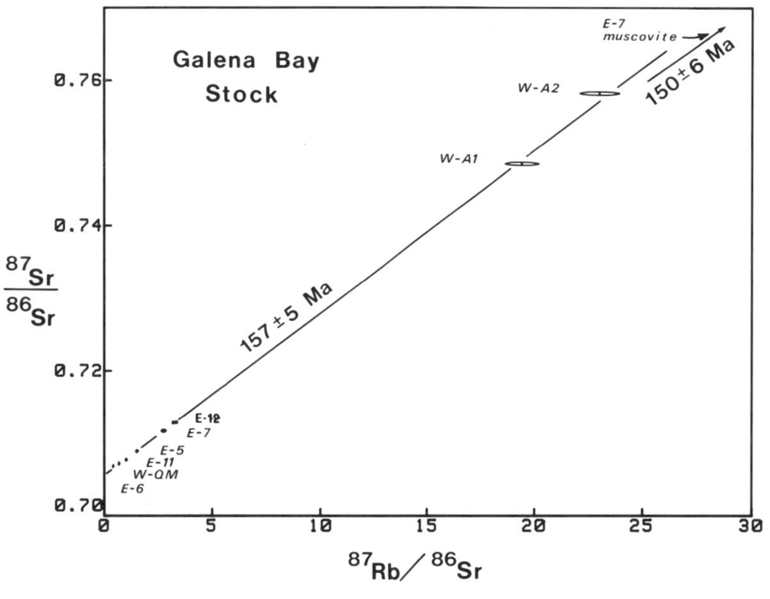

Figure 3. Rb-Sr evolution diagram for whole-rocks and muscovite from the Galena Bay stock. See Table 1 for analytical data. Errors plotted reflect two standard errors of the mean.

Table 1. Rb-Sr data, Galena Bay stock

Sample, description	ppm Sr	ppm Rb	$^{87}Rb/^{86}Sr$	$^{87}Sr/^{86}Sr$
Galena Bay stock, west shore Upper Arrow Lake, Shelter Bay, near ferry landing, 50°38′06″N, 117°55′24″W				
GB-W-QM, medium grained biotite-muscovite quartz monzonite	656	154	0.679	0.70715
GB-W-A1, aplite dyke	41.7	278	19.4	0.7485
GB-W-A2, aplite dyke	32.6	258	23.0	0.7582
Galena Bay stock, east shore Upper Arrow Lake, ferry landing, picnic area, 50°36′35″N, 117°52′44″W				
GB-E-6, medium grained biotite-muscovite quartz monzonite	726	102	0.407	0.7068
GB-E-11, muscovite-bearing leucocratic dyke	423	147	1.008	0.7077
GB-E-5, medium-to coarse-grained garnet-mica quartz monzonite	308	160	1.50	0.7089
GB-E-12, muscovite-garnet-feldspar-quartz pegmatite	182	202	3.21	0.7126
GB-E-7, leucocratic coarse grained granite pegmatite	217	206	2.74	0.7117
GB-E-7M, muscovite from pegmatite	20.4	889	129.0	0.9812

Calculated dates with two sigma erros:

Whole-rocks, n = 8, MSWD−1.2, age = 156.7 ± 4.8 Ma $^{@87}Sr/^{86}Sr_1$ = 0.7056 ± 2
GB-E muscovite-whole-rock age = 150 ± 6 Ma $^{@87}Sr/^{86}Sr_1$ = 0.7058 ±2

Analytical procedures: Rb and Sr concentrations were determined by replicate analysis of pressed powder pellets using X-ray fluorescence. U. S. Geological Survey rock standards were used for calibration; mass absorption coefficients were obtained from Mo K-alpha Compton scattering measurements. Rb/Sr ratios have a precision of 2% (1 sigma) and concentrations a precision of 5% (1 sigma). Sr isotopic composition was measured on unspiked samples prepared using standard ion exchange techniques. A modified U. S. National Bureau of Standards design, 60° sector, 30 cm radius mass spectrometer and a VG Isomass 54R mass spectrometer, both with digitized and automated data acquisition were used. A value of 0.1194 for $^{86}Sr/^{88}Sr$ was used for normalization and ratios were adjusted so that the NBS SRM 987 and E and A standards give $^{87}Sr/^{86}Sr$ values of 0.71020 ± 2 and 0.70800 ± 2, respectively. The precision of a single $^{87}Sr/^{86}Sr$ ratio is better than 0.00013 (one sigma). Rb-Sr dates are based on a decay constant of 1.42×10^{-11}/yr. The regressions are calculated according to York (1969).

Table 2. U-Pb zircon and monazite data, Galena Bay stock

Fraction, size[1]	wt, (mg)	U, ppm	Pb^2, ppm	$^{206}Pb^3/^{204}Pb$	$Pb_c{}^4$, (pg)	$^{208}Pb^2$ (%)	$^{206}Pb/^{238}U$ ±1SEM%[5]	$^{207}Pb/^{235}U$ ±1SEM%[5]	$^{207}Pb/^{206}Pb$ ±1SEM%[5]	$^{207}Pb/^{206}Pb$ age, error (Ma)[6]
PCA-86-322, biotite-muscovite quartz monzonite, massive; west shore Upper Arrow Lake; 50°37'58″N, 117°56'05″W.										
Zircon:										
ab, clr, ndles	0.026	992.4	24.18	582	73	6.1	0.02537 (.10)	0.1726 (.29)	0.04933 (.23)	163.6 (5.3)
ab, clr, frag	0.076	1524	36.69	8160	23	4.9	0.02538 (.06)	0.1731 (.10)	0.04945 (.05)	169.0 (1.0)
Monazite:										
+149, clr	0.3555	5613	864.0	7690	419	85	0.02565 (.08)	0.1731 (.10)	0.04893 (.05)	144.6 (1.0)
+149, clr	0.556	5647	816.2	13610	370	84	0.02546 (.26)	0.1719 (.27)	0.04897 (.04)	146.4 (1.0)

Notes: [1]size (i.e. +105) refer to length aspect of zircons in microns; ab = abraded, clr = extremely clear, ndles = needles, frag = fragments of needles; [2]radiogenic Pb; [3]measured ratio, corrected for spike and fractionation; [4]total common Pb in analysis corrected for fractionation and spike; [5]corrected for blank Pb and U, common Pb, errors quoted are one sigma in percent; [6]corrected for blank and common Pb, errors are one sigma in Ma; decay constants used are those of Steiger and Jäger (1977); for analytical details see Parrish et al. (1987).

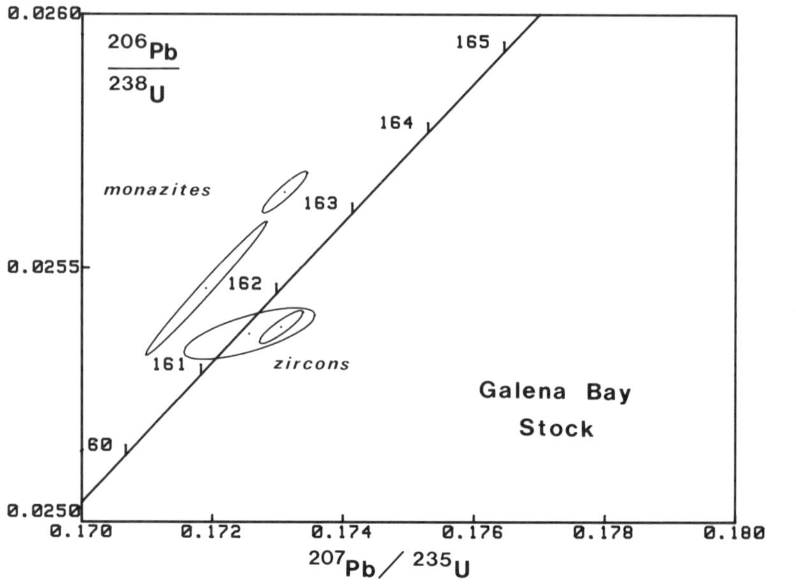

Figure 4. U-Pb concordia diagram of zircon and monazite data from the Galena Bay stock. Error ellipses reflect two standard errors of the mean. The age estimate of the stock is considered to be 161.6 ± 0.5 Ma. See text for discussion.

Zircons by and large contain visibly distinct xenocrystic cores, but there were a number of crystals which were absolutely free of both cores and other non-zircon visible inclusions. Zircons chosen for analysis were very clear, elongate (aspect ratios of 5:1 to 10:1), free of any visible material and sharply terminated. One fraction contained whole crystals and the other had fragments of similar type. Both were strongly abraded by the method of Krogh (1982) to ensure that surface-correlated Pb loss would be removed if it was present. Very clear, euhedral monazite fractions were chosen for analysis, but were not abraded.

The U-Pb techniques utilize a mixed ^{205}Pb-^{233}U-^{235}U tracer (Parrish and Krogh, 1987), multicollector mass spectrometry for U and Pb (Roddick et al., 1987), and were characterized by Pb and U analytical blanks of 20 and 0-2pg, respectively. When plotted on the concordia diagram of Figure 4, one of the zircons is concordant at 161.5 Ma, while the other is very slightly discordant as a result of inheritance. Monazites plot above concordia, with $^{207}Pb/^{235}U$ ages of 161.1 to 162.1 Ma. Our interpretation of the reversely discordant monazite analyses follows Schärer (1984) who showed that reversely discordant monazites can be interpreted as the result of incorporation of ^{230}Th upon crystallization that subsequently decays to ^{206}Pb and becomes unsupported because that part of the ^{206}Pb was not produced by in situ decay of ^{238}U. Because of this problem, only the $^{207}Pb/^{235}U$ ages are interpretable in terms of age or thermal significance.

The best estimate of age for the Galena Bay stock is considered to be the average and two standard errors of the concordant zircon analysis (161.5 Ma) and the $^{207}Pb/^{235}U$ ages of the two monazite fractions, that is 161.6 ± 0.5 Ma.

DISCUSSION

The contrast in field relationships between the pre- or synkinematic 173 ± 5 Ma Kuskanax batholith (Parrish and Wheeler, 1983) and the post-tectonic 161.6 Ma Galena Bay stock brackets the age of deformation, metamorphism, and plutonism within the Kootenay Arc of southeast British Columbia. The Galena Bay stock cooled relatively rapidly because of the close agreement between monazite and zircon ages and the slightly younger ca.150 Ma Rb-Sr muscovite date.

This chronology is in distinct contrast to the cooling history of the metamorphic complexes of southeastern British Columbia which are bounded by Eocene normal faults (Fig. 1) which cooled much later (see Price, 1981b; Ewing, 1981; Mathews, 1981; Parrish and Carr, 1986; Carr et al., 1987). The explanation envisioned for this difference is that the metamorphic core complexes (Monashee, Valhalla, Okanagan, Kettle-Grand Forks, Priest River; Fig. 1) were tectonically exhumed during the Eocene as footwall terranes beneath important normal faults during a period of crustal extension (Price, 1981b, 1985; Ewing, 1981; Parrish et al., 1984; Parrish and Carr, 1986). The field relations of the Columbia River fault adjacent to the Galena Bay stock as described in this paper are consistent with this model, and they refute the hypothesis of Read (1979) and Read and Brown (1981) that the Galena Bay stock intrudes metamorphic rocks of the Monashee complex. The field and chronological data clearly show that the stock does not constrain the age of deformation, metamorphism, or development of mylonite within the Monashee complex.

ACKNOWLEDGMENTS

We thank S. D. Carr and D. L. Parkinson for their contributions to mapping during 1985 and 1986, respectively. We thank D. W. Klepacki and S. D. Carr for discussion of the regional geology and S. D. Carr for a review which improved the manuscript. I. J. Duncan collected samples used for Rb-Sr analysis. The Rb-Sr data were generated at the University of British Columbia which is supported by NSERC grant A8841. The staff of the Geochronology Section of the Geological Survey of Canada is thanked for assistance in generating U-Pb data.

REFERENCES

Berthé, D., Choudroune, P., and Jegouzo, P.
1979: Orthogneiss, mylonite and non-coaxial deformation of granites: The example of the South American shear zone; Journal of Structural Geology, v. 1, p. 31-42.

Brown, R.L. and Read, P.B.
1983: Shuswap terrane of British Columbia: A Mesozoic ''core complex''; Geology, v. 11, p. 164-168.

Brown, R.L. and Journeay, J.M.
1987: Tectonic denudation of the Shuswap metamorphic terrane of southeastern British Columbia; Geology, v. 15, p. 142-146.

Brown, R.L., Journeay, J.M., Lane, L.S., Murphy, D.C., and Rees, C.J.
1986: Obduction, backfolding and piggyback thrusting in the metamorphic hinterland of the southeastern Canadian Cordillera; Journal of Structural Geology, v. 8, p. 255-268.

Carr, S.D., Parrish, R. R., and Brown, R.L.
1987: Eocene structural development of the Valhalla complex, southeastern British Columbia; Tectonics, v. 6, p. 175-196.

Duncan, I.J.
1984: Structural evolution of the Thor-Odin gneiss dome; Tectonophysics, v. 101, p. 87-130.

Ewing, T.E.
1981: Paleogene tectonic evolution of the Pacific Northwest; Journal of Geology, v. 88, p. 619-638.

Journeay J.M., and Brown, R.L.
1986: Major tectonic boundaries of the Omineca belt in southern British Columbia: a progress report; *in* Current Research, Part A, Geological Survey of Canada, Paper 86-1A, p. 81-88.

Klepacki, D.W.
1986: Stratigraphy and structural geology of the Goat Range area, southeastern British Columbia; unpublished Ph.D. thesis, Massachusetts Institute of Technology, Cambridge, Massachusettes, 268 p.

Klepacki, D.W. and Wheeler, J.O.
1985: Stratigraphic and structural relations of the Milford, Kaslo and Slocan groups, Goat Range, Lardeau and Nelson map areas, British Columbia; *in* Current Research Part A, Geological Survey of Canada, Paper 85-1A, p. 277-286.

Krogh, T. E.
1982: Improved accuracy of U-Pb ages by the creation of more concordant systems using an air abrasion technique; Geochimica et Cosmochimica Acta, v. 46, p. 637-649.

Lane, L.S.
1984: Brittle deformation in the Columbia River fault zone near Revelstoke, southeastern British Columbia; Canadian Journal of Earth Sciences, v. 1, p. 584-598.

Lister, G.S. and Snoke, A.W.
1984: S — C Mylonites; Journal of Structural Geology, v. 6, p. 617-638.

Mathews, W.H.
1981: Early Cenozoic resetting of potassium-argon dates and geothermal history of north Okanagan area, British Columbia; Canadian Journal of Earth Sciences, v. 18, p. 1310-1319.

Okulitch, A.V.
1984: The role of the Shuswap metamorphic complex in Cordilleran tectonism: a review; Canadian Journal of Earth Sciences, v. 21, p. 1171-1193.
1985: Paleozoic plutonism in southeastern British Columbia; Canadian Journal of Earth Sciences, v. 22, p. 1409-1424.

Parrish, R. R. and Wheeler, J. O.
1983: U - Pb zircon age of the Kuskanax batholith, southeastern British Columbia; Canadian Journal of Earth Sciences, v. 20, p. 1751-1756.

Parrish, R., Carr, S.D. and Parkinson, D.
1985: Metamorphic complexes and extensional tectonics, southern Shuswap complex, southeastern British Columbia; *in* Field Guides to Geology and Mineral deposits in the Southern Canadian Cordillera, Geological Society of America, Cordilleran Section Meeting, p. 12.1-12.15.

Parrish, R. and Carr, S.D.
1986: Extensional tectonics of southeastern British Columbia: new data and interpretations; Geological Association of Canada, Program with Abstracts, v. 11, p. 112.

Parrish, R. R. and Krogh, T. E.
1987: Synthesis and purification of ^{205}Pb for U-Pb geochronology; Isotope Geoscience, v. 66, p. 103-110.

Parrish, R. R., Roddick, J. C., Loveridge, W. D. and Sullivan, R. W.
1987: Uranium-lead analytical techniques at the geochronology laboratory, Geological Survey of Canada; *in* Radiogenic Age and Isotopic Studies Geological Survey of Canada, Paper 87-2.

Price, R.A.
1981a: The Cordilleran forelnd thrust and fold belt in the southern Canadian Rocky Mountains; *in* Thrust and Nappe Tectonics, ed. K.R. McClay and N.J. Price; Geological Society of London, Special Publication, 9, p. 427-488.
1981b: Eocene stretching and necking of the crust and tectonic unroofing of the Cordilleran metamorphic infastructure, southeastern British Columbia and adjacent Washington and Idaho; Geological Association of Canada, Program and Abstracts, p. A47.

Price, R.A.
1985: Metamorphic core complexes of the first and second kind in the Cordillera of southern Canada and northern U.S.A.; Geological Society of America, Program with Abstracts, v. 17, p. 401.

Read, P. B.
1973: Petrology and structure of Poplar Creek map-area, British Columbia; Geological Survey of Canada, Bulletin 193.
1979: Relationship between the Shuswap metamorphic complex and Kootenay Arc, Vernon east-half, southern British Columbia; *in* Current Research, Part A, Geological Survey of Canada, Paper 79-1A, p. 37-40.

Read, P. B. and Brown, R. L.
1981: Columbia River fault zone: southeastern margin of the Shuswap and Monashee complexes, southern British Columbia; Canadian Journal of Earth Sciences, v. 18, p. 1127-1145.

Read, P. B. and Klepacki, D.
1981: Stratigraphy and structure: northern half of Thor-Odin nappe, Vernon east-half map-area, southern British Columbia; *in* Current Research, Part A, Geological Survey of Canada, Paper 81-1A, p. 169-173.

Read, P.B. and Wheeler, J.O.
1976: Geology of Lardeau west-half map area; Geological Survey of Canada, Open File 288.

Roddick, J. C., Loveridge, W. D. and Parrish, R. R.
1987: Precise U/Pb dating of zircon at the subnanogram Pb level; Isotope Geoscience, v. 66, p. 111-121.

Schärer, U.
1984: The effect of initial ^{230}Th equilibrium on young U-Pb ages: the Makalu case, Himalaya; Earth and Planetary Science Letters, v. 67, p. 191-204.

Steiger, R. H. and Jäger, E.
1977: Subcommission on geochronology: convention on the use of decay constants in geo- and cosmochronology; Earth and Planetary Science Letters, v. 36, p. 359-362.

York, D.
1969: Least squares fitting of a straight line with correlated errors; Earth and Planetary Science Letters, v. 5, p. 320-324.

Age of the Ice River complex, southeastern British Columbia

R. R. Parrish[1], S. Heinrich[2] and D. Archibald[2]

Parrish, R.R., Heinrich, S., and Archibald, D., Age of the Ice River complex, southeastern British Columbia; in Radiogenic Age and Isotopic Studies: Report 1, Geological Survey of Canada, Paper 87-2, p. 33-37, 1987.

Résumé

Le complexe d'Ice River est un complexe intrusif de nature alcaline, situé dans la partie ouest du chaînon principal (Main Ranges) des Rocheuses, en Colombie-Britannique. L'âge exact du complexe demeure incertain en raison du manque de concordance entre les dates obtenues par la méthode classique de datation radiométrique au K-Ar sur des minéraux magmatiques provenant de roches intrusives.

Dans la présente étude, les datations sur du zircon et du sphène par la méthode U-Pb et l'analyse des minéraux réalisée à l'aide de la méthode Rb-Sr, effectuées, outre un spectre d'âge détaillé obtenu par la méthode $^{40}Ar/^{39}Ar$ sur la hornblende (tous ces minéraux proviennent d'une syénite néphélinique), indiquent que tous les systèmes isotopiques sont faussés par la perte de Pb dans le zircon et dans le sphène et par un excès d'argon dans la hornblende. Malgré ces complications, les données indiquent que l'âge du complexe varie entre 356 et 372 Ma, mais que l'âge le plus probable est de 368 ± 4 Ma.

Abstract

The Ice River complex is an alkaline intrusive complex in the western Main Ranges of the Rocky Mountains, British Columbia. The exact age of the complex has been uncertain because of a lack of agreement of conventional K-Ar dates on magmatic minerals from intrusive rocks.

In this study U-Pb zircon and sphene age determinations, Rb-Sr analysis of minerals, and a detailed $^{40}Ar/^{39}Ar$ age spectrum on hornblende, all from nepheline syenite, indicate that all isotopic systems are disturbed with Pb loss in zircon and sphene and excess argon in hornblende. Despite these complications, the data indicate that the age of the complex falls between 356 and 372 Ma, with an estimate of 368 ± 4 Ma as most probable.

INTRODUCTION

The Ice River complex (Allan, 1914) is an alkaline igneous complex in the western Rocky Mountains of Yoho and Kootenay National Parks of southeastern British Columbia.

Although the complex has been claimed to represent Precambrian basement (Gussow, 1977), other workers (*see* Currie, 1975 for other references) clearly agree that the complex intrudes country rocks of Cambrian and Ordovician age. Currie (1975) described in detail the igneous petrology of the complex and its contact relationships, and these aspects will not be dealt with further.

There has been a lack of agreement between previous attempts to date the complex; K-Ar dates have varied from as young as about 245 Ma (Currie, 1975) to as old as 488 Ma (Hunt and Roddick, 1987). Excess argon, thermal resetting, and fluid alteration all may have contributed to the overall discordance of dates.

This study was undertaken to solve the problem of the age of the complex utilizing U-Pb methods on zircon and sphene and the $^{40}Ar/^{39}Ar$ method on hornblende. Only these methods have the potential for resolving disturbances and excess argon (Faure, 1986).

ANALYTICAL METHODS

The sample chosen for study was collected by W. C. Gussow from the nepheline syenite phase of the complex. Its location is 51°08'N, 116°26'W. The constituent phases include K-feldspar, nepheline, hornblende, sphene, apatite, and very minor zircon. Pure mineral separates of hornblende, sphene and zircon were made along with somewhat impure K-feldspar and nepheline concentrates. After analysis of sphene, K-feldspar, and nepheline for Rb and Sr contents and Sr isotopic composition, purer separates were prepared and were given a brief ultrasonic leach in weak HCl and reanalyzed (labeled II in Table 2). Analytical methods for U-Pb

[1] Geological Survey of Canada, 601 Booth Street, Ottawa, Ontario K1A 0E8
[2] Department of Geological Sciences, Queen's University, Kingston, Ontario

and Rb-Sr dating are outlined by Parrish et al. (1987, U-Pb) and below (Rb-Sr). ^{40}Ar-^{39}Ar methods are similar to those published in McBride et al. (1987).

Rb-Sr ANALYTICAL METHODS

Mineral separates were weighed and spiked with a mixed ^{85}Rb-^{84}Sr tracer prior to HF dissolution and cation exchange column HCl chemistry. Isotopic analyses for Rb and Sr were performed on a MAT 261 mass spectrometer equipped with Faraday cups for simultaneous multicollection of ion signals. Two standard errors of the mean in $^{87}Rb/^{86}Sr$ and $^{87}Sr/^{86}Sr$ are estimated as $\pm$ 1.0 % and 0.00010, respectively, unless otherwise noted.

RESULTS

U-Pb

Zircons recovered from the sample were very sparse (less than 3mg/10kg sample), cloudy, high in uranium (1800-4200 ppm) and of poor quality. They consisted of subhedral to anhedral cloudy brownish fragments; only three analyses were performed before virtually all of the acceptable crystals were expended. All three analyses are discordant and cluster in a small region opposite 350 Ma (Fig. 1). The $^{207}Pb/^{206}Pb$ dates range from 360 to 369 Ma (Table 1), and represent minimum ages in this case because there is no evidence of inheritance, and because the zircons are considered to be igneous on the basis of their morphology.

Abundant very pale sphene of good quality was analyzed to supplement the zircon data and assist in its interpretation. Although of low uranium content (10 ppm), the common Pb content was very low (0.2 ppm) and allowed good analyses with small common Pb corrections (Table 1). Two of the sphene analyses are slightly more concordant than zircons, and the sphenes have similar $^{207}Pb/^{206}Pb$ dates of 357 to 369 Ma, although the errors are relatively large. The minimum age for the complex is given by the most concordant sphene age of about 356 Ma; the most likely age, however, is close to the maximum $^{207}Pb/^{206}Pb$ ages for sphene and zircon which are about 368-370 Ma.

Table 1. U-Pb zircon and sphene data, Ice River complex

Fraction, size[1]	wt, (mg)	U, ppm	Pb[2], ppm	$^{206}Pb/^{204}Pb$ [3]	Pb_c[4], (pg)	^{208}Pb[2] (%)	$^{206}Pb/^{238}U$ ±SEM %[5]	$^{207}Pb/^{235}U$ ±SEM %[5]	$^{207}Pb/^{206}Pb$ ±SEM %[5]	$^{207}Pb/^{206}Pb$ age, error (Ma)[6]
zircon:										
euhedral pink, clr	0.209	1815	103.7	2366	571	10.9	0.05619 (.25)	0.4163 (.30)	0.05373 (.10)	359.9 (2.2)
fragments pink, cdy	0.040	4232	226.3	2131	279	5.8	0.05559 (.25)	0.4124 (.30)	0.05381 (.10)	363.1 (2.2)
annhedral clr, pink	0.750	3367	186.1	3050	2787	8.6	0.05573 (.25)	0.4138 (.30)	0.05385 (.10)	364.9 (2.2)
sphene #1	3.541	10.05	1.337	217	635	61.3	0.05659 (.07)	0.4209 (.36)	0.05395 (.31)	368.8 (7.0)
sphene #2	6.255	11.38	1.450	155	1794	60.8	0.05483 (.12)	0.4058 (.44)	0.05367 (.37)	357.3 (8.4)
sphene #3	5.062	11.03	1.435	150	1494	60.9	0.05582 (.13)	0.4138 (.47)	0.05376 (.40)	361.2 (9.0)
sphene #4	2.613	10.67	1.428	107	1119	61.3	0.05683 (.20)	0.4204 (.83)	0.05366 (.71)	356.6 (16.1)

Notes: [1]sizes (i.e. + 105) refer to length aspect of zircons in microns, ab = abraded, cdy = cloudy, clr = clear; [2]radiogenic Pb; [3]measured ratio, corrected for spike and fractionation; [4]total common Pb in analysis corrected for fractionation and spike; [5]corrected for blank Pb and U, common Pb, errors quoted are 1 standard error of the mean in per cent; [6]corrected for blank and common Pb, errors are 1 standard error of the mean in Ma; the Pb blank for sphene analyses 1, 2 and 3 was 0.66 ng.

Table 2. Rb-Sr data, Ice River complex nepheline syenite

Sample, description	ppm Sr	ppm Rb	$^{87}Rb/^{86}Sr$	$^{87}Sr/^{86}Sr$ [1]
whole-rock	1362	214.1	0.455	0.70586 (10)
K-feldspar I	1480	238.1	0.465	0.70565 (10)
K-feldspar II	1743	280.7	0.465	0.70576 (10)
nepheline I	2227	184.3	0.239	0.70555 (10)
nepheline II	3058	178.6	0.169	0.70540 (20)
sphene I	396.6	4.495	0.0327	0.70417 (10)
sphene II	336.1	36.01	0.310	0.70540 (30)
hornblende	383.6	30.85	0.233	0.70465 (10)

Note: [1]numbers in parentheses are 2 standard errors in the measurement of $^{87}Sr/^{86}Sr$.

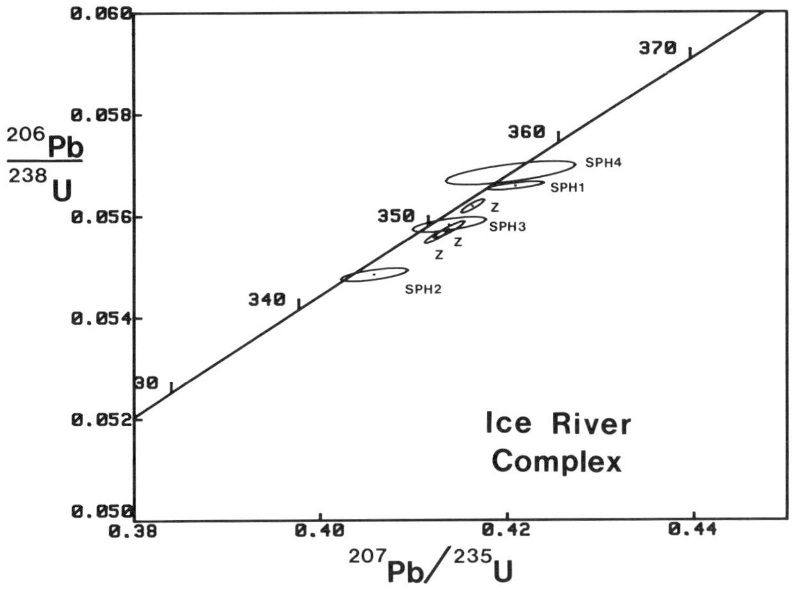

Figure 1. U-Pb concordia diagram with analyses of zircon and sphene plotted with their respective two standard errors of the mean error ellipses. See Table 1 for analytical details.

Rb-Sr

Rb-Sr analyses on whole-rock, sphene, feldspar, nepheline, and hornblende show a great deal of scatter and implied disturbance (Table 2, Fig. 2). A 370 Ma reference line passing through an initial $^{87}Sr/^{86}Sr$ of 0.704 is shown. Obviously there is no possibility of gaining any estimate of the age of the rock using this method. It is presented mainly to illustrate that the mineral Rb-Sr systematics are highly disturbed. The initial $^{87}Sr/^{86}Sr$ for the sample is estimated to be 0.7035 to 0.704.

K-Ar and $^{40}Ar/^{39}Ar$

The hornblende separate from the nepheline syenite is of good quality and yielded a conventional age of 488 Ma (Hunt and Roddick, 1987). This is older than but reasonably similar to, a sample of hornblende dated earlier from the same lithology and locality at 420 Ma (Stevens et al., 1982). The older age suggests considerable excess argon relative to the U-Pb indications of a 360-370 Ma age. The $^{40}Ar/^{39}Ar$ age spectrum shown in Figure 3 and tabulated in Table 3 confirms this with initial gas ages in excess of 1 Ga. The integrated age for all of the gas, interpreted as being equivalent to a K-Ar age on this same aliquot of hornblende, is 444 Ma. The difference between this and the 488 K-Ar age of a different aliquot of this same separate is probably due to highly variable amounts of the extraneous Ar component in a nonuniform mineral separate. The hornblende exhibits a saddle shaped age spectrum which can be attribued to excess argon trapped in two sites of different retentivity. These two types of sites degassed in the furnace at different temperatures, one being in the 500-650°C range and the other above 950°C.

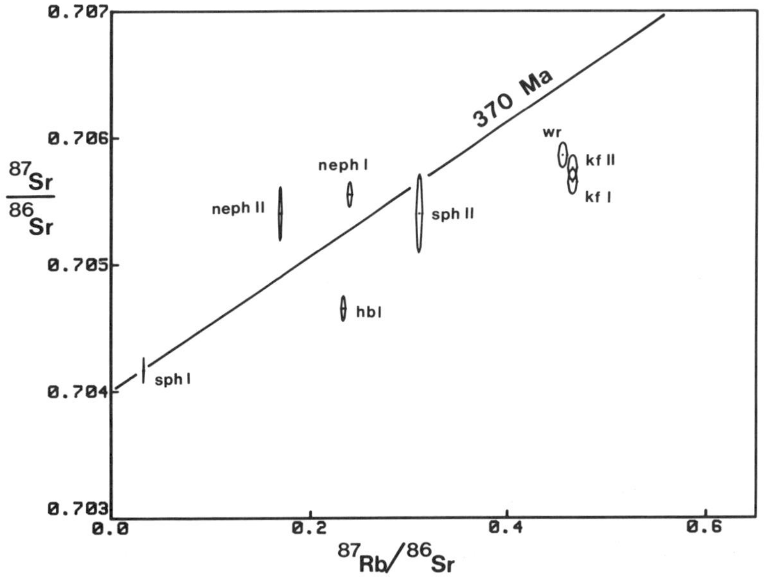

Figure 2. Rb-Sr evolution diagram for minerals from nepheline syenite of the Ice River complex. A 370 Ma reference line is shown. The large scatter in the data indicates a severe isotopic disturbance to the sample.

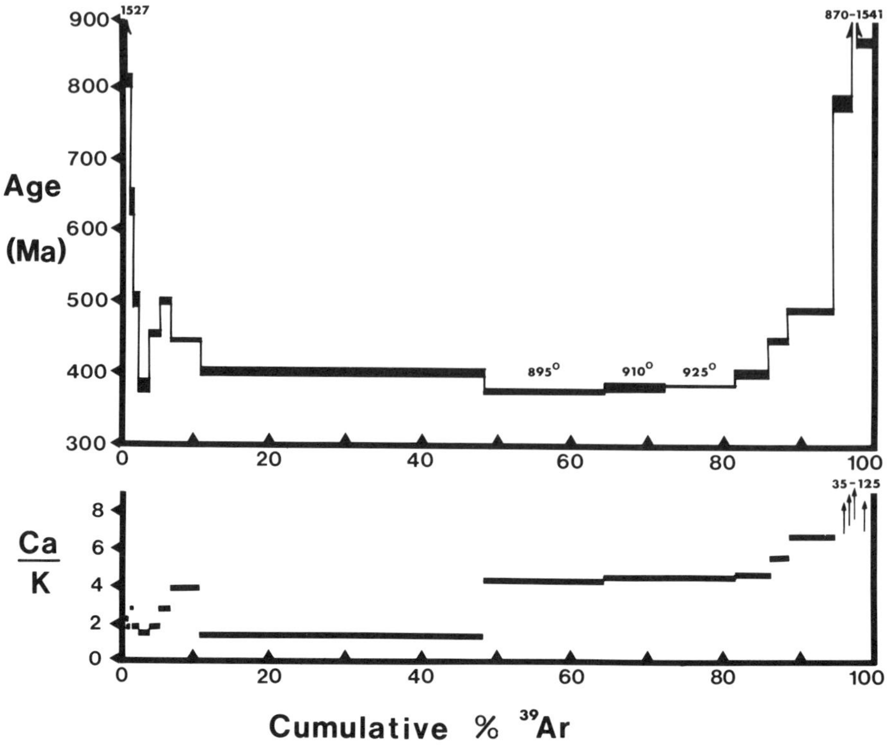

Figure 3. $^{40}Ar/^{39}Ar$ age spectrum with accompanying Ca/K plot of hornblende illustrating the two components of excess Ar. The bottom of the saddle is a 372 ± 4 Ma step at 895°, which is interpreted as a maximum age of the hornblende. The ages of several steps are off scale and are listed above vertical arrows. The four steps at the highest temperatures have Ca/K ratios which are off scale and range from 35 to 125, as indicated above vertical arrows. See text for discussion and Table 3 for analytical details of the $^{40}Ar/^{39}Ar$ step-heating analysis.

Table 3. ^{40}Ar-^{39}Ar data for Ice River complex hornblende

Temp. °C	$^{40}Ar^{1}/^{39}Ar$	$^{36}Ar^{1}/^{39}Ar$	$^{37}Ar^{2}/^{39}Ar$	Vol. $^{39}Ar^{3}$ 10^{-6} cc	%^{40}Ar rad.	Age[4] ±	2 SEM
500	246.2	0.267	1.207	0.001526	68.01	1527.6	28.6
550	81.94	0.037	1.028	0.001022	86.89	809.7	9.6
600	61.02	0.026	1.559	0.000893	87.36	637.9	17.2
650	45.61	0.019	1.065	0.001251	88.13	500.6	8.9
700	33.92	0.016	0.897	0.002308	85.99	376.5	5.9
750	38.40	0.00925	1.051	0.002153	93.05	451.4	3.6
790	41.87	0.00722	1.601	0.002967	95.15	496.9	3.2
830	35.82	0.00308	2.142	0.006634	97.86	444.0	1.5
870	31.80	0.00214	0.690	0.0630	98.13	399.9	6.4
895	29.23	0.00184	2.315	0.0262	98.68	372.9	4.0
910	29.65	0.00165	2.415	0.0136	98.92	378.7	6.4
925	29.88	0.00174	2.431	0.0156	98.84	381.1	1.4
940	31.16	0.00112	2.603	0.00747	99.51	398.2	6.6
990	39.59	0.00290	3.686	0.00983	98.49	488.1	3.4
1030	70.53	0.0151	19.017	0.00434	95.63	782.3	12.0
1080	131.6	0.0548	55.012	0.00700	90.72	1235.3	15.2
1120	204.7	0.1589	68.01	0.00152	79.44	1541.2	54.0
1200	79.14	0.0147	25.135	0.00450	96.87	869.8	5.7

J value = 7.950×10^{-3} for this irradiation

Notes: [1]True ratios corrected for fractionation and discrimination ($^{40}Ar/^{36}Ar$ atmospheric = 295.5), ratios are not corrected for system blank Ar but vol. of blank ^{40}Ar is 1×10^{-8}cc STP for 500°-1050°; [2] the $^{37}Ar/^{39}Ar$ ratio is corrected for the decay of ^{37}Ar during and after irradiation ($^{37}Ar = 1.975 \times 10^{-2}$/day); K/Ca = 1.83 × $^{37}Ar/^{39}Ar$; [3]volume of ^{39}Ar determined using the equilibrium peak height and mass spectrometer sensitivity; [4]ages calculated using constants of Steiger and Jager (1977); errors represent the analytical precision only, not including error in J values; Flux monitors used were LP-6 biotite (128.5 Ma) and Da-83-48-BB biotite (97.5 Ma) referenced to HB3gr and mmHb-1 hornblendes.

As can be seen from Table 3 and the Ca/K plot of Figure 3, the second, higher temperature site is rich in calcium relative to potassium suggesting that clinopyroxene may represent the second reservoir of excess Ar. Two steps at 910° and 925° have a total of 17.3 % of the gas and agree within error at 380 ± 1.2 Ma. These steps, despite their agreement, may still contain excess argon because the previous 895° step at 372.9 ± 4.0 Ma (15.5 % of the gas) may be closer to the botton of the saddle shaped spectra. In view of the U-Pb data, we feel that the most reasonable interpretation is that none of the age spectra steps is demonstrably free of excess argon and that the 372 ± 4 Ma age of the 895° step is the maximum age of the hornblende.

DISCUSSION

U-Pb and $^{40}Ar/^{39}Ar$ methods indicate that the age of the Ice River complex is between 356 and 372 Ma; we choose a value of 368 ± 4 Ma to encompass the most likely range of age and error. The explanation for the disturbed Rb-Sr systematics and the considerable excess Ar in hornblende is uncertain, but may relate to the later deformation and low grade metamorphism of the complex during the Mesozoic.

ACKNOWLEDGMENTS

Dale Loveridge of the Geological Survey of Canada is thanked for his assistance in Rb-Sr analyses. W.C. Gussow is thanked for suggesting this project and collecting the sample for U-Pb dating.

REFERENCES

Allan, J. A.
1914: Geology of the Field map-area, British Columbia and Alberta; Geological Survey of Canada, Memoir 55.

Currie, K. L.
1975: The geology and petrology of the Ice River alkaline complex, British Columbia. Geological Survey of Canada, Bulletin 245, 68 p.

Faure, G.
1986: Principles of Isotope Geology, second edition; John Wiley and Sons, New York, 589 p.

Gussow, W. C.
1977: The Ice River complex, British Columbia, is Precambrian basement; Bulletin of Canadian Petroleum Geology, v. 25, p. 505-517.

Hunt, P. A. and Roddick, J. C.
1987: A compilation of K-Ar ages *in* Radiogenic Age and Isotopic Studies, Geological Survey of Canada, Paper 87-2.

McBride, S. L., Clark, A. H., Farrar, E., and Archibald, D. A.
1987: Delimitation of a cryptic Eocene tectono-thermal domain in the Eastern Cordillera of the Bolivian Andes through K-Ar dating and ^{40}Ar-^{39}Ar step-heating; of the Geological Society of London, Journal, v. 144, pp. 243-255.

Parrish, R. R., Roddick, J. C., Loveridge, W. D., and Sullivan, R. W.
1987: Uranium-lead analytical techniques at the geochronology laboratory, Geological Survey of Canada; *in* Radiogenic Age and Isotopic Studies, Geological Survey of Canada, Paper 87-2.

Steiger, R. H. and Jager, E.
1977: Subcommission on geochronology: convention on the use of decay constants in geo- and cosmochronology. Earth and Planetary Science Letters, v. 36, p. 359-362.

Stevens, R. D., Delabio, R. N., and Lachance, G. R.
1982: Age determinations and geological studies, K-Ar isotopic ages, report 16; Geological Survey of Canada, Paper 82-2.

Age of the NOR breccia pipe, Wernecke Supergroup, Yukon Territory

R. R. Parrish[1] and R. T. Bell[1]

Parrish, R.R. and Bell, R.T., Age of the NOR breccia pipe, Wernecke Supergroup, Yukon Territory; in Radiogenic Age and Isotopic Studies: Report 1, Geological Survey of Canada, Paper 87-2, p. 39-42, 1987.

Abstract

The NOR megabreccia body occurs within the Wernecke Supergroup, south Richardson Mountains, Yukon Territory. A sample of brannerite-hematite-calcite-monazite occurring in the breccia body has yielded ca.1270 ± 40 Ma reversely discordant U-Pb ages on monazite which we interpret as the approximate age of formation of the breccia. Previous U-Pb dates on uranium minerals are much younger, generally disturbed, and only provide minimum ages for mineralizing events. The ca.1.3 Ga age of the breccia is consistent with geological relationships for this and other bodies and demonstrates that the Wernecke Supergroup is older than 1.3 Ga.

Résumé

Le massif mégabréchique de NOR se trouve à l'intérieur du supergroupe de Wernecke dans le sud des monts Richardson, dans le territoire du Yukon. Un échantillon contenant de la brannérite, de l'hématite, de la calcite et de la monazite, prélevé dans le massif bréchique, a été daté à environ 1270 ± 40 Ma par la méthode U-Pb appliquée à la monazite; cet âge, qui présente une discordance inverse, est interprété comme étant l'âge approximatif de la formation bréchique. Des datations antérieures par la méthode U-Pb appliquée à des minéraux uranifères indiquent des âges beaucoup plus récents. Ces datations montrent des distorsions, et n'indiquent que l'âge minimum des épisodes de minéralisation. L'âge approximatif de la brèche (1,3Ga) concorde avec les corrélations géologiques appliquées à ce massif et à d'autres, et démontre que le supergroupe de Wernecke est plus ancien que la massif en question.

INTRODUCTION AND GEOLOGICAL SETTING

The Wernecke breccia pipes are large breccia bodies which occur exclusively in the Wernecke Supergroup of the northern Canadian Cordillera in Yukon Territory. The age of the breccias has been the subject of several papers; geological evidence (Bell 1978, 1986; Bell and Delaney, 1977) indicates that the breccias in the eastern Wernecke Mountains and south Richardson Mountains are all intrusive into the three lowermost Proterozoic units that together make up the Wernecke Supergroup. To the west in the Coal Creek Dome of the Ogilvie Mountains, many of the breccias may be sedimentary (Thompson, 1986; Mercier, 1985) but others are intrusive or tectonic (Lefebvre, 1980; G. Abbott, personal communication, 1986). Most if not all of these breccias appear to have formed at the end of deposition of the Wernecke Supergroup or immediately afterwards, but before the deposition of the younger Mackenzie Mountains Supergroup.

The breccias are large in extent and often contain kilometre-size fragments. There is evidence of multiple brecciation. Some have been severely and multiply metasomatized both during the phase of main breccia formation and later when the breccias appear to have mainly provided conduits for mineralizing fluids.

NOR BRECCIA

The NOR breccia pipe is an irregularly shaped megabreccia body about 800 by 1 600m. The host rocks are greenish-grey phyllitic, calcareous and chloritic siltstone and mudstone of the Fairchild Lake Group, the lowest exposed unit of the Wernecke Supergroup. Most of the material is poorly exposed and frost-heaved, but it is clear that the outer borders are gradational into country rock. Templeman-Kluit (1981) described an irregular central part about 300m across, of mainly fluxion-textured material interfingering with the breccia. He described an earlier phase of massive hematite-magnetite-jasper-quartz with disseminated chalcopyrite, and a second, later phase including coarsely crystalline brannerite with quartz and pink feldspar at the east edge of the fluxion textured core. Templeman-Kluit interpreted the breccia as a diatreme.

[1] Geological Survey of Canada, 601 Booth Street, Ottawa, Ontario K1A 0E8

One of us (RTB) visited the property in 1978 and sampled the breccia. The second, brannerite-bearing, zone contains abundant pink calcite and is cut by hematite-quartz veins similar to those in the massive hematite zones. Therefore, there are more than one of these hematite-rich phases. It is unwise to use the term diatreme for the NOR body, originally applied by Bell and Delaney (1977) to the breccias to the south but later discounted (*see* Bell 1978, 1986; Delaney et al., 1982). The gradational contacts with the Fairchild Lake Group suggest otherwise.

A sample of the NOR body was collected for U/Pb dating of brannerite, the results of which have been presented by Archer et al. (1986). This sample is dominated by brannerite and calcite but was later found to contain considerable monazite which is also amenable to U-Pb dating but is known to be much more resistant to Pb loss and remobilization of uranium.

PREVIOUS GEOCHRONOLOGICAL STUDIES

Recent U/Pb dating results of Archer et al. (1986) were confined to uraniferous whole-rocks and uranium minerals including brannerite and pitchblende. Their attention was directed primarily to determining the age of uranium mineralization. These results are by and large much younger than 1 200 Ma, highly discordant, and difficult to interpret because of the effects of loss or gain of either U parent or Pb daughter isotopes, or because of multiple mineralizing events. The oldest date they obtained approached 1 200 Ma, but it was clearly anomalous with respect to other samples. A number of explanations were invoked to explain the pattern of data, but none of these data clearly reflect the age of the development of the breccias. The tendency of uranium and uranium-rich minerals to be mobilized by fluids is well known and implicit in the interpretation of their data. It seems unlikely that U-Pb dating of brannerite or pitchblende, or particularly whole-rocks enriched in these minerals will provide a clear-cut age of the breccia bodies.

MONAZITE FROM THE NOR BODY

The discovery of substantial primary monazite in a sample of the NOR breccia provides an alternative to the dating of only U-rich phases. Monazite is clearly capable of acting as a closed mineral system for long periods of time and is very resistant to thermal resetting. It seemed appropriate to use it as a chronometer in this study.

The dated sample contained major brannerite, hematite, calcite, and monazite. The sample was processed by crushing and magnetic separation, followed by a brief weak HNO3 leach to remove calcite. This resulted in a pure, pale yellow monazite separate. Four fractions of this separate were analyzed using analytical techniques outlined by Parrish et al. (1987) and the results are presented in Table 1 and Figure 1.

Uranium concentrations in monazite are unusually low (about 120ppm), this being attributable to the tendency of U to partition into the brannerite. One of the fractions was abraded by the method of Krogh (1982). All of the non-abraded fractions plot above concordia with $^{207}Pb/^{235}U$ ages of 1214 to 1268 Ma and $^{207}Pb/^{206}Pb$ ages of 1103 to 1216 Ma.

The explanation we favour for the reverse discordance is that offered by Schärer (1984); briefly, ^{230}Th which is produced as an intermediate daughter of ^{238}U occurs in low concentrations in equilibrium with ^{238}U and will be incorporated into Th-rich minerals such as monazite at the time of their crystallization. The ^{230}Th subsequently decays to ^{206}Pb and this part of the ^{206}Pb becomes unsupported because it was not produced by in situ decay of ^{238}U, yielding a $^{206}Pb/^{238}U$ ratio above concordia. The $^{207}Pb/^{235}U$ ages are uninfluenced by this effect and are more easily interpreted as having age significance. In the case of the NOR monazites we accept the $^{207}Pb/^{235}U$ ages of 1214 to 1268 Ma as approximately correct. These ages are minimum because of a likely small degree of Pb loss which is difficult to estimate because of excess ^{206}Pb at the time of formation. This is plausible because the environment of formation of monazite within the breccia is one of very high uranium contents which would have been in equilibrium with relatively large concentrations of ^{230}Th; also the thorium content of most of these breccias is much lower than uranium, compounding this effect. In addition, it is possible that some of the scatter in the data is related to incorporation of common Pb with a composition deviating significantly from the model age for upper crustal lead (Stacey and Kramers, 1975), particularly in view of the U-rich geochemical environment of the breccia.

Table 1. U-Pb monazite data, NOR breccia

Fraction, size[1]	Wt. (mg)	U, ppm	Pb[2], ppm	$^{206}Pb/^{204}Pb$[3]	Pb_c[4], (pg)	^{208}Pb[2] (%)	$^{206}Pb/^{238}U$ ± 1 SEM%[5]	$^{207}Pb/^{235}U$ ± 1 SEM%[5]	$^{207}Pb/^{206}Pb$ ± 1 SEM%[5]	$^{207}Pb/^{206}Pb$ age, error (Ma)[6]
+149,pale	0.137	112.8	101.61	1756	120	77.7	0.2151 (.21)	2.364 (.23)	0.07971 (.08)	1189.9 (1.6)
+149,pale	0.118	125.6	99.12	1651	125	74.2	0.2191 (.21)	2.305 (.24)	0.07630 (.09)	1102.8 (1.8)
−149+74,pale	0.151	106.6	102.6	2007	113	78.3	0.2234 (.20)	2.487 (.22)	0.08076 (.07)	1215.7 (1.4)
−149+74,abr	0.121	310.4	94.58	2654	67	77.0	0.07465 (.21)	0.8524 (.23)	0.08281 (.06)	1264.7 (1.2)

Note: [1] sizes (i.e. + 105) refer to length aspect of monazites in microns; abr = abraded; [2]radiogenic Pb; [3]measured ratio, corrected for spike and fractionation; [4]total common Pb in analysis corrected for fractionation and spike; [5]corrected for blank Pb and U, common Pb, errors quoted are 1 standard error of the mean in per cent; [6]corrected for blank and common Pb, errors are 1 standard error of the mean in Ma.

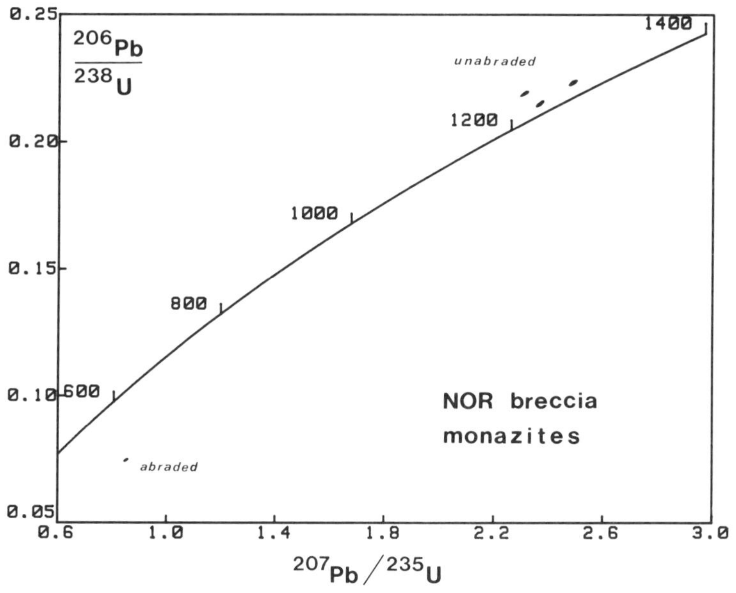

Figure 1. U-Pb concordia diagram of monazite analyses from the NOR breccia, Yukon Territory. The $^{207}Pb/^{235}U$ ages of 1214 to 1268 are interpreted to be close but minimum estimates of the age of primary monazite in the sample. They are reversely discordant due to reasons discussed in the text. For the sake of discussion, we choose a value of 1270 ± 40 Ma for the age of the monazite.

The abraded fraction is very discordant with a $^{207}Pb/^{206}Pb$ age of 1265 Ma, and its uranium content is 310ppm. The explanation of its discordance is uncertain; we speculate that the cores of most of the monazite crystals are enriched in uranium (supported by the abraded monazite's higher U content), and that some of the radiogenic Pb produced by subsequent decay diffused or was redistributed toward the margins of grains relatively recently. Abrasion of the outer rim would tend to remove the redistributed radiogenic Pb leaving the core relatively depleted with a low Pb/U ratio. This could produce the large discordance of the abraded point.

These considerations suggest that the age of monazite in the NOR breccia is at least 1268 Ma, and without additional data, a precise age cannot be assigned. For the sake of discussion, we choose 1270 ± 40 Ma.

DISCUSSION

The ca. 1.3 Ga age, presumed to date initial development of monazite syngenetic to the formation of the breccia, is similar to $^{207}Pb/^{206}Pb$ model ages of 1238 ± 21 Ma (model of Cumming and Richards, 1975) or 1288 ± 85 Ma (model of Stacey and Kramers, 1975) on galena from the Hart River deposit within the upper part of the Werneke Supergroup (Morin, 1979). This age was interpreted to suggest that deposition of the upper Wernecke Supergroup occurred at that time. It is clearly older than either of the Mackenzie Mountains or Windermere supergroups (Eisbacher, 1981), and is consistent with the breccias forming during, or soon after, deposition of the Wernecke Supergroup. Abbott (personal communication, 1986) suggested that some of the breccia bodies within the Wernecke Supergroup are spatially related to mafic volcanics and dykes, and has implied that they may be genetically related. If so, the two may have a common tectonic cause. The ca. 1.3 Ga age of the NOR breccia presented in this paper implies that dates on uranium minerals such as brannerite or pitchblende are clearly secondary and are not reliable in determining the age of formation of the breccia. Monazite age determinations on other breccias should be undertaken to test the potential contemporaneity of breccia bodies from disparate localities; this type of information may shed light on their underlying tectonic cause.

REFERENCES

Archer, A., R. T. Bell, and R. I. Thorpe
1986: Age relationships from U-Th-Pb isotope studies of uranium mineralization in Wernecke breccias, Yukon Territory; *in* Current Research, Part A, Geological Survey of Canada, Paper 86-1A, p. 385-391.

Bell, R. T.
1978: Breccias and uranium mineralization in the Wernecke Mountains, Yukon — a progress report; *in* Current Research, Part A, Geological Survey of Canada, Paper 78-1A, p. 317-322.
1986: Megabreccias in northeastern Wernecke Mountains, Yukon Territory; *in* Current Research, Part A, Geological Survey of Canada, Paper 86-1A, p. 375-384.

Bell, R. T. and G. D. Delaney
1977: Geology of some uranium occurrences in Yukon Territory; *in* Current Research, Part A, Geological Survey of Canada, Paper 77-1A, p. 33-37.

Cumming, G. L. and J. R. Richards
1975: Ore lead isotope ratios in a continuouisly changing earth; Earth and Planetary Science Letters, v. 28, p. 155-171.

Delaney, G. D., Jefferson, C. W., Yeo, G. M., McLennan, S. M., Aitken, J. D. and Bell, R. T.
1982: Some Proterozoic sediment-hosted metal occurrences of the northeastern Canadian Cordillera. Idaho Bureau of Mines and Geology, Bulletin 24, pp. 97-116.

Eisbacher, G. H.
1981: Sedimentary tectonics and glacial record in the Windermere Supergroup, Mackenzie Mountains, northwestern Canada; Geological Survey of Canada, Paper 80-27, 40 p.

Krogh, T. E.
1982: Improved accuracy of U-Pb ages by the creation of more concordant systems using an air abrasion technique; Geochimica et Cosmochimica Acta, v. 46, p. 637-649.

Lefebvre, J.-J.
1980: À propos de l'existence d'un ''wildflysch katangien''. Annales de la Société Géologique de Belgique, v. 103, p. 1-13.

Mercier, E.
1985: Précambrien de "Coal Creek dome" (montagnes Ogilvie, Yukon, Canada); these troisième cycle en géologie, non publiée, Université des sciences et techniques de Lille, France, 246 p.

Morin, J.
1979: A preliminary report on Hart River — a Proterozoic massive sulphide deposit, Yukon Territory; Mineral Industry Report, Department of Indian and Northern Affairs, 1979-9, p. 22-24.

Parrish, R. R., Roddick, J. C., Loveridge, W. D. and Sullivan, R. W.
1987: Uranium-Lead analytical techniques at the geochronology laboratory, Geological Survey of Canada; *in* Radiogenic Age and Isotopic Studies: Report 1, Geological Survey of Canada, Paper 87-2.

Schärer, U.
1984: The effect of initial ^{230}Th equilibrium on young U-Pb ages: the Makalu case, Himalaya; Earth and Planetary Science Letters v. 67, p. 191-204.

Stacey, J. S. and Kramers, J. D.
1975: Approximation of terrestrial lead isotope evolution by a two-stage model; Earth and Planetary Science Letters, v. 26, p. 207-221.

Templeman-Kluit, D. J.
1981: NOR, summary of assessment work and description of mineral properties; Yukon, Geology and Exploration 1979-1980, Department of Indian and Northern Affairs, p. 300-301.

Thompson, R.
1986: Repeated extension on the Proto-Pacific margin, west-central Yukon; Geological Survey of Canada, Paper 86-8, 11 p.

Pb isotopic ratios of Paleozoic granitoids from the Miramichi terrane, New Brunswick, and implications for the nature and age of the basement rocks[1]

Mary Lou Bevier[2]

Bevier, M.L., Pb isotopic ratios of Paleozoic granitoids from the Miramichi terrane, New Brunswick, and implications for the nature and age of the basement rocks; in Radiogenic Age and Isotopic Studies: Report 1, Geological Survey of Canada, Paper 87-2, p. 43-50, 1987.

Abstract

New Pb isotopic data for Ordovician and Devonian granitoids from the Miramichi terrane, northern New Brunswick, provide the first evidence of a Precambrian upper crustal component in the source area for these plutonic bodies. This Precambrian crustal component has a range of Pb isotopic compositions similar to the range displayed by Precambrian rock units from Avalonian terrane in southern New Brunswick, thus raising the possibility that Avalonian basement is much more extensive in the subsurface than previously recognized. To the north of Miramichi terrane, Devonian granitoids within Elmtree terrane have fundamentally different, less radiogenic Pb isotopic compositions than Miramichi granitoids, and therefore Elmtree terrane does not contain Avalonian basement. Viable tectonic models for Miramichi terrane are thus restricted to those in which the terrane had a continental basement by Ordovician time, i.e. a back-arc basin or intracontinental rift setting.

Résumé

De nouvelle données sur les isotopes du Pb présents dans des granitoïdes ordoviciens et dévoniens du terrain de Miramichi, dans le nord du Nouveau-Brunswick, fournissent la première preuve de la présence d'une composante crustale supérieure d'âge précambrien dans la zone source de ces corps plutoniques. Cette composante crustale d'âge précambrien présente une plage de compositions isotopiques du Pb semblable à celle d'unités précambriennes du terrain avalonien situé dans le sud du Nouveau Brunswick; ce phénomène semble indiquer que le socle avalonien pourrait être beaucoup plus vaste en subsurface qu'on ne l'avait supposé. Au nord du terrain de Miramichi, dans des granitoïdes dévoniens du terrain d'Elmtree, les compositions isotopiques du Pb sont fondamentalement différentes et cet élément est moins radiogénique que dans les granitoïdes de Miramichi, donc le terrain d'Elmtree ne contiendrait pas un socle avalonien. Des modèles tectoniques viables du terrain de Miramichi se limitent donc à ceux où le terrain avait un socle continental durant l'Ordovicien, c'est-à-dire constituait un bassin dans la convexité d'un arc insulaire ou faisait partie d'un rift intracontinental.

INTRODUCTION

Granitoid rocks in New Brunswick range in age from Proterozoic to Late Paleozoic (Pajari et al., 1974; Fyffe et al., 1977, 1981; Fyffe and Cormier, 1979; Poole, 1980; Hauseaux, 1980, 1982; McCutcheon et al., 1981; Fyffe, 1982; Olszewski and Gaudette, 1982; Fyffe and Pronk, 1985; MacLellan et al., 1986). Previous studies have documented the diversity among these plutons and a preliminary attempt to classify them by tectonic setting of origin (Ruitenberg and Fyffe, 1982) has been made. Proterozoic granitoids are confined to Avalonian terrane in southern New Brunswick whereas Paleozoic granitoids intrude the Miramichi, Elmtree, St. Croix, and Mascarene terranes (Fig. 1).

This study measured initial Pb isotopic compositions of Paleozoic granitoids from Miramichi terrane to discern the nature and age(s) of the source for these granitoids, because nothing is known about the basement to the Miramichi terrane. Additional samples of granitoid from the Elmtree terrane and granitoid and basement gneiss from Avalonian terrane were examined for comparison. These data permit an assessment

---

[1] Contribution to Canada-New Brunswick Mineral Development Agreement 1984-1989. Project carried by Geological Survey of Canada.

[2] Geological Survey of Canada, 601 Booth Street, Ottawa, Ontario K1A 0E8

of possible melt versus source rock relationships between the Paleozoic granitoid rocks and the Proterozoic granitoids and gneisses from Avalonian terrane.

Pb isotopes have been used elsewhere with success in determining the age and type of source material for granitoid rocks (c.f. Zartman, 1974; Doe and Delevaux, 1980; Gariépy and Allègre, 1985; Mukasa, 1986; Ayuso, 1986). The data and interpretations presented here are consistent with Ayuso's (1986; personal communication, 1987) conclusions regarding different crustal sources for Paleozoic granitoids in three tectonic belts in Maine.

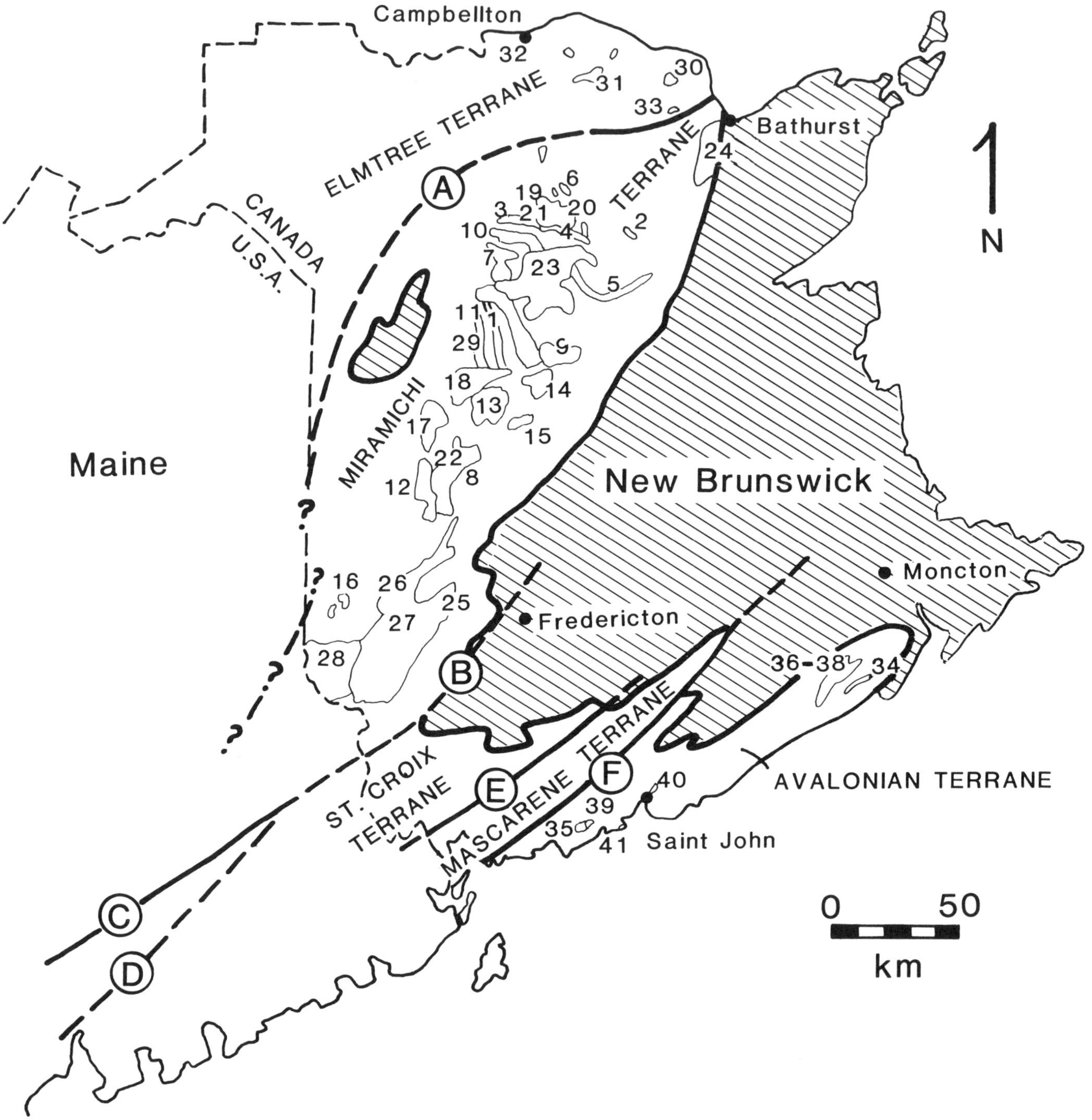

Figure 1. Location map showing granitic and gneissic units sampled for this study, and terrane boundaries and designations after Fyffe and Fricker (1987). Where dashed, terrane boundaries are covered by younger rock units. Lined area- Carboniferous cover; A-Rocky Brook-Millstream fault; B- Fredericton fault; C- Norumbega fault; D- Turtlehead fault; E- Pendar Brook fault; F- Belleisle fault. Numbers refer to localities listed in Table 1.

GENERAL GEOLOGY

Granitoid and basement gneiss units sampled for this study are listed by terrane and age in Table 1 and located in Figure 1. Within the Miramichi terrane, Ordovician and Devonian granitoid bodies intrude sedimentary rocks of the (?)Cambro-Ordovician Tetagouche Group. Proposed tectonic models for the Tetagouche Group include an island-arc setting (Rast and Stringer, 1974; Pajari et al., 1977), an intracontinental rift (Janes, 1976), an off-axis rift environment (Whitehead and Goodfellow, 1978), and a back-arc basin (van Staal, 1987). The Ordovician granitoid bodies were emplaced during the Taconic orogeny and are pervasively deformed (Ruitenberg and Fyffe, 1982). The largest volume of felsic material in Miramichi terrane was emplaced during the Devonian in a number of polyphase granitoid suites (North Pole Stream, Mount Elizabeth, Pokiok; *see* Fig. 1), and these large-volume intrusive suites only in part have a fabric (Ruitenberg and Fyffe, 1982). The Devonian plutons were intruded after Acadian deformation in the area and after considerable movement on faults that bound and segment the Miramichi terrane (Fyffe et al., 1981). A detailed description of Ordovician and Devonian granitoids in the northern part of the Miramichi terrane is provided by Whalen (1987).

Within Elmtree terrane, granitoid bodies are commonly small stocks and plutons that range from granodiorite to granite. Some granitoids outcrop in an area known to be underlain by ophiolitic basement (Pajari et al., 1977; Rast and Stringer, 1980) whereas others outcrop through the Late Ordovician and younger Matapedia cover sequence which presumably overlies Elmtree terrane basement (Fyffe and Fricker, 1987).

Granitoids and gneisses from Avalonian terrane were sampled for this study because they represent a possible upper crustal source for granitoids in the Miramichi and Elmtree terranes. Avalonian terrane is one of the closest occurrences of known Precambrian basement to the Miramichi and Elmtree terrane granitoids. Proterozoic basement gneisses of Avalonian terrane in the region around Saint John (Fig. 1) include those of the Green Head Group (metasedimentary rocks), Brookville tonalite gneiss (Olszewski and Gaudette, 1982), and Coldbrook Group (metavolcanic and minor metasedimentary rocks; Currie, 1987). Associated with the Coldbrook Group are cogenetic granitoid bodies including the Hansen Stream and Prince of Wales plutons (Fig. 1). In southeastern New Brunswick, granitoid plutons of Late Proterozoic age (Barr and White, 1986; Barr, 1987) intrude rocks of the Coldbrook Group, and several samples of these from the Point Wolfe River and the Fortyfive River plutons (Fig. 1) were included in this study.

RESULTS

Analytical techniques

Feldspars are U-poor and Pb-rich minerals; therefore, their $^{238}U/^{204}Pb$ values are very low, and changes in Pb isotopic composition as a function of age are usually very small. Hence measured Pb isotopic ratios on feldspars can be interpreted as initial ratios.

A clean K feldspar or plagioclase fraction was separated from each granitic or gneissic sample. After initial handpicking, the feldspar fraction was ground with a boron carbide mortar and pestle and sieved to separate a 75-150μm split,

Table 1. Granitic and gneissic rock units sampled for Pb isotopic analysis.

MIRAMICHI TERRANE		
Ordovician	Devonian	
1. Fox Ridge	12. Becaguimec Lake	21. Mount Elizabeth syenite
2. Heath Steele	13. Burnthill	22. Nashwaak
3. Meridian Brook	14. Dungarvon	23. North Pole Stream
4. Mount Johnson	15. Trout Brook	24. Pabineau Falls
5. Mullin Stream Lake	16. Gibson	25. Allandale
6. Popple Depot	17. Juniper Barrens	26. Hartfield
7. Serpentine River	18. Lost Lake	27. Hawkshaw
8. Sisson Brook	19. Mount Elizabeth alkali granite	28. Skiff Lake
9. South Renous	20. Mount Elizabeth biotite granite (n=2)	29. Redstone Mountain
10. Sweat Hill		
11. Trousers Lake		
ELMTREE TERRANE	*AVALONIAN TERRANE*	
Devonian	Late Proterozoic	
30. Antinouri Lake	34. Fortyfive River	39. Prince of Wales
31. Benjamin River	35. Hansen Stream	40. Brookville gneiss
32. Campbellton	36. Point Wolfe River granite (n=4)	41. Green Head gneiss
33. Nicholas Denys	37. Point Wolfe River monzonite	
	38. Point Wolfe River quartz monzonite	

Terrane designation after Fyffe and Fricker (1987).

which was then subjected to a severe acid-leaching procedure in order to remove grain-boundary contaminants, following Gariépy and Allègre (1985).

Approximately 15 mg of the powdered sample was digested with a mixture of high-purity 48 % HF and 70 % HNO_3 in clean, covered teflon beakers in a clean air environment. Lead was separated by anion exchange chromatography using HCl and HBr, following procedures described in Barreiro (1982). Purified Pb samples were loaded onto previously outgassed Re filaments using the silica gel-phosphoric acid technique. Analyses were performed on a Finnigan MAT 261 mass spectrometer equipped with fully adjustable multiple collectors. Measurements were made in static multicollection mode with four Faraday collectors positioned to simultaneously collect ^{208}Pb, ^{207}Pb, ^{206}Pb, and ^{204}Pb. Measured Pb isotopic compositions were corrected for mass fractionation of 0.12 ± 0.04 % per a.m.u., determined by the average deviation of NBS 981 from its absolute value (Todt et al., 1984). Within-run precision averages 0.03 % (1σ) for all measured Pb isotopic ratios. Six Pb procedural blanks averaged 317 picograms and represent less than 0.21 % of the estimated total Pb in the samples; therefore no blank correction was made. Analyses of Proterozoic samples from Avalonian terrane were corrected (for additional decay) to 460 Ma (approximate age of Ordovician plutons from Miramichi terrane). Because U, Th, and Pb

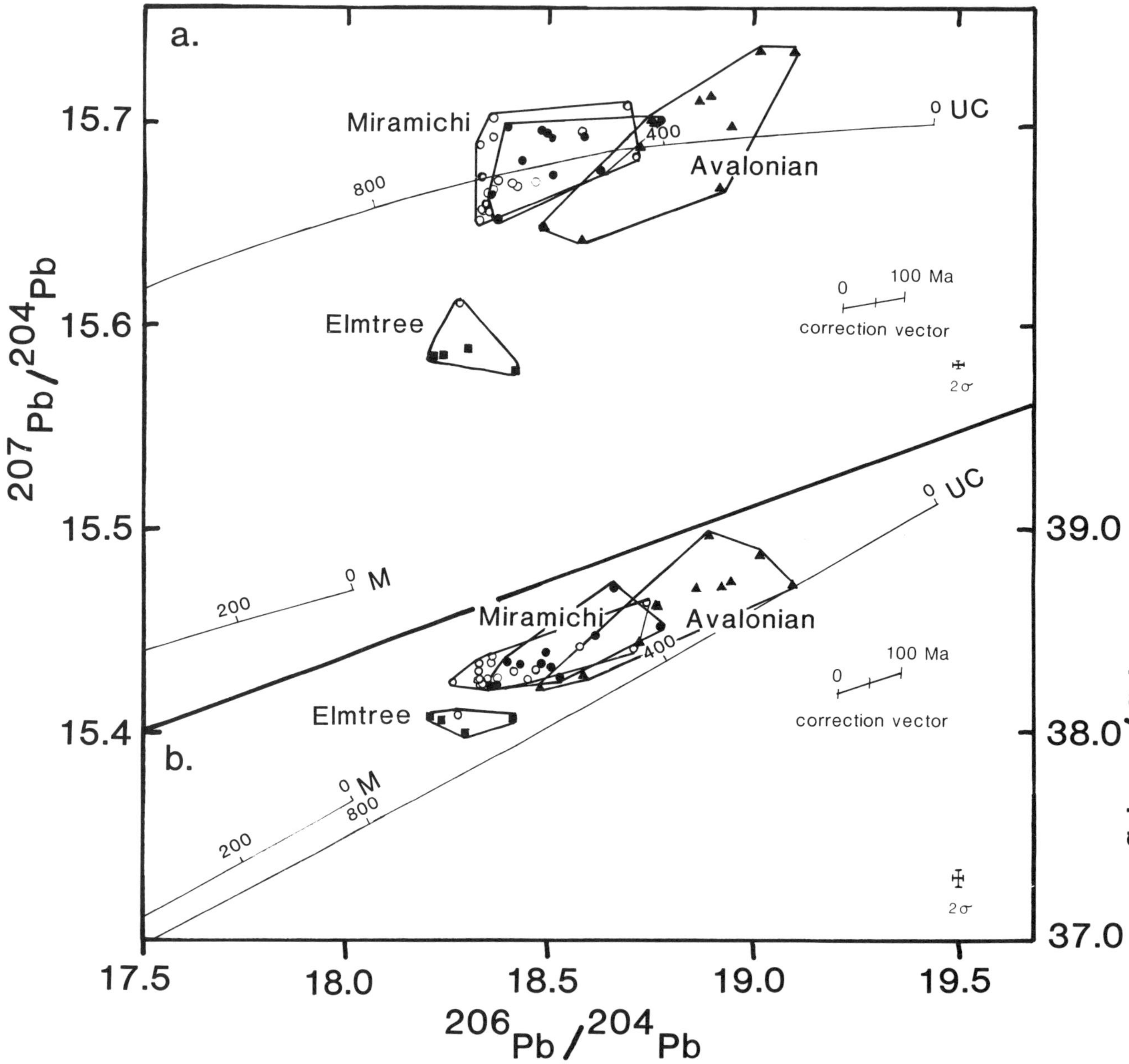

Figure 2. Pb-Pb correlation diagrams for New Brunswick granitoids and gneisses. Filled circles- Ordovician granitoids, Miramichi terrane; open circles- Devonian granitoids, Miramichi terrane; filled squares- Devonian granitoids, Elmtree terrane; filled triangles- Proterozoic granitoids and gneisses, Avalonian terrane. Also shown for comparison are Pb isotope evolution curves for upper crust (UC), and mantle (M) (Zartman and Doe, 1981).

Table 2. Pb isotopic ratios, New Brunswick granitoids and gneisses.

Unit number from Table 1	$^{206}Pb/^{204}Pb$	$^{207}Pb/^{204}Pb$	$^{208}Pb/^{204}Pb$
Miramichi terrane — Ordovician			
1.	18.397	15.699	38.348
2.	18.658	15.694	38.707
3.	18.491	15.696	38.398
4.	18.618	15.677	38.467
5.	18.428	15.682	38.332
6.	18.770	15.702	38.528
7.	18.501	15.693	38.306
8.	18.370	15.654	38.226
9.	18.479	15.697	38.342
10.	18.353	15.675	38.232
11.	18.506	15.675	38.333
Miramichi terrane — Devonian			
12.	18.353	15.702	38.360
13.	18.327	15.657	38.261
14.	18.329	15.652	38.252
15.	18.356	15.693	38.377
16.	18.738	15.709	38.631
17.	18.348	15.665	38.247
18.	18.336	15.660	38.263
19.	18.574	15.696	38.415
20.	18.411	15.669	38.303
	18.399	15.670	38.294
21.	18.467	15.672	38.311
22.	18.326	15.690	38.340
23.	18.539	15.665	38.250
24.	18.276	15.612	38.089
25.	18.369	15.672	38.270
26.	18.349	15.658	38.224
27.	18.329	15.658	38.229
28.	18.330	15.674	38.304
29.	18.708	15.684	38.420
Elmtree terrane — Devonian			
30.	18.212	15.585	38.075
31.	18.410	15.578	38.073
32.	18.293	15.589	38.005
33.	18.236	15.586	38.044
Avalonian terrane — late Proterozoic[1]			
34.	18.930	15.670	38.766
35.	18.592	15.644	38.333
36.	18.711	15.705	38.682
	18.769	15.706	38.674
	18.904	15.714	39.007
	18.995	15.699	38.791
37.	18.874	15.712	38.762
38.	19.105	15.736	38.774
39.	18.490	15.649	38.278
40.	19.027	15.737	38.911
41.	18.731	15.689	38.485

[1]Avalonian Pb isotopic ratios are recalculated to 460 Ma (*see* **Analytical Techniques** in text).

concentration data were not available, the data for Proterozoic samples were corrected using average crustal values of $\mu=9.5$ (Zartman and Doe, 1981) and Th/U=4 (Clark et al., 1966).

Pb isotopic data

Results from 45 samples of granitoid rock and basement gneiss are shown diagramatically in Figure 2 and listed in Table 2. In order for the Pb isotopic ratios of the Proterozoic and Ordovician samples to be directly comparable with the Pb isotopic data of the Devonian samples, an additional ≈ 80 Ma of radiogenic Pb should be added to the isotopic ratios for 460 Ma. Approximate correction vectors (calculated using $\mu=9.5$ and Th/U=4) are shown in Figure 2.

Pb isotopic analyses of Miramichi granites of Ordovician and Devonian age form overlapping clusters of points on the Pb-Pb correlation diagrams. Four samples of Devonian granitoids from Elmtree terrane are distinctly less radiogenic than samples from the Miramichi or Avalonian terranes. One sample from the northeast-most locality in the Miramichi terrane, Pabineau Falls granite (Fig. 1), has a Pb isotopic composition that is widely different from all other Miramichi granitoids but quite close to the isotopic composition of Elmtree granitoids. It plots within the field of Elmtree granitoids (Fig. 2). Samples of granitic rocks and gneisses in Avalonian terrane, corrected to an age of 460 Ma, form a cluster of points that partially overlap with the fields of Ordovician and Devonian granitoids from the Miramichi terrane.

DISCUSSION

All Proterozoic and Paleozoic granitoids that were analyzed for this study have an upper crustal Pb isotopic signature, significantly more radiogenic than that displayed by mantle leads of Proterozoic or Paleozoic age (Zartman and Doe, 1981; Fig. 2). Ordovician and Devonian granitoid suites from Miramichi terrane show no or little difference in Pb isotopic composition, suggesting that melting of essentially the same upper crustal source rocks occurred in both Ordovician and Devonian time. U-Pb zircon systematics for several Middle Ordovician granitic orthogneisses (Fox Ridge, Trousers Lake, South Renous) in Miramichi terrane indicate that a Precambrian (Middle to Early Proterozoic?) inherited component is present in some if not all of the Ordovician granitoids (M. L. Bevier, unpublished data, 1987). The geochronological and isotopic tracer data taken together imply that Precambrian upper crustal rocks must exist in the subsurface of Miramichi terrane, although there is no surface outcrop of a unit older than (?) Cambrian (sedimentary rocks of the lower Tetagouche Group; van Staal, 1987).

The closest candidates for Precambrian basement to the Miramichi terrane are Grenville-age crust (approximately 1.1 Ga old) of the North American Craton, exposed in southern Quebec, the allochthonous (approximately 1.5 Ga old) Chain Lakes massif of western Maine (Naylor et al., 1973), and late Proterozoic rocks of Avalonian terrane in southern New Brunswick. No evidence of a subcontinental mantle or a Grenvillian basement component such as Ayuso (1986) found for Paleozoic plutons in northern Maine is found in Paleozoic granitoids of the Miramichi or Elmtree terranes, and no Pb isotopic data exist for the Chain Lakes massif for comparison. Proterozoic granitoids and gneisses of Avalonian terrane, however, have Pb isotopic compositions at 460 Ma that partially overlap with those of Miramichi granitoids (Fig. 2), and these data are consistent with the possibility that Avalonian-age upper crustal rocks were the source for Paleozoic granitoids of Miramichi terrane. Thus, Pb isotopic data provide the first evidence to suggest that Avalonian basement is present in the subsurface of Miramichi terrane.

Devonian granitoids of the Elmtree terrane have a fundamentally different, less radiogenic upper crustal source than granitoids of Miramichi terrane; the source contains Pb which evolved in a lower μ and $^{232}Th/^{238}U$ environment. Because of the pronounced isotopic difference between Miramichi and Elmtree granitoids, the upper crustal sources for the two terranes must have evolved along distinct μ paths for long intervals. The Elmtree crustal source volume could be derived through mixing of either mantle or a lower crustal component with an upper crustal component. Ophiolitic basement of the Elmtree terrane is a likely candidate for a mantle or lower crustal end member for the granitoid source in the New Brunswick part of Elmtree terrane. Ayuso (1986) reported Pb isotopic data from Devonian plutons in Merrimack terrane (in part equivalent to the Elmtree terrane) in Maine that are consistent with those reported here from New Brunswick, and he suggested that isotopic and geochemical features of Devonian felsic rocks in Merrimack terrane are compatible with their origin from a mixture of immature geosynclinal sediments and more mafic rocks.

There is no overlap in Pb isotopic composition between granitoids of Elmtree terrane and the Precambrian samples from Avalonian terrane. If there is Precambrian upper crust at depth in Elmtree terrane, it is extremely unlikely that it is Avalonian in age. Based on data from Ayuso (1986) for northern Maine, it is also unlikely that Grenvillian basement is a component of Elmtree terrane in northern New Brunswick. However, Pb isotopic data for the McGerrigle intrusive complex in the Gaspésie (Whalen and Gariépy, 1986), which may be indicative of a Grenvillian component in the source, overlap the field for Elmtree granitoids and permit a Grenvillian upper crustal component in Elmtree terrane. More data are needed to resolve this problem.

Pb isotopic data from the Devonian Pabineau Falls granite are problematic because they suggest that the magma was derived from an upper crustal source like that present at depth in Elmtree terrane and distinctly unlike that inferred for other Miramichi granitoids. Because Pb isotopic ratios so precisely delineate granitoids of the Miramichi and Elmtree terranes elsewhere in New Brunswick, it is tempting to speculate that a fundamental crustal break must occur between the Pabineau Falls pluton and the closest other pluton in Miramichi terrane, the Heath Steele pluton, which outcrops 30 km to the southwest (Fig. 1). Additional Pb isotopic analyses of feldspar porphyry dykes that intrude the Tetagouche Group in the area between the Pabineau Falls and the Heath Steele plutons may help resolve this problem.

Ayuso (1986) reported no granitoids from Maine that have Pb isotopic signatures like those reported herein for Miramichi granitoids; however, Ayuso (personal communication, 1987) now confirms the existence of Devonian granitoid rocks with Pb isotopic ratios like those of Miramichi granitoids in a narrow belt in coastal Maine, bounded by the Norumbega and Turtlehead faults (Fig. 1). This is consistent with Zen et al.'s (1986) recognition of the continuation of Miramichi terrane into Maine as being confined between these two faults. Two Pb isotopic analyses of samples of gneiss (age unknown but probably >550 Ma; R. A. Ayuso, personal communication, 1987) from the "coastal belt" in Maine (Ayuso, 1986) are consistent with analyses of basement gneiss from Avalonian terrane in southern New Brunswick. Perhaps the Miramichi terrane in southern Maine does contain outcrops of Avalonian basement, unlike Miramichi terrane in New Brunswick.

In summary, Pb isotopic data provide the first evidence indicative of a Precambrian upper crustal source for Ordovician and Devonian granitoids of the Miramichi terrane. This Precambrian basement source is similar in isotopic composition to rock units of Avalonian terrane in southern New Brunswick, thus raising the possibility that Avalonian terrane is much more extensive in the subsurface than previously recognized. A fundamental difference exists between the upper crustal sources for Paleozoic granitoids of the Miramichi and Elmtree terranes in both New Brunswick and Maine. The presence of Avalonian basement in the subsurface of Miramichi terrane but not in Elmtree terrane is a plausible reason for this difference. The boundary separating granitoids with a Miramichi-type source from those with a Elmtree-type source can be traced from New Brunswick into Maine.

The presence of Avalonian basement constrains tectonic models of origin for Miramichi terrane to those in which continental crustal basement was present by Middle Ordovician time. Thus formation of the Tetagouche Group in an island-arc setting is negated. The Pb isotopic data are, however, compatible with either an intracontinental rift or a back-arc basin model.

ACKNOWLEDGMENTS

I thank J. B. Whalen, L. R. Fyffe, C. R. van Staal, and J. K. Mortensen for helpful discussions of the regional geology and the isotopic data. J. B. Whalen critically reviewed the manuscript.

REFERENCES

Ayuso, R.A.
1986: Lead-isotopic evidence for distinct sources of granite and for distinct basements in the northern Appalachians, Maine; Geology, v. 14, p. 322-325.

Barr, S.M.
1987: Field relations, petrology and age of plutonic and associated metavolcanic and metasedimentary rocks, Fundy National Park area, New Brunswick; *in* Current Research, Part A, Geological Survey of Canada, Paper 87-1A, p. 263-280.

Barr, S.M. and White, C.E.
1986: Field relations, petrology, and age of granitoid rocks of the Point Wolfe pluton (eastern half) and associated metavolcanic and metasedimentary rocks, Caledonian Highlands, New Brunswick; *in* Eleventh Annual Review of Activities, Mineral Resources Division, ed. S.A. Abbott; New Brunswick Department of Natural Resources and Energy, Information Circular 86-2, p. 110-111.

Barreiro, B.A.
1982: Lead isotope evidence for crust-mantle interaction during magmagenesis in the South Sandwich island arc and in the Andes of South America; unpublished Ph.D. thesis, University of California, Santa Barbara, 173 p.

Clark, S.P., Jr., Peterman, Z.E., and Heier, K.S.
1966: Abundances of uranium, thorium, and potassium; *in* Handbook of Physical Constants — revised edition, ed.S.P. Clark; Geological Society of America, Memoir 97, p. 521-541.

Currie, K.L.
1987: Late Precambrian igneous activity and its tectonic implications, Musquash-Loch Alva region, southern New Brunswick; *in* Current Research, Part A, Geological Survey of Canada, Paper 87-1A, p. 663-671.

Doe, B.R. and Delevaux, M.H.
1980: Lead isotope investigations in the Minnesota River Valley — Late-tectonic and posttectonic granites; *in* Selected Studies of Archean Gneisses and Lower Proterozoic Rocks, Southern Canadian Shield, ed.G.B. Morey, and G.N. Hansen; Geological Society of America, Special Paper 182, p. 105-112.

Fyffe, L.R.
1982: Geology in the vicinity of the 1982 Miramichi earthquake, Northumberland County, New Brunswick; Mineral Resources Division, New Brunswick Department of Natural Resources, Open File Report 82-27, 37 p.

Fyffe, L.R. and Cormier, R.F.
1979: The significance of radiometric ages from the Gulquac Lake area of New Brunswick; Canadian Journal of Earth Sciences, v. 16, p. 2046-2052.

Fyffe, L.R. and Fricker, A.
1987: Tectonostratigraphic terrane analysis of New Brunswick; Maritime Sediments and Atlantic Geology, v. 23.

Fyffe, L.R. and Pronk, A.G.
1985: Bedrock and surficial geology — rock and till geochemistry in the Trousers Lake area, Victoria County, New Brunswick; Mineral Resources Division, New Brunswick Department of Natural Resources, Report of Investigation 20, 74 p.

Fyffe, L.R., Irrinki, R.R., and Cormier, R.F.
1977: A radiometric age of deformed granitic rocks in north-central New Brunswick; Canadian Journal of Earth Sciences, v. 14, p. 1687-1689.

Fyffe, L.R., Pajari, G.E., and Cherry, M.E.
1981: The Acadian plutonic rocks of New Brunswick; Maritime Sediments and Atlantic Geology, v. 17, p. 23-36.

Gariépy, C. and Allègre, C.J.
1985: The lead isotope geochemistry and geochronology of late kinematic intrusives from the Abitibi greenstone belt, and the implications for late Archaean tectonic crustal evolution; Geochimica et Cosmochimica Acta, v. 49, p. 2371-2383.

Hauseaux, M.
1980: Diamond drilling, Summer 1980. Long Claim Group; Canadian Occidental Petroleum, 298 p.
1982: Diamond drilling, Winter 1982. Long Lake, Long South, Shawn, Shawn North, and Swamp Claims; Canadian Occidental Petroleum, 350 p.

Janes, P.A.
1976: Geochemistry of altered volcanic and intrusive rocks of the Bathurst district, New Brunswick; unpublished M.Sc. thesis, University of Montreal, Montreal, 110 p.

MacLellan, H.E., Taylor, R.P., and Lux, D.R.
1986: Geology and geochronologic investigations of the Burnthill granite; *in* Eleventh Annual Review of Activities, Mineral Resources Division, ed. S.A. Abbott; New Brunswick Department of Natural Resources and Energy, Information Circular 86-2, p. 48-56.

McCutcheon, S., Lutes, G., Gauthier, G., and Brooks, C.
1981: The Pokiok batholith: a contaminated Acadian intrusion with an anomalous Rb/Sr age; Canadian Journal of Earth Sciences, v. 18, p. 910-918.

Mukasa, S.B.
1986: Common Pb isotopic compositions of the Lima, Arequipa, and Toquepala segments in the Coastal Batholith, Peru: implications for magmagenesis; Geochimica et Cosmochimica Acta, v. 50, p. 771-782.

Naylor, R.S., Boone, G.M., Boudette, E.L., Ashenden, D.D., and Robinson, P.
1973: Pre-Ordovician rocks in the Bronson Hill and Boundary Mountain anticlinoria, New England; EOS, v. 54, p. 495.

Olszewski, Jr., W.J. and Gaudette, H.E.
1982: Age of the Brookville Gneiss and associated rocks, southeastern New Brunswick; Canadian Journal of Earth Sciences, v. 19, p. 2158-2166.

Pajari, G.E., Trembath, L.T., Cormier, R.F., and Fyffe, L.R.
1974: The age of Acadian deformation in southwestern New Brunswick; Canadian Journal of Earth Sciences, v. 11, p. 1309-1313.

Pajari, G.E., Rast, N., and Stringer, P.
1977: Paleozoic volcanicity along the Bathurst-Dalhousie geotraverse, New Brunswick and its relation to structure; *in* Volcanic Regimes in Canada, ed. W.R.A. Baragar, L.C. Coleman, and Hall, J.M.; Geological Association of Canada, Special Paper 16, p. 111-124.

Poole. W.H.
1980: Rb-Sr ages of the "sugar" granite and Lost Lake granite, Miramichi Anticlinorium, New Brunswick; *in* Current Research, Part C, Geological Survey of Canada, Paper 80-1C, p. 174-180.

Rast, N. and Stringer, P.
1974: Recent advances and the interpretation of the geological structure of New Brunswick; Geoscience Canada, v. 1, p. 15-25.
1980: A geotraverse across a deformed Ordovician ophiolite and its Silurian cover, northern New Brunswick, Canada; Tectonophysics, v. 69, p. 221-245.

Ruitenberg, A.A. and Fyffe, L.R.
1982: Mineral deposits associated with granitoid intrusions and related subvolcanic stocks in New Brunswick and their relationship to Appalachian tectonic evolution; Canadian Mining and Metallurgical Bulletin, v. 75, p. 1-15.

Todt, W., Cliff, R.A., Hansen, A., and Hofmann, A.W.
1984: ^{202}Pb + ^{205}Pb double spike for Pb isotopic analyses; Terra Cognita, v. 4, p. 209.

van Staal, C.R.
1987: Tectonic setting of the Tetagouche Group in northern New Brunswick: implications for plate tectonic models of the northern Appalachians; Canadian Journal of Earth Sciences, v. 24, p. 1329-1351.

Whalen, J.B.
1987: Geology of a northern portion of the Central Plutonic Belt, New Brunswick; *in* Current Research, Part A, Geological Survey of Canada, Paper 87-1A, p. 209-217.

Whalen, J.B. and Gariépy, C.
1986: Petrogenesis of the McGerrigle plutonic complex, Gaspé, Québec: a preliminary report; *in* Current Research, Part A, Geological Survey of Canada, Paper 86-1A, p. 265-274.

Whitehead, R.E.S. and Goodfellow, W.D.
1978: Geochemistry of volcanic rocks from the Tetagoughe Group, Bathurst, New Brunswick, Canada; Canadian Journal of Earth Sciences, v. 15, p. 207-219.

Zartman, R.E.
1974: Lead isotopic provinces in the Cordillera of the western United States and their geologic significance; Economic Geology, v. 69, p. 792-803.

Zartman, R.E. and Doe, B.R.
1981: Plumbotectonics — The Model; Tectonophysics, v. 75, p. 135-162.

Zen, E-an, Stewart, D.B., and Fyffe, L.R.
1986: Paleozoic tectonostratigraphic terranes and their boundaries in the mainland northern Appalachians; Geological Society of America, Abstracts with Programs, v. 18, p. 800.

The age of deformation related to the emplacement of the Elzevir Batholith and its regional implications, Grenville Province, southeastern Ontario

J.N. Connelly[1], O. van Breemen[2], and S.K. Hanmer[2]

Connelly, J.N., van Breemen, O., and Hanmer, S.K., The age of deformation related to the emplacement of the Elzevir Batholith and its regional implications, Grenville Province, southeastern Ontario; in Radiogenic Age and Isotopic Studies: Report 1, Geological Survey of Canada, Paper 87-2, p. 51-54, 1987.

Abstract

Strongly developed planar and linear fabrics within the margin of the Elzevir batholith, previously dated at 1226 ± 25 Ma, and surrounding metavolcanic rocks of the Grenville Supergroup have been attributed to diapiric emplacement of the batholith. A weakly foliated dyke which crosscuts the fabrics within the metavolcanic rocks proximal to the Elzevir batholith has yielded an U-Pb date of 1229 ± 11/4 Ma. Assuming this is not an inherited age, this confirms that the age of deformation around the batholith is synchronous with the emplacement of the batholith. Structural evidence further suggests that the regional deformation is coeval with the batholith emplacement. The new data raise the possibility that the Flinton Group is older than the Elzevir batholith.

Résumé

Des fabriques linéaires et planaires bien visibles, à la périphérie du batholite d'Elzevir, autrefois daté à 1226 ± 25 Ma, et dans les roches métavolcaniques environnantes du supergroupe de Grenville ont été attribuées à la mise en place du batholite à la façon d'un diapir. Un dyke faiblement feuilleté, qui traverse les fabriques à l'intérieur des roches volcaniques à proximité du batholite d'Elzevir, a été daté par la méthode U-Pb à 1229 ± 11/4 Ma. Cet âge, si l'on admet qu'il n'est pas acquis, confirme que la déformation du pourtour du batholite est synchrone à la mise en place de ce dernier. D'autres indices structuraux semblent indiquer en outre que la déformation régionale est contemporaine de la mise en place du batholite. Les nouvelles données viennent appuyer l'hypothèse selon laquelle le groupe de Flinton serait plus ancien que le batholite d'Elzevir.

INTRODUCTION

The study area is located in southeastern Ontario approximately 150 km west of Ottawa within the Hastings Basin of the Central Metasedimentary Belt in the Grenville Province (Fig. 1). The Hastings Basin is characterized by lower average metamorphic grade than the rest of the southwestern Grenville Province and a weak north-northeast-trending structural grain which tends locally to parallel major intrusive contacts. The supracrustal rocks are host to subcircular plutonic bodies ranging in composition from gabbro to granite.

PREVIOUS WORK

The Elzevir Batholith has yielded an U-Pb zircon date of 1226 ± 25 Ma (Silver and Lumbers, 1966; recalculated according to constants of Steiger and Jäger, 1977). Thompson (1972) and Moore and Thompson (1980) suggested that the batholith was emplaced at this time into rocks of the Grenville Supergroup either during or after a regional deformation event which they termed the Elzevirian Orogeny. This is supported by misoriented foliated xenoliths within the margins of the batholith and Elzevir dykes which crosscut the foliation within the surrounding host metavolcanic rocks of the Grenville Supergroup.

Moore and Thompson (1980) suggested that subsequent regional uplift and erosion unroofed the Elzevir batholith providing basement for the deposition of the Flinton Group sediments which are devoid of Elzevir dykes. They noted pebbles of plutonic rock lithologically similar to the Elzevir batholith and Addington Granite and argued for a correlation between facies changes within the Flinton Group and changes in lithology of the underlying basement. They suggested the age of the Flinton Group is restricted by 1030 ± 20 Ma (zircon date; Silver and Lumbers, 1966)

[1] Mineral Development Division, Newfoundland Department of Mines, P.O. Box 4750, St. John's, Newfoundland A1C 5T7
[2] Geological Survey of Canada, 601 Booth Street, Ottawa, Ontario K1A 0E8

crosscutting pegmatite dykes and the underlying Addington Granite, which was previously dated by Rb-Sr methods at 1060 ± 30 Ma (Bell and Blenkinsop, 1980). It should be noted that U-Pb zircon evidence (unpublished data, O. van Breemen) indicates a minimum age of 1.2 Ga for the Addington Granite. Moore and Thompson (1980) attributed the bulk of the deformation and metamorphism within the Grenville Supergroup and Flinton Group to a post-Flinton Group, Grenville deformation event.

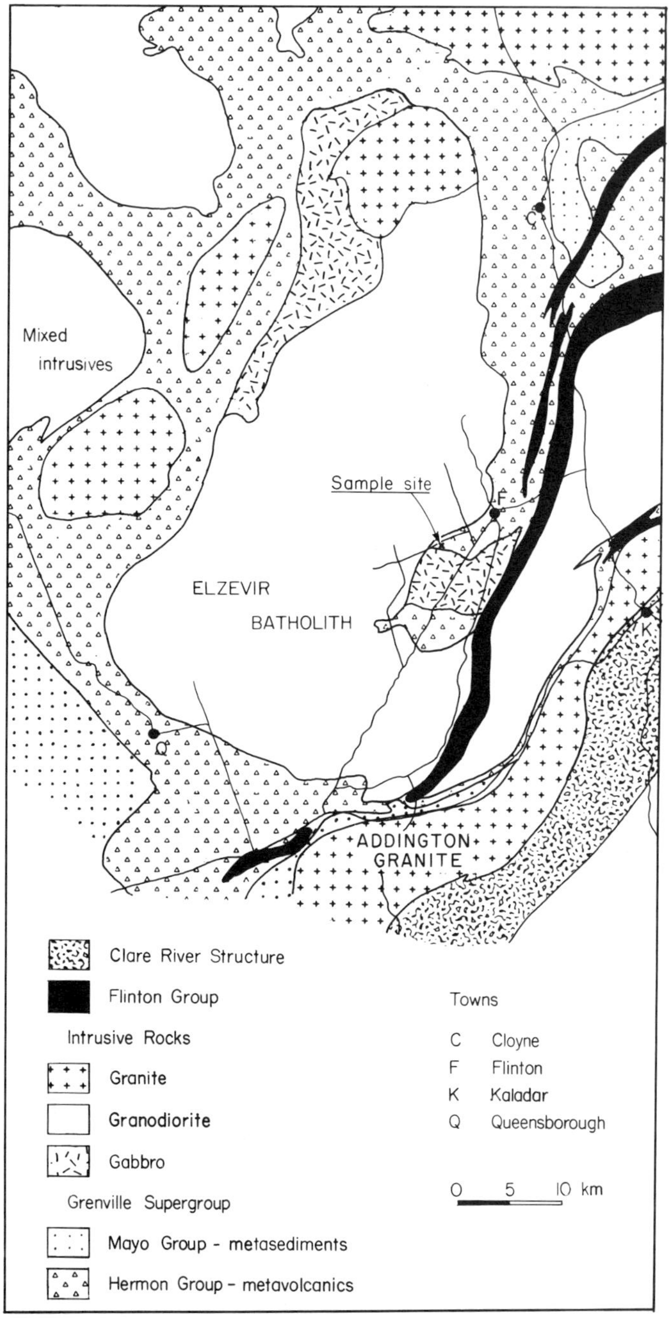

Figure 1. Simplified geological map of the study area showing U-Pb sample site southwest of the town of Flinton.

RECENT WORK

Detailed field studies do not support a differentiation between pre- and post-Flinton Group deformation and metamorphism (Connelly, 1985,1986). Although the Elzevir batholith interior is massive to weakly foliated, well developed planar fabrics within the margin of the Elzevir batholith and in the surrounding metavolcanic rocks are consistently parallel to the batholith contact (Fig. 2). Linear fabrics within this domain are foliation parallel and near vertical. The presence of deformed and undeformed dykes crosscutting fabrics within the metavolcanic rocks suggest the deformation was synchronous with the emplacement of the Elzevir batholith.

The regional post-Flinton Group fabrics do not overprint the fabrics proximal to the Elzevir batholith; instead the planar fabrics merge with one another and form foliation triple junctions where the two fabrics meet at an appropriate angle (Fig. 2). Strain within the rocks bounded by the foliation triple junctions is mainly constrictional. Away from the

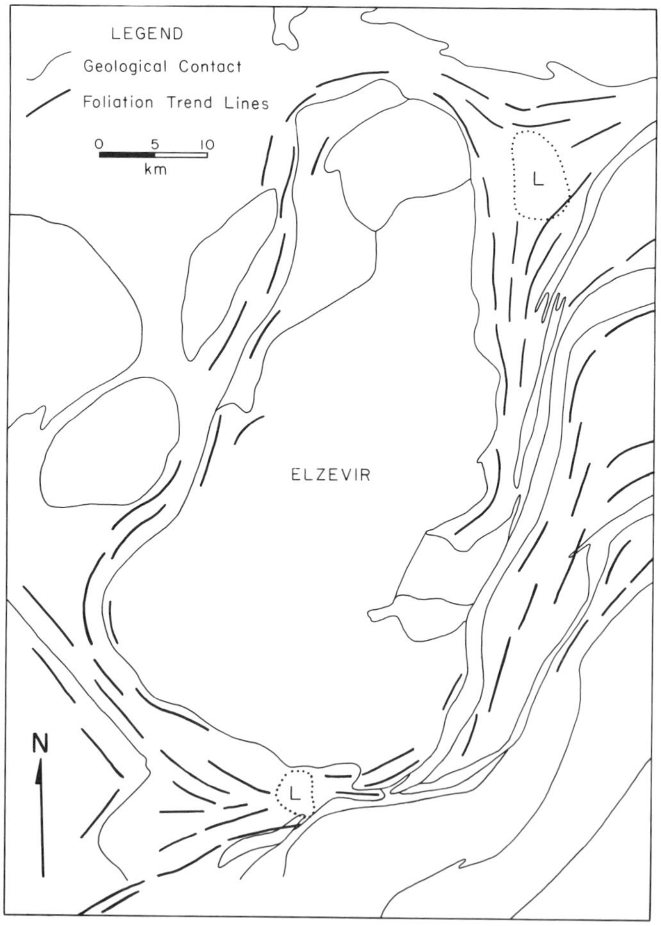

Figure 2. Structural trendline map; geological contacts correspond to Figure 1. Note the concentric cleavage elements proximal and parallel to the Elzevir batholith boundary, the tangential cleavage elements (NE-SW) and the triangular strain interference patterns (foliation triple points in the NE and SW (?) quadrants. Note also areas of constrictional strain (L).

batholith regional planar fabrics within the Grenville Supergroup and overlying Flinton Group trend approximately 060° (Fig. 2); fabrics within these two groups are very similar in orientation and intensity.

Metamorphic grade associated with the regional deformation ranges from greenschist to lower amphibolite facies; metamorphic grade within the Grenville Supergroup is very similar to that in the Flinton group. The metavolcanic rocks proximal to the batholith contain amphibolite facies mineral assemblages which formed by contact metamorphism during the emplacement of the Elzevir batholith. This contact metamorphic aureole is not overprinted by the regional post-Flinton Group metamorphism; in one instance a regional northeast-trending isograd defining the first appearance of hornblende-quartz merges with a similar contact aureole isograd which is north-trending, parallel to the batholith contact.

INTERPRETATION

The presence of strongly developed, contact parallel planar and linear fabrics within the margins of the otherwise massive to weakly foliated Elzevir batholith and in the proximal metavolcanic rocks, suggests that the batholith may have been emplaced as a radially expanding diapir into the Grenville Supergroup. The lack of overprinting of the regional fabrics and isograds on the batholith related fabrics and contact aureole, and the preservation of the foliation triple junctions suggest that the diapiric emplacement occurred during the post-Flinton Group tectonothermal event. This is in contradiction with previous models which envisioned the Elzevirian deformation to be earlier than the unconformable Flinton Group and would require either: 1) the Flinton Group is older than the Elzevirian deformation; 2) the ca. 1226 Ma old Elzevir batholith was remobilized and emplaced diapirically during the post-Flinton Group Grenville deformation; or 3) that most of the deformation within the Grenville Supergroup is Elzevirian and that the Flinton Group is allochthonous, thrust over previously deformed rocks during the Grenville deformation, which had only a minor effect on the underlying rocks.

Given the similarities in fabric orientations and the metamorphic grade between the Flinton Group and underlying rocks, scenario three is considered unlikely. A second diapiric remobilization (scenario two) is not supported by the presence of dykes apparently related to the Elzevir batholith. To account for the lack of Elzevir dykes in the Flinton Group, scenerio one may require that the Flinton Group was faulted against the Elzevir batholith late in the deformation event.

GEOCHRONOLOGY

In an attempt to test models 1 and 2 listed above, a weakly foliated Elzevir granitoid dyke which crosscuts a strong planar and linear fabric within the metavolcanic rocks was dated by U-Pb methods (Parrish et al., 1987) to determine the age of the deformation within the metavolcanic rocks. The sample site is located about 5 km southwest of the town of Flinton in a field along the south side of the road (Fig. 1). Three zircon fractions and two monazite fractions were analyzed (Table 1). Zircons were scarce and of poor quality possibly reflecting their high uranium content. Three data points plot close to a line between 300 and 1300 Ma (Fig. 3). In view of the heterogeneous nature of the zircon population, the upper intercept age may reflect a component of older inherited zircon. The monazites yielded $^{207}Pb/^{206}Pb$ ages of 1228 ± 2 Ma and 1229 ± 2 Ma and these data are tentatively extrapolated to a maximum age of 1240 Ma assuming a line from the lower intercept (300 Ma) of the strongly discordant zircon fractions through the monazite points to concordia (Fig. 3). This 1229 ± 11/4 Ma age is in agreement with the U-Pb age of 1226 ± 25 Ma determined by Silver and Lumbers (1966) for the Elzevir batholith.

Table 1. U.Pb isotopic data on zircon and monazite

Mineral and size fraction (microns)	Weight mg	U ppm	Pb ppm	Measured $^{206}Pb/^{204}Pb$	Isotopic abundance ^{206}Pb = 100			Isotopic ratios		Age, Ma
					^{204}Pb	^{207}Pb	^{208}Pb	$^{206}Pb/^{238}U$	$^{207}Pb/^{235}U$	$^{207}Pb/^{207}Pb$
a, zircon, +74	0.12	845.2	137.4	3131	.222	83.82	198.9	.15760	1.7533	1213.8
b, zircon, − 74 + 44	0.10	2524	307.3	1577	.582	84.91	133.2	.11504	1.2158	1112.1
c, zircon, − 74 + 44	0.19	3302	347.5	1318	.734	84.04	128.6	.099596	1.0106	1030.3
d, monazite	0.13	2219	2259	13739	.046	81.92	4754	.20128	2.2555	1228.3
e, monazite	0.12	2126	2370	10191	.067	82.27	5349	.19996	2.2421	1229.3

Zircons are non-magnetic on Frantz isodynamic separator model L-1, side slope 10° and current 1.8 amps.
All zircon fractions were abraded.

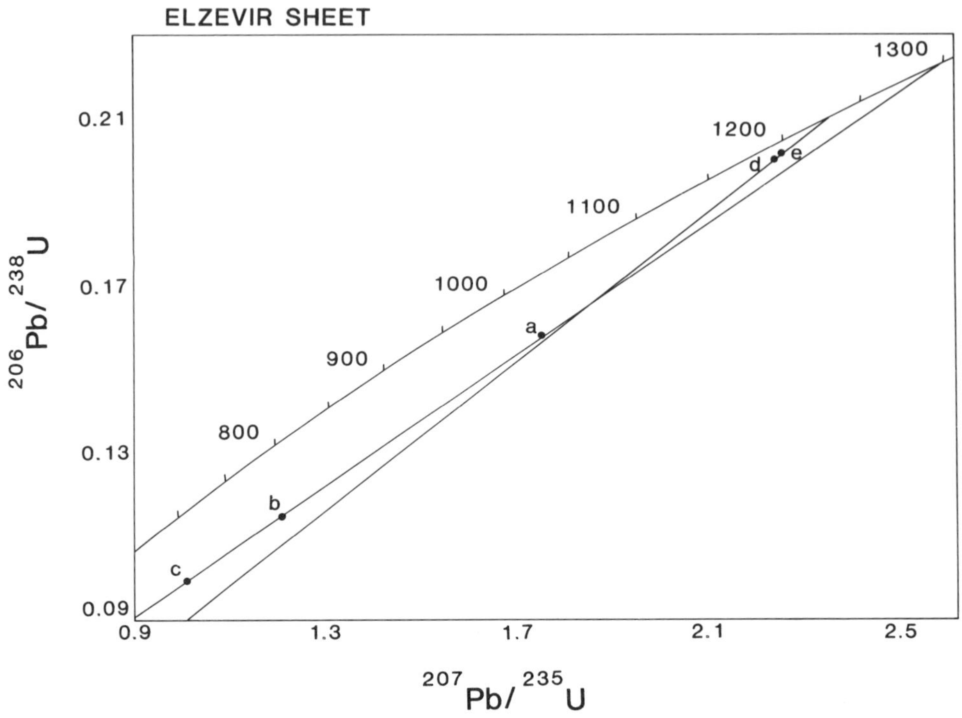

Figure 3. Concordia diagram for the Elzevir dyke. Data for fractions a to e are listed in Table 1. Line through zircon data points a, b, and c passes through upper and lower concordia intercepts of ca. 1.3 Ga and 0.3 Ga. A hypothetical Pb loss line passing through monazite data points d and e has been drawn through 0 Ma.

Given the evidence for the possibility of a weathering-related origin for monazite in some S-type granites, the possibility that the monazite age is inherited (Sawka et al., 1986) cannot be ruled out. However the zircon data are also consistent with the dyke being of Elzevir batholith age.

DISCUSSION

The age data suggest that the deformation which affected the rocks proximal to the Elzevir batholith, and by extension the regional deformation, occurred at the time of emplacement at ca.1226 Ma. If this interpretation is correct, and if the Flinton is not significantly allochthonous with respect to the underlying rocks, then the Flinton Group must be older than the Elzevir batholith. Moore and Thompson (1980) suggested the maximim age of the Flinton Group is constrained by the Addington granite; which, however, has recently yielded an age similar to the Elzevir batholith (see above).

Given the respective uncertainties for ages of the Elzevir batholith and Addington Granite, an age older than the Elzevir batholith may be permissible for the Flinton Group.

ACKNOWLEDGMENTS

This manuscript was critically read by R.R. Parrish.

REFERENCES

Bell, K. and Blenkinsop, J.
1980: Whole rock Rb/Sr studies in the Grenville Province of southeastern Ontario and western Quebec — a summary report: *in* Current Research, Part C, Geological Survey Canada, Paper 80-1C, p. 152-154.

Connelly, J.N.
1985: The Elzevir batholith: emplacement history with respect to the Grenville Supergroup and Flinton Group, southeastern Ontario; *in* Current Research, Part B, Geological Survey of Canada, Paper 85-1B, p. 161-167.

1986: The emplacement history of the Elzivir batholith relative to the regional deformation of the supracrustal rocks in the Central Metasedimentary Belt, Grenville Province, southeastern Ontario; unpublished M.Sc. thesis, Queens University at Kingston, Kingston, Ontario, 152 p.

Moore, J.M. and Thompson, P.H.
1980: The Flinton Group: a late Precambrian metasedimentary succession in the Grenville Province of eastern Ontario: Canadian Journal of Earth Sciences, v. 17, p. 1685-1707.

Parrish, R.R., Roddick, J.C., Loveridge, D.W. and Sullivan, R.W.
1987: Uranium-lead analytical techniques at the geochronology laboratory, Geological Survey of Canada; *in* Radiogenic Age and Isotopic Studies, Geological Survey of Canada, Paper 87-2, p. 3-7.

Sawka, W.N., Banfield, J.F. and Chappell, B.W.
1986: A weathering-related origin of widespread monazite in S-type granites; Geochimica et Cosmochimica Acta, v. 50, p 171- 174.

Silver, L.T. and Lumbers, S.B.
1966: Geochronologic studies in the Bancroft-Madoc area of the Grenville Province, Ontario, Canada; Geological Society of America Special Paper 87, p. 156 (Abstract).

Steiger, R.H. and Jäger, E.
1977: Subcommission on geochronology: convention on the use of decay constants in geo- and cosmochronology; Earth and Planetary Science Letters, v. 36, p. 359-362.

Thompson, P.H.
1972: Stratigraphy, structure and metamorphism of the Flinton Group in Bishop Corners-Madoc area, Grenville Province; Unpublished Ph.D. thesis, Carleton University, Ottawa, Ontario, 268 p.

Grenvillian plutonism in the eastern Grenville province

C.F. Gower[1] and W.D. Loveridge[2]

Gower, C.F. and Loveridge, W.D., Grenvillian plutonism in the eastern Grenville province; in Radiogenic Age and Isotopic Studies: Report 1, Geological Survey of Canada, Paper 87-2, p. 55-58, 1987.

Abstract

This study tests the hypothesis that circular magnetic anomalies in southernmost Labrador and adjacent Quebec can serve to identify Grenvillian plutons. A U-Pb age of 966 ± 3 Ma was obtained on zircon from a monzonite pluton in southeastern Labrador characterized by a small well developed magnetic anomaly. The Grenvillian age confirms the hypothesis in this instance.

Résumé

Dans la présente étude, on cherche à vérifier l'hypothèse selon laquelle les anomalies magnétiques circulaires observées dans la région la plus méridionale du Labrador, contiguë au Québec, peuvent être utilisées pour identifier les plutons grenvilliens. On a daté à 966 ± 3 Ma, par la méthode U-Pb employée sur du zircon, un pluton monzonotique du sud-est du Labrador, caractérisé par une petite anomalie magnétique très évidente. L'âge grenvillien confirme l'hypothèse formulée dans ce cas.

INTRODUCTION

Although considerable evidence exists for Grenvillian resetting of older rocks in the eastern Grenville Province (Schärer et al., 1986; Schärer and Gower, in press; Thomas et al., 1986) only limited information is available documenting Grenvillian plutonism. Figure 1 and Table 1 include all data known to the authors that can be interpreted as pluton emplacement ages in the eastern Grenville Province. The reason for the relative dearth of evidence for Grenvillian plutonism can be attributed, in part, to the fact that most geochronology in the section of the Grenville Province outlined in Figure 1 has been carried out within 150 km of the Grenville Front. Numerous high-precision U-Pb zircon dates have demonstrated that this region is largely underlain by granitoid rocks emplaced during the 1710-1650 Ma Labradorian Orogeny (op. cit.). Gower and Owen (1984) and Gower and Ryan (1986) noted, however, that, by analogy with the Sveconorwegian Orogenic Belt (the Scandinavian counterpart of the Grenville Province), Grenvillian plutonism can be anticipated to be more extensive, than presently indicated, in the interior of the Grenville Province.

This study was largely initiated as a result of an observation by Loveridge (1986) that a pluton in the Lac de Morhiban area, dated at 993 ± 3 Ma (U-Pb, zircon) was characterized by a strong, positive, circular aeromagnetic signature. Similar anomalies are common throughout southernmost Labrador and adjacent Quebec, and although it has been surmised informally for sometime that these probably indicate post-tectonic (with respect to the Grenvillian Orogeny) granitic plutons, there was no proof that this was so. Hence, this study set out to test the hypothesis that circular magnetic anomalies can serve to identify Grenvillian plutons.

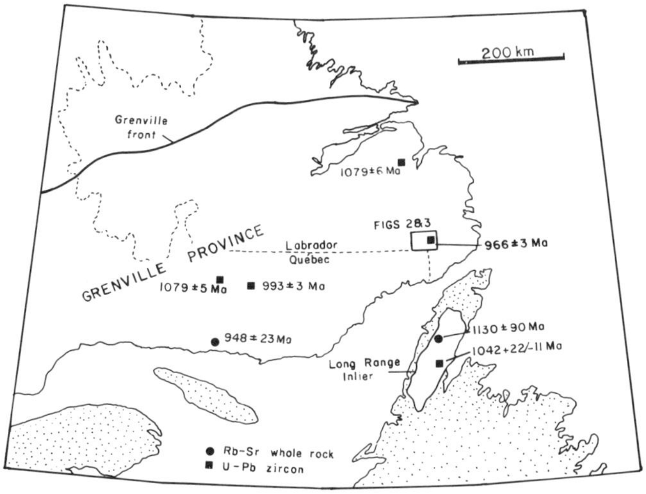

Figure 1. The eastern Grenville Province showing the study area and distribution and ages of dated Grenvillian plutons. Stippled areas are underlain by Phanerozoic rocks.

[1] Mineral Development Division, Newfoundland Department of Mines and Energy, P.O. Box 4750, St. John's, Newfoundland A1C 5T7
[2] Geological Survey of Canadam 601, Booth Street, Ottawam, Ontario K1A 0E8

AREA OF STUDY

The area selected (Fig. 2) contains three circular aeromagnetic anomalies, and was choosen because the magnetic signatures are distinctive and nearly circular, contrasting with the prevailing southeast structural trend in the region. In detail the anomalies have a doughnut form, with magnetically high rims and low interiors (Fig. 3). Such anomalies are typical of high-level plutons, and can be attributed to higher oxygen fugacity, magnetite-forming, thermal aureole conditions in the outer part of the pluton and its country rock envelope.

The geology of the area is poorly known. It is included in the 1:500 000 map of Eade (1962) who depicted the region as being underlain by massive to poorly foliated granite to granodiorite (porphyritic in part), associated with some granitic gneiss and minor syenite. A strip east of 57°00'W will be included in a 1:100 000 mapping project by the Newfoundland Department of Mines in 1987.

The smallest and best-defined anomaly was initially selected for sampling (CG-84-195) and was subsequently dated (this report). Samples were later collected from bedrock underlying the other two anomalies and geochronological studies are planned for 1987-8.

SAMPLE DESCRIPTION

Modal analyses of samples collected from the four sites indicated in Figure 2 are given in Table 2.

Sample CG-84-195 is a medium grained, allotriomorphic granular monzonite comprising plagioclase, alkali feldspar and amphibole, with lesser clinopyroxene, biotite, opaque minerals and quartz, and traces of zircon, sphene and apatite. Plagioclase (moderately calcic) forms well-twinned, lightly sericitized, anhedral grains, locally containing inclusions of perthitic alkali feldspar. Alkali feldspar also occurs as separate anhedral grains of stringlet perthite. Clinopyroxene occurs as cores to mafic grains having amphibole and quartz symplectite rims. Biotite occurs as discrete orange-brown flakes. Two opaque minerals, probably ilmenomagnetite and pyrite, are present. A whole-rock, major and trace element analysis is given in Table 2.

Samples CG-86-697 and CG-86-698, are both coarse grained, massive, pink- weathering rocks that are interpreted

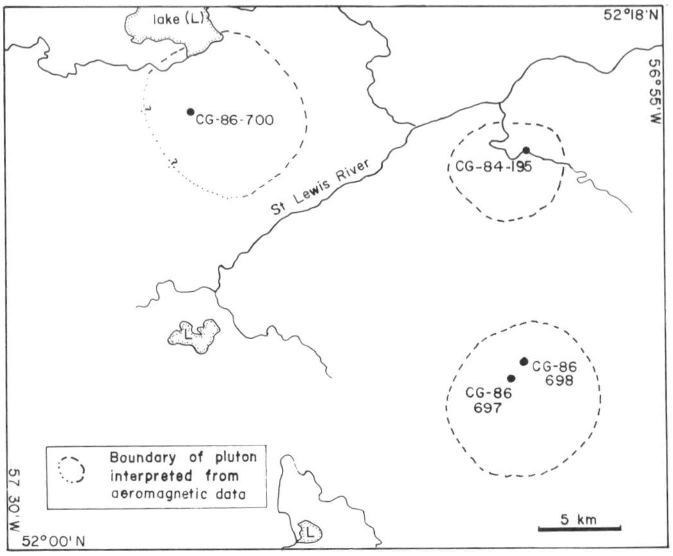

Figure 2. Outline of Grenvillian plutons in the area interpreted from aeromagnetic data. Sample sites are indicated.

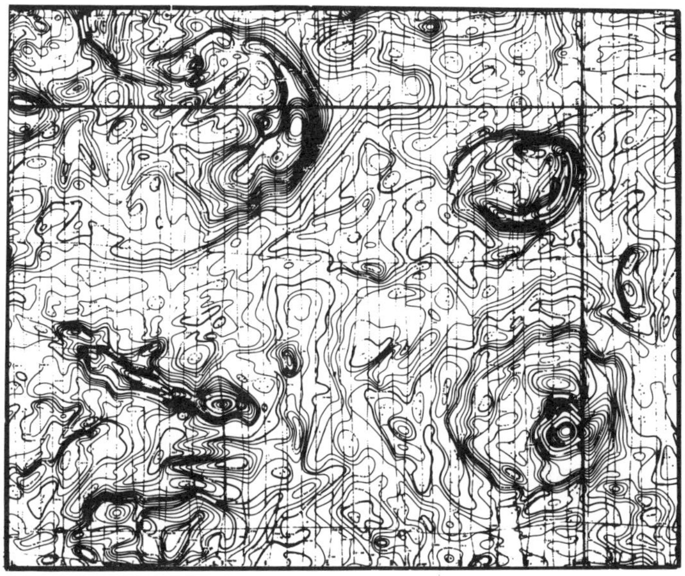

Figure 3. Aeromagnetic patterns in the area outlined in Figure 2. Data from Geological Survey of Canada (1974).

Table 1. Summary of geochronological data for Grenvillian plutons in the eastern Grenville Province

Rock unit	Age (Ma)	Method	Reference
Portland Creek Pond and Hawke's Bay granites (Lake Michael pluton of Bostock et al., 1983)*	1130 ± 90	Rb-Sr w.r.	Pringle et al. (1971)
Southwest Brook granite	1079 ± 6	U-Pb zircon	Schärer and Gower (in press)
Foliated monzogranite	1079 ± 5	U-Pb zircon	Loveridge (1986)
Granite to granodiorite*	1042 + 22/−11	U-Pb zircon	Erdmer (1986)
Monzogranite	993 ± 3	U-Pb zircon	Loveridge (1986)
Monzonite	966 ± 3	U-Pb zircon	this study
Turgeon Lake granite	948 ± 23	Rb-Sr w.r.	Fowler and Doig (1983)

* pooled data from more than one pluton

to be from the same pluton, but they have contrasting compositions. CG-86-697 is granite, whereas CG-86-698 is alkali-feldspar quartz syenite. The fourth sample, CG-86-700, from the third circular anomaly is a coarse grained, massive, pink-weathering granite.

ANALYTICAL PROCEDURES AND RESULTS

Analytical procedures are described by Roddick et al., 1987. Error analysis is according to Roddick (1987). Zircon grains were all broken, probably during routine crushing and mineral separation procedures. Zircon shards from crystal interiors were hand picked for analysis to avoid external surfaces most prone to lead loss. Shards are colourless to pale yellow, clear and generally featureless except for occasional red speck inclusions. No fractures, cores or zoning were noted. All fractions analyzed were non-magnetic at a Frantz Isodynamic Separator slope of $-2°$; none were abraded.

Results of analyses of 4 zircon fractions from CG-84-195 are presented in Table 3 and displayed on a concordia diagram (Fig. 4). The 4 data points have overlapping error enve-

Table 2. Modal analyses and one chemical analysis of samples from sites indicated in Figure 2.

	Modal analyses				Chemical analysis			
	CG-84 195	CG-86 697	CG-86 698	CG-86 700		CG-84-195 wt.%		ppm
Plagioclase	46.1	12.5	6.1	26.5	SiO_2	55.75	F	861
K-feldspar	37.5	48.6	75.5	34.5	Al_2O_3	18.60	Rb	39
Quartz	1.2	35.5	12.8	33.0	Fe_2O_3	2.69	Sr	1134
Biotite	3.3	1.6	1.1	2.7	FeO	3.30	Y	23
Amphibole	5.0	tr	0.6	—	MgO	1.76	Zr	164
Clinopyroxene	2.6	—	1.1	—	CaO	5.00	Nb	14
Apatite	1.0	tr	tr	tr	Na_2O	4.95	Ga	31
Zircon	tr	tr	tr	tr	K_2O	3.70	Ba	6369
Sphene	tr	tr	tr	1.3	TiO_2	1.27	V	22
White mica	—	0.3	—	0.5	MnO	0.16	Ce	115
Opaques	3.1	1.3	2.5	1.3	P_2O_5	0.60	Th	nd
					LOI	0.63	U	0.1
					Total	98.41		

tr = trace amounts
— = not present
nd = not detected

Table 3. Analytical data, zircon fractions from sample CG-84-195

Fraction no. and grain size (μm)	Weight (mg)	U (ppm)	Pb (ppm)	Measured $^{206}Pb/^{204}Pb$	Isotopic abundance ^{206}Pb = 100 ^{204}Pb	^{207}Pb	^{208}Pb	$^{206}Pb/^{238}U$	$^{207}Pb/^{235}U$	$^{207}Pb/^{206}Pb$ age (Ma)
1 + 149	0.349	52.85	10.53	1529	0.0423	7.738	37.64	0.16114	1.5850	967.2
2 + 149	0.195	46.35	9.39	774	0.0830	8.308	41.36	0.16108	1.5816	963.4
3 + 149	0.120	75.45	15.08	1523	0.0294	7.557	37.65	0.16129	1.5872	968.2
4 − 149 + 105	0.072	59.73	12.10	761	0.0558	7.925	40.56	0.16101	1.5822	965.1

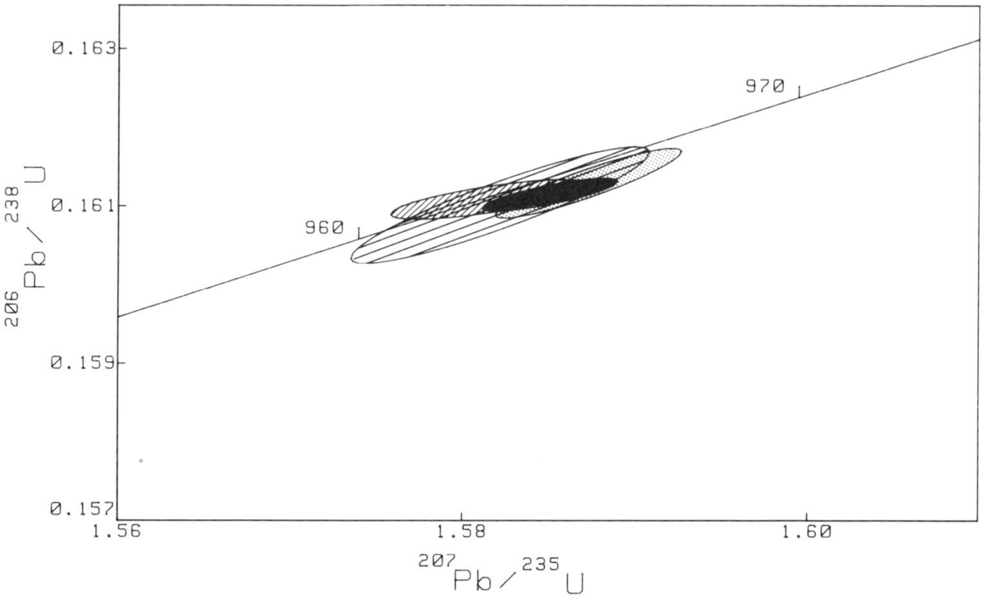

Figure 4. Concordia diagram showing U-Pb isotopic data from 4 zircon fractions, CG-84-195.

lopes. Two of these are concordant within analytical uncertainty and the other two almost concordant. This is in accordance with the low measured U contents, 46 to 75 ppm, and suggests a simple history for these zircons: zircon crystallized at 966 ± 3 Ma and lost minor amounts of Pb at or near 0 Ma. The age quoted is the average of the $^{207}Pb/^{206}Pb$ ages and the uncertainty (95% confidence) is derived from the standard error of the mean. The data points are sufficiently concordant that even if the lead loss occurred at 500 Ma rather than 0 Ma (an unlikely event considering the low U contents) the calculated age would only increase by 2 Ma.

DISCUSSION

The data presented here support the hypothesis that circular, positive magnetic anomalies in the southeast Grenville Province delineate Grenvillian plutons. Comparison of magnetic data with plutons that have been dated already show, however, that not all Grenvillian plutons are necessarily associated with obvious magnetic anomalies.

A summary of dated Grenvillian plutons in the eastern Grenville Province is given in Table 1. It should be noted that these include a fairly broad range of rock types. The dated units are too widespread geographically, and insufficient compositional data are available to make comments on their possible co-sanguinuity or mode of origin. An observation of potential economic significance is that at least two of the units, the Turgeon Lake granite and the Southwest Brook granite west of Paradise River are associated with U enrichment. This is indicated by known U mineralization in the case of the Turgeon Lake granite (Hauseux, 1977) and by anomalous radioactivity and high, lake sediment U anomalies in the case of the Southwest Brook granite (Gower et al., 1985).

REFERENCES

Bostock H.H., Cumming, L.M., Williams, H. and Smyth, W.R.
1983: Geology of the Strait of Belle Isle area, northwestern insular Newfoundland, southern Labrador, and adjacent Quebec; Geological Survey of Canada, Memoir 400, 145 p.

Eade, K.E.
1962: Geology, Battle Harbour — Cartwright, Labrador; Geological Survey of Canada, Map 22-1962.

Erdmer, P.
1986: Geology of the Long Range Inlier in the Sandy Lake map area, western Newfoundland; *in* Current Research, Part B, Geological Survey of Canada, Paper 86-1B, p. 19-27.

Fowler, A.D. and Doig, R.
1983: The age and origin of Grenville Province uraniferous granites and pegmatites; Canadian Journal of Earth Sciences, v. 20, p. 92-104.

Geological Survey of Canada
1974: Aeromagnetic Map 13A and 3D, Battle Harbour; Geological Survey of Canada, Map 7377G, scale 1:250 000.

Gower, C.F. and Owen, V.
1984: Pre-Grenvillian and Grenvillian lithotectonic regions in eastern Labrador — correlations with the Sveconorwegian Orogenic Belt in Sweden; Canadian Journal of Earth Sciences, v. 21, p. 678-693.

Gower, C.F., Noel, N. and Van Nostrand, T.
1985: Geology of the Paradise River region, Grenville Province, Labrador; *in* Current Research, Mineral Development Division, Newfoundland Department of Mines and Energy, Report 85-1, p. 19-32.

Gower, C.F. and Ryan, B.
1986: Proterozoic evolution of the Grenville Province and adjacent Makkovik Province in eastern-central Labrador; *in* The Grenville Province, Geological Association of Canada, Special Paper 31, p. 281-296.

Hauseux, M.A.
1977: Mode of uranium occurrence in a migmatitic granite terrain, Baie Johan Beetz, Quebec; Canadian Mining and Metallurgical Bulletin, v. 70, p. 110-116.

Loveridge, W.D.
1986: U-Pb ages on zircon from rocks of the Lac de Morhiban map area, Quebec; *in* Current Research, Part A, Geological Survey of Canada, Paper 86-1A, p. 523-530.

Pringle, I.R., Miller, J.A. and Warrell, D.M.
1971: Radiometric age determinations from the Long Range Mountains, Newfoundland; Canadian Journal of Earth Sciences, v. 8, p. 1325-1330.

Roddick, J.C.
1987: Generalized numerical error analysis with applications to geochronology and thermodynamics; Geochimica et Cosmochimica Acta, v. 51, p. 2129-2135.

Roddick, J.C., Loveridge, W.D. and Parrish, R.R.
1987: Precise U/Pb dating of zircon at the subnanogram Pb level; Isotope Geoscience, v. 66, p. 111-121.

Schärer, U., Krogh, T.E. and Gower, C.F.
1986: Age and evolution of the Grenville Province in eastern Labrador from U-Pb systematics in accessory minerals; Contributions to Mineralogy and Petrology, v. 94, p. 438-451.

Schärer, U. and Gower, C.F.
— Crustal evolution in eastern Labrador: constraints from precise U-Pb ages; Precambrian Research, (in press).

Thomas, A., Nunn, G.A.G. and Krogh, T.E.
1986: The Labradorian Orogeny: evidence for a newly identified 1600 to 1700 Ma orogenic event in Grenville Province crystalline rocks from central Labrador; *in* The Grenville Province, Geological Association of Canada, Special Paper 31.

U-Pb ages on zircon from the Maggo Gneiss, the Kanairiktok Plutonic Suite and the Island Harbour Plutonic Suite, coast of Labrador, Newfoundland

W.D. Loveridge[1], I. F. Ermanovics[1] and R.W. Sullivan[1]

Loveridge, W.D., Ermanovics, I.F., and Sullivan, R.W., U-Pb ages on zircon from the Maggo Gneiss, the Kanairiktok Plutonic Suite and the Island Harbour Plutonic Suite, coast of Labrador, Newfoundland; in Radiogenic Age and Isotopic Studies: Report 1, Geological Survey of Canada, Paper 87-2, p. 59-65, 1987.

Abstract

A U-Pb age of 3105 + 6/−9 Ma was determined on zircon from the Maggo Gneiss, dating the time of emplacement of an igneous precursor to the gneiss. An even older component of zircon within the gneiss was recognized but not dated.

U-Pb measurement on zircon provided a minimum age of emplacement for the Kanairiktok Plutonic Suite of 2838 Ma. Previous Rb-Sr studies by others indicated a minimum age of about 2750 Ma for the time of epidote-amphibolite facies metamorphism.

An age of 1805 ± 5 Ma (U-Pb zircon) was obtained for the Island Harbour Plutonic Suite, providing minimum ages for the Moran Lake and possibly Aillik groups as well as defining a minimum age for metamorphism and deformation of the northwestern Makkovik Province. The age is in agreement with a 1802 + 13/−7 Ma U-Pb age obtained by other workers on zircon from the Long Island granodiorite in Kaipokok Bay.

Résumé

On a daté par la méthode U-Pb appliquée au zircon, à 3105 + 6/−9 Ma le gneiss de Maggo, donc de la mise en place d'un précurseur igné du gneiss. On a même identifié une composante plus ancienne du zircon au sein du gneiss, mais on ne l'a pas daté.

La datation par la méthode U-Pb effectuée sur du zircon a indiqué un âge minimum de 2838 pour la mise en place de la série plutonique de Karnairiktok. Des datations antérieures réalisées par d'autres chercheurs à l'aide de la méthode Rb-Sr avaient permis d'établir un âge d'environ 2750 Ma pour le métamorphisme du faciès à épidote et amphibolite.

On a déterminé un âge de 1805 ± 5 Ma (méthode U-Pb sur du zircon) pour la série plutonique d'Island Harbour, datation qui fournit par le même fait l'âge minimum du groupe de Moran Lake et peut-être du groupe d'Aillik, en même temps que l'âge minimum du métamorphisme et de la déformation du nord-ouest de la province de Makkovik. Cet âge concorde avec l'âge de 1802 + 13/−7 Ma obtenu par d'autres chercheurs à l'aide de la méthode U-Pb sur du zircon provenant de la granodiorite de Long Island, dans la baie de Kaipokok.

GEOLOGICAL SETTING

A brief description of geological setting and relationships is presented for the 3 units studied. The reader is referred to the listed references for a more complete discussion. Sample locations which are described in the Appendix may be found on the geological sketch map, Figure 14.1, of Grant et al. (1983).

Maggo Gneiss

The gneiss, at the sample locality, is generally a fine grained, granoblastic, subtly layered as shades of grey, biotite and hornblende (± garnet) bearing, leucocratic orthogneiss of tonalitic composition. A younger (?) light grey, medium-to fine-grained, porphyritic granodiorite phase of the gneiss appears to grade imperceptibly in and out of the earlier grey

[1] Geological Survey of Canada, 601, Booth Street, Ottawa, Ontario K1A 0E8

phase. The Maggo Gneiss (Ermanovics et al., 1982) hosts mafic gneisses, amphibolites, ultramafics and rare pelitic remnants that indicate, commonly upper amphibolite facies, possibly retrogressed granulite facies, and high-pressure amphibolite facies (staurolite-garnet-cordiorite-kyanite-biotite-rutile). These assemblages accompany a pronounced NNW-trending planar and moderate SE-plunging linear structural fabric (Hopedalian deformation) that appears to be superimposed on previously polydeformed rocks. The Hopedalian structures were subsequently overprinted to produce NE-trending and mainly NE-plunging planar and linear fabrics at epidote- amphibolite grade during ductile shearing (Fiordian deformation).

A phase of Maggo Gneiss intrudes the Hunt River Group that comprises mafic volcanics and ultramafic rocks, and lesser amounts of greywacke, pelite and calcareous rocks. Some of the amphibolitic and ultramafic relics in Maggo Gneiss, including the kyanite rock, were probably derived from Hunt Group. Accordingly, the age of Maggo Gneiss can also be considered as a minimum age for Hunt Group.

Several Rb-Sr whole-rock determinations of various phases of Maggo Gneiss have yielded ages in the range 3200 to 2900 Ma and are considered to be 'igneous' ages modified by metamorphism (Grant et al., 1983; G. Finn, personal communication 1986). A Rb-Sr whole-rock analysis from the present sample locality yielded an age of 3011 $\pm$ 100 Ma, initial ratio 0.7019 $\pm$ 0.0007, (Voner, 1985).

Maggo Gneiss at the present locality has not been affected by Fiordian deformation, although retrograde metamorphism (chlorite, epidote, fibroblastic amphibole), in keeping with this deformation, is evident. The sample is fine grained (<0.5 mm), grey, granoblastic and comprises poikiloblastic andesine, 10-15% biotite + poikiloblastic hornblende $\pm$ skeletal garnet, and about 10-15% quartz and rare grains of microcline. Flattened, cigar-shaped concentrations of biotite in quartz-rich matrices range in length from a few centimetres to several metres and accentuate the linear, dominant fabric of the gneiss at this locality.

Kanairiktok Plutonic Suite

The Kanairiktok Plutonic Suite (Ermanovics et al., 1982), occurs as a group of small, elongate plutonic bodies of tonalite and granodiorite in the region of Kijuksuatalik Island (Diorite Island, Grant et al., 1983) and extends, as a large batholith, southwestward to underlie large areas of NTS map sheet 13K. The suite contains trains of cognate diorite inclusions and occasional net-veined diorite bodies, suggesting contemporaneous intrusion of acid and basic magma. Locally, reactions between acid and basic components form orbicular diorites which have been stretched by the prevailing Fiordian deformation at epidote-amphibolite grade metamorphism.

The suite intrudes Florence Lake Group supracrustal rocks and consequently the age of the suite also serves as a minumum age for Florence Lake Group. At the present locality the tonalite is a mesocratic, coarse grained, zoned andesine, hornblende-bearing, weakly foliate, rock in which the Fiordian structural fabric is only weakly developed. Nevertheless, metamorphism accompanying this deformation has here produced fibroblastic green amphibole, chlorite, quartz-epidote symplektites and epidote veins.

A Rb-Sr whole-rock determination at this locality yields an age of 2831 $\pm$ 178 Ma, initial ratio 0.7017 $\pm$ 0.0004 (Voner, 1985). The time of the epidote-amphibolite facies metamorphism is determined from a number of Rb-Sr whole-rock age determinations to be about 2750 Ma (Grant et al., 1983; G. Finn, personal communciation, 1986).

Island Harbour Plutonic Suite

The Island Harbour Plutonic Suite (Ryan et al., 1983) intrudes gneisses of Makkovik Province that were derived from adjacent Nain Province (cf. Maggo Gneiss and Kanairiktok Plutonic Suite) during Proterozoic deformation and metamorphism. The suite consists of an early phase of foliate granodiorite and a younger massive porphyritic phase of granite and granodiorite sampled in this study. The map dimensions of the body are 70 $\times$ 25 km, yet despite this size the suite is thought to be a relatively thin sheet like body of granodiorite, capped by porphyritic and megacrystic granite.

Although rocks of the Island Harbour Plutonic Suite do not intrude the metamorphosed Moran and Aillik groups, adjacent massive rocks of the suite are thought to be younger than these supracrustals because they are not metamorphosed.

The sample contain 10% biotite, epidote and lesser muscovite, and equal amounts of zoned oligoclase and porphyritic-poikilitic microcline. Trace amounts of sphene, allanite and occasionally fluorite are also present. Three Rb-Sr whole-rock determinations of the suite yielded the following ages (Grant et al., 1983):

granite 1805 $\pm$ 42 Ma, initial ratio 0.7020 $\pm$ 0.0015
granodiorite 1843 $\pm$ 90 Ma, initial ratio 0.7044 $\pm$ 0.0003,
and 1794 $\pm$ 71 Ma, initial ratio 0.7044 $\pm$ 0.0003

ANALYTICAL PROCEDURES AND RESULTS

Results of U-Pb analyses on 9 zircon fractions from the Maggo Gneiss, 6 from the Kanairiktok Plutonic Suite and 5 from the Island Harbour Plutonic Suite are presented in Table 1 and displayed on concordia diagrams, Figures 1 and 2. Analyses were carried out on these samples at 3 different times; Table 2 relates fraction numbers to time periods and provides references for analytical techniques.

Zircon morphologies and sample locations are given in the Appendix and individual fraction characteristics are listed in Table 1. All fractions were hand picked to ensure 100% zircon composition. Linear regression of U-Pb data and estimates of uncertainty in age are based on Davis (1982). Analytical uncertainties are estimated for the 1980-84 measurements but precisely calculated for the 1986-87 results (Parrish et al., 1987).

Maggo Gneiss

U-Pb analysis of zircon from the Maggo Gneiss was initiated in 1980 (M5 to M9) but the results obtained were not readily interpretable. More recent measurements (1986, 7: M1 to M4) using smaller, more carefully selected zircon fractions which were abraded prior to dissolution, have clarified the systematics to a certain extent.

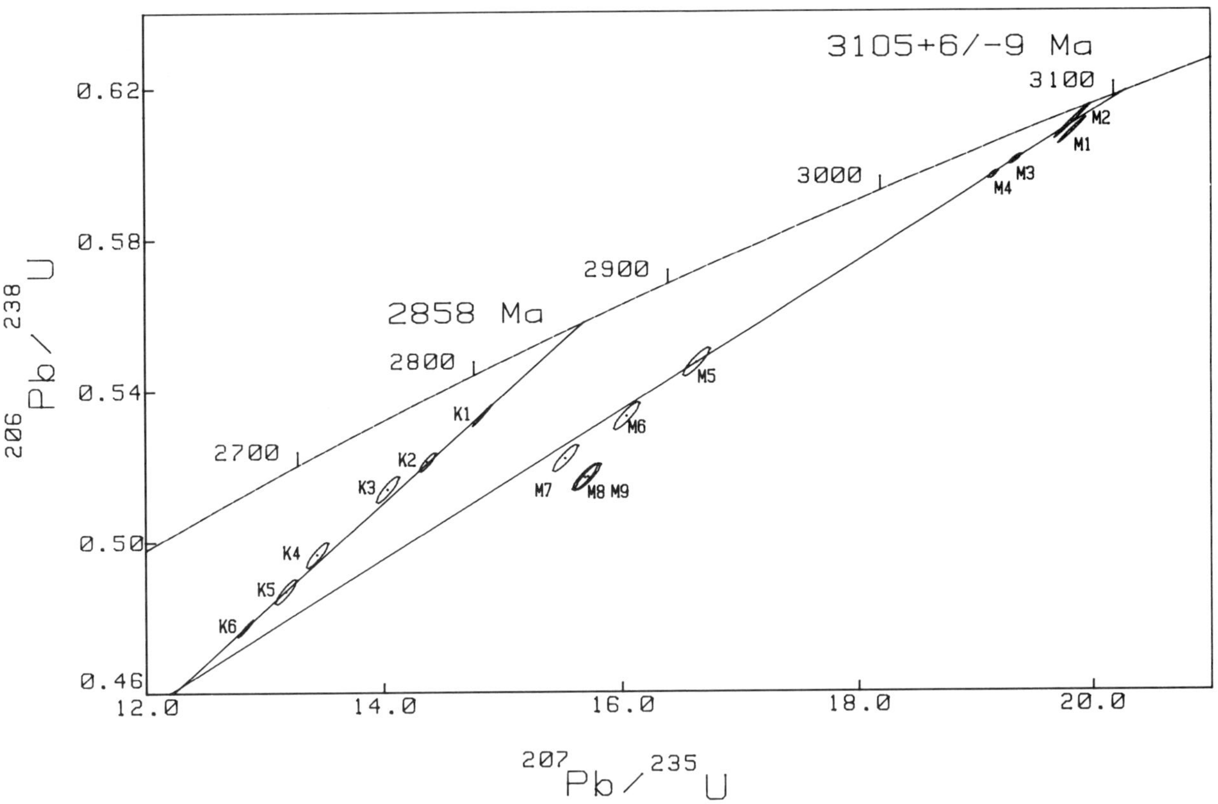

Figure 1. Concordia diagram showing the results of U-Pb analyses of zircon from the Maggo Gneiss (M1 to M9) and the Kanairiktok Plutonic Suite (K1 to K6).

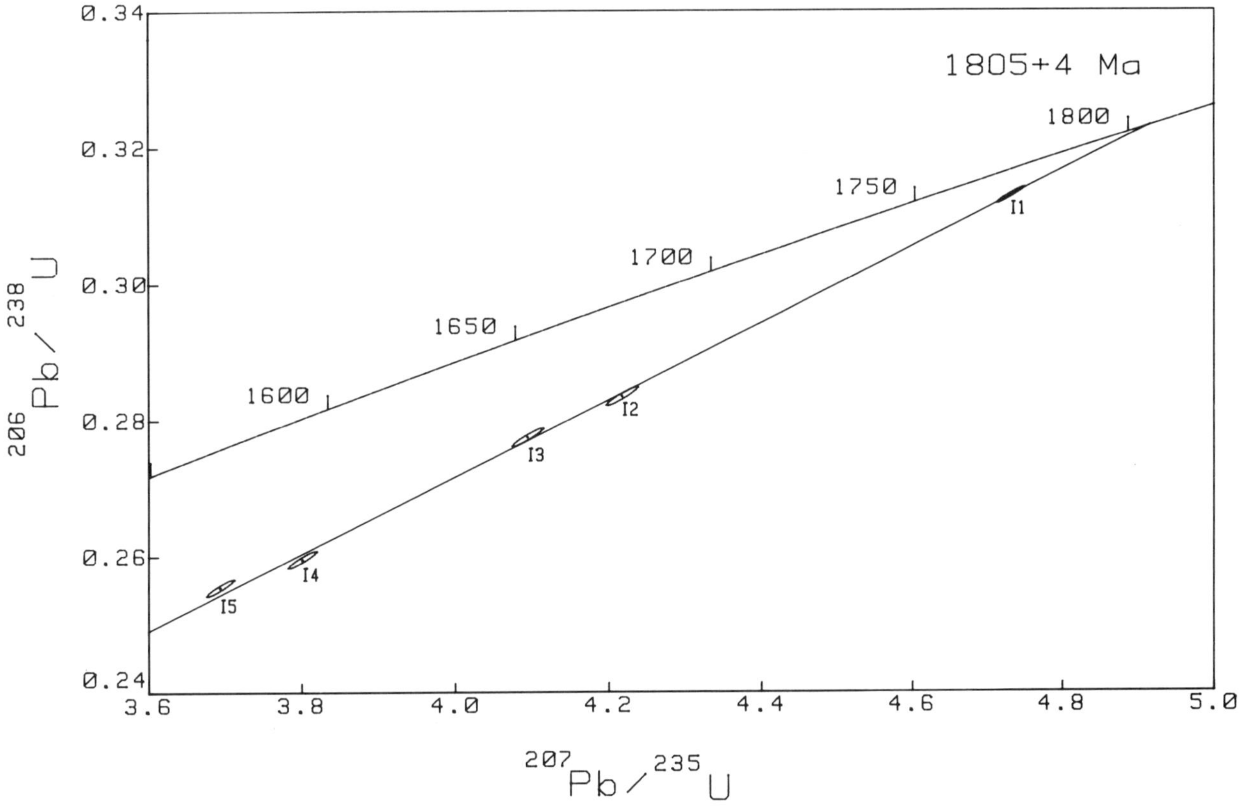

Figure 2. Concordia diagram showing the results of U-Pb analyses on zircon from the Island Harbour Plutonic Suite.

The analytical results must be considered in light of the zircon morphology (see Appendix). The euhedral zoning which is visible throughout the crystal interior suggests a magmatic origin for the precursor to the gneiss (Silver, 1969) while the surface embayments and necking indicate chemical corrosion during the gneiss forming event (Jocelyn and Pidgeon, 1974). A thin layer of surficial zircon, coating most grains, indicates minor zircon growth following resorption.

Four zircon fractions (M1 to M4) were heavily abraded prior to analysis, removing any surficial zircon layer present. The clearest zircon was selected for M1 and M2 without regard to shape, while fractions M3 and M4 sampled clear members of the long prismatic and rounded ovaloid populations respectively. M1 was a more magnetic fraction (Frantz slope = 1°); M2, M3, and M4 were less magnetic fractions at the same slope.

Fractions M2, M3 and M4 yielded collinear data points whereas the M1 point fell distinctly to the right of the other three (Fig. 1). Regression of the three collinear results yields concordia intercept ages 3098 + 25/ −7 and 1543 Ma. There is no apparent difference between the results from the long prismatic (M3) vs. rounded (M4) zircons except that the former have a higher U content.

Table 1. Analytical data, U-Pb measurements on zircon

Sample fraction	Weight (mg)	U (ppm)	Pb (ppm)	Measured $^{206}Pb/^{204}Pb$	Isotopic abundance, $^{206}Pb = 100$ ^{204}Pb	^{207}Pb	^{208}Pb	$^{206}Pb/^{238}U$	$^{207}Pb/^{235}U$	$^{207}Pb/^{206}Pb$ age (Ma)
Maggo Gneiss										
M1. IV m1°	0.017	250.6	176.0	3760	0.0087	23.75	10.88	0.60802	19.829	3097
M2. II n1°A	0.011	280.2	197.8	2521	0.0156	23.74	11.28	0.61050	19.831	3091
M3. II n1°lpA	0.004	458.0	320.0	1466	0.0255	23.66	12.63	0.60052	19.353	3078
M4. II n1°rA	0.028	333.7	227.5	5696	0.0083	23.39	9.81	0.59649	19.161	3073
M5. IV n-2°	3.72	395.4	239.5	7647	0.0069	22.12	6.82	0.54726	16.634	2984
M6. I	0.28	502.2	295.6	485	0.0139	21.99	6.99	0.53300	16.043	2968
M7. I	3.32	500.1	286.9	10360	0.0036	21.62	6.32	0.52191	15.527	2949
M8. IV m-2°	2.01	458.2	262.4	3399	0.0130	22.16	7.16	0.51687	15.691	2982
M9. V	2.45	441.2	251.0	9354	0.0015	22.06	5.92	0.51704	15.716	2984
Kanairiktok Plutonic Suite										
K1. II n1°A	0.024	201.6	118.7	2804	0.0159	20.36	8.66	0.53327	14.829	2840
K2. II n-2°	3.49	215.9	124.6	4096	0.0198	20.24	8.15	0.52099	14.370	2827
K3. VII	3.42	285.4	167.3	2532	0.0248	20.10	13.54	0.51389	14.034	2810
K4. VII ce	1.58	156.6	85.01	1916	0.0145	17.17	7.89	0.49656	13.444	2796
K5. I m-2°	4.41	257.1	137.7	2391	0.0296	19.98	9.19	0.48688	13.179	2796
K6. III n1°A	0.017	295.4	155.3	2501	0.0196	19.74	9.14	0.47725	12.838	2786
Island Harbour Plutonic Suite										
I1. VI N0°A	0.022	347.3	113.2	2735	0.0157	11.183	10.70	0.31281	4.7313	1794
I2. VI n0°	3.26	554.2	163.2	1738	0.0565	11.562	11.92	0.28341	4.2175	1765
I3. VIII n+2°A	1.19	895.0	257.5	1354	0.0718	11.682	12.32	0.27739	4.0945	1750
I4. VI m0°,n1°	3.45	582.8	157.4	933	0.1060	12.066	13.95	0.25953	3.8008	1735
I5. VIII n+2°	1.16	914.3	243.4	858	0.1147	12.054	14.56	0.25538	3.6932	1712

Size fractions (μm)

I	−62
II	−62 + 44
III	+62
IV	−74 + 62
V	+74
VI	−105 + 74
VII	+105
VIII	−149 + 105

Pb ppm: radiogenic Pb
Measured $^{206}Pb/^{204}Pb$: spike subtracted
Isotopic abundance, $^{206}Pb = 100$: corrected for fractionation and Pb blank subtracted

m = magnetic at given angle
n = nonmagnetic at given angle
A = abraded
lp = long prismatic
r = rounded
ce = clear ends

Table 2. Periods of time of U-Pb analyses of zircon fractions and references to analytical techniques.

	1986-7	1983-4	1980-1
Maggo Gneiss	M1-M4		M5-M9
Kanairiktok P.S.	K1, K6	K2	K3-K5
Island Harbour P.S.	I 1	I 2-I 4	

References
1980 — 1 Sullivan and Loveridge (1980)
1983 — 4 Van Breemen et al. (1986)
1986 — 7 Parrish et al. (1987)

Fraction M5 and M8 constituted a less magnetic-more magnetic pair, with the data point for M8, the more magnetic fraction, falling to the right (as did M1, also a more magnetic fraction). Data point M5 is collinear with the other 3 non-magnetic fractions, M2 to M4; regression yields concordia intercept ages of 3105 + 6/ −5 and 1740 Ma.

Conversely, data points M6 and M7 also show excellent collinearity with points M2 to M4, yielding ages 3100 ± 4 and 1609 Ma. Since the possible validity of these results cannot be excluded, the uncertainty associated with the above (M2-M5) age is increased to 3105 + 6/ −9 Ma thereby incorporating the 3100 ± 4 Ma result.

A minimum age may be obtained by regressing points M8 and M9 with M2. Since the resultant age, 3096 Ma is encompassed by the 3105 + 6/ −9 result, the error limits are accepted as realistic.

The lower concordia intercept age estimates, 1268 Ma (M1, M8, M9) to 1740 Ma (M2-M5), are higher than that indicated by the Kanairiktok Plutonic Suite results, c. 950 Ma. This is attributed to the higher U content in the Maggo Gneiss zircons and the greater age of the Maggo Gneiss. It is recognized that the thin surficial zircon coating on the Maggo Gneiss zircon may be of apprecably younger age than 3105 Ma. However, the possible effect this would have on U-Pb systematics, increasing calculated lower intercept ages by shifting the more discordant unabraded zircon data points to the left, is discounted as minimal due to the small proportion and low U content of the surficial zircon.

We interpret these results as defining the age of a primary igneous precursor to the Maggo Gneiss. A second, older, zircon component is also present, separable by its more magnetic character (i.e. M1 and M8). Since the data point for M9 is essentially superimposed on that for M8, M9 must also include some of the secondary component. These 3 fractions are probably hybrid in nature consisting mostly of primary zircon with a small constituent of older more magnetic zircon. The older zircon may be present in either of two modes, as zircon grains from a second, older, igneous precursor to the gneiss, or as an inherited component within the zircon of the primary igneous precursor. The presence of metasedimentary inclusions within the gneiss provides a possible source for older zircon grains.

Kanairiktok Plutonic Suite

The 6 data points for the suite form a linear trend (Fig. 1). These results must be interpreted in light of the hybrid nature of the zircon population analyzed (zircon morphology in Appendix). Grains typically consist of a relatively high U, euhedrally zoned, euhedral zircon core enclosed in a thick, lower U, zircon overgrowth.

The euhedral zoning and crystal shape of the core zircons indicate growth in a magmatic environment (Silver, 1969) without major metamorphic rounding or resorption. The thick, rounded, lower U surficial zircon, which shows indistinct crystal faces at the terminations, is interpreted as a metamorphic overgrowth (Pidgeon and Aftalion, 1972).

The Rb-Sr age associated with the epidote-amphibolite facies metamorphism is about 2750 Ma. One would expect a slightly older age for the zircon overgrowths as there probably would have been some period of time between zircon growth and final closure of the Rb-Sr systems.

The locations of the data points in Figure 2 are attributable to 3 phenomena:

1) mixture between magmatic and metamorphic zircon
2) primary lead loss (ca. 1 Ga?)
3) secondary (0 Ma) lead loss.

It is impossible to completely unravel the influence of each of these factors but some inferences can be made.

A minimum age for the magmatic zircon component can be obtained from the highest $^{207}Pb/^{206}Pb$ age obtained, 2840 ± 2 Ma on fraction K1. This establishes a minimum age of emplacement for the Kanairiktok Plutonic Suite at 2838 Ma. Fractions K1 and K6 were abraded, maximizing the ratio of core to surficial zircon. Both data points fall slightly to the right of the general trend, also indicating an enhanced component of magmatic zircon. A chord through these points intersects concordia at 2860 and 958 Ma. These two fractions are relatively non-magnetic, and yield data points collinear within analytical uncertainty with the data points from the other two non-magnetic fractions, K2 and K5. The chord through the 4 points, illustrated in Figure 1, yields concordia intercept ages of 2858 + 4/ −3 and 946 Ma. We suggest that the 2858 Ma age may be a closer approximation to the age of emplacement of the Kanairiktok Plutonic Suite than the 2838 Ma age previously discussed.

Conversely, the two larger size fractions, K3 and K4 yield data points that fall to the left of the linear trend. One of these, K4, was hand picked to include grains with clear ends maximizing the content of metamorphic zircon. The low U content of K4 (157 ppm) also attests to the high proportion of low U, metamorphic zircon. The $^{207}Pb/^{206}Pb$ age from this fraction, 2796 ± 2 Ma, may provide a reasonable maximum for the age of the metamorphic component, for which a minimum of ca. 2750 Ma is given by Rb-Sr measurements.

Both of these latter dates, although appearing to provide reasonable limits, must be considered speculative due to the unknown effects of two stages of lead loss. The linearity of data points suggests primary lead loss at about 950 Ma; some secondary (0 Ma) lead loss has probably also occurred and the combination of the two could effect each point differently.

Island Harbour Plutonic Suite

Data points from the five zircon fractions analyzed form a linear trend (Fig. 2). Fractions I2 to I5, analyzed in 1984, were selected from the clearest most euhedral zircon grains and fragments available, and I3 was heavily abraded in anticipation of achieving a nearly concordant data point. However, since this was not obtained, an additional fraction, I1, was analyzed in 1986 when improved techniques allowed a decrease in sample size by a factor of 100. The very clearest, most non-magnetic zircon was selected and heavily abraded, resulting in a much more concordant point.

The original rock sample was a composite of 7, one kilogram samples of similar pink granite collected from 7 localities along a bay 4 km south of Arachat Tickle on the Labrador coast. Although the rocks are probably of the same age, they would not be expected to have precisely the same lead loss systematics. Thus the deviation from collinearity of the data points probably results from minor variations in time and amount of lead loss in the zircon analyzed. This lead loss systematic is comparable with the model for lead loss proposed by Davis (1982) in his method for regression of nonlinear U/Pb data.

Using this method, the concordia intercept ages obtained by regressing data from the five analyses are 1805 ± 5 and 522 Ma. The euhedral zircon morphology and zoning indicate a magmatic origin for the zircon. The 1805 ± 5 Ma result is therefore interpreted as the time of emplacement of this member of the Island Harbour Plutonic Suite.

DISCUSSION AND CONCLUSIONS

Maggo Gneiss

An age of 3105 + 6/ −9 Ma has been determined for a primary magmatic precursor to the Maggo Gneiss. A smaller component of older zircon is present; this may either be inherited zircon present within the individual grains of the primary magmatic zircon, or it may constitute a second component derived, perhaps from metasedimentary inclusions within the gneiss. A surficial veneer of low U zircon indicates a third period of zircon growth at, or more recent than, 3105 +6/ −9 Ma.

This age agrees with the less precise Rb-Sr age of Voner (1985) of 3011 ± 100 Ma on rocks from the same locality.

Kanairiktok Plutonic Suite

Due to the hybrid nature of the zircon analyzed (central magmatic zircon overgrown by younger metamorphic zircon) only a minimum age, 2838 Ma, was obtained for the age of emplacement of the Kanairiktok Plutonic Suite. Previous Rb-Sr whole-rock studies yielded a minimum age of about 2750 Ma for the time of epidote-amphibolite facies metamorphism (Grant et al., 1983; G. Finn, personal communication, 1986).

This minimum age of emplacement is in general agreement with the poorly defined Rb−Sr age of Voner (1985) of 2831 ± 178 Ma for rocks collected from the same location.

Island Harbour Plutonic Suite

An age of 1805 ± 5 Ma has been obtained for the Island Harbour Plutonic Suite. The age of the suite serves as a minimum age for Moran Lake and possibly Aillik groups, as well as indicating the minimum time of metamorphism and deformation of northwesten Makkovik Province.

An early granodioritic phase of the Suite (*see* photo in Korstgård and Ermanovics, 1984, p. 308) is similar to the Long Island grandiorite in Kaipokok Bay. The latter, which is a discordant intrusion into folded and faulted Aillik Group and its basement (Gandhi et al., 1969), has yielded a K-Ar date of 1832 ± 58 Ma on hornblende plus biotite (Gandhi et al., 1969) and recently obtained U-Pb ages of 1802 + 13/ −7 Ma on zircon and 1746 ± 2 Ma on titanite (S.S. Gandhi, T.E. Krogh and F. Corfu, personal communication, 1987). The titanite date is interpreted as a cooling age following an episode of Makkovikian metamorphism. It is thus apparent that the Island Harbour Plutonic Suite and Long Island granodioritic intrusion are coeval, and are possibly cogenetic intrusions that are syn- to post-kinematic in relation to a late phase of Makkovikian deformation.

REFERENCES

Davis, D.W.
1982: Optimum linear regression and error estimation applied to U-Pb data; Canadian Journal of Earth Sciences, v. 19, no. 11, p. 2141-2149

Ermanovics, I.F., Korstgård, J.A., and Bridgwater, D.
1982: Structural and lithological chronology of the Archean Hopedale block and the adjacent Proterozoic Makkovik Subprovince, Labrador: Report 4; *in* Current Research, Part B, Geological Survey of Canada, Paper 82-1B, p. 153-165.

Gandhi, S.S., Grasty, R.L., and Grieve, R.A.F.
1969: The geology and geochronology of the Makkovik Bay area, Labrador; Canadian Journal of Earth Sciences, v. 6, no. 5, p. 1019-1035.

Grant, N.K., Voner, F.R., Marzono, M.C., Hickman, M.H., and Ermanovics, I.F.
1983: A summary of Rb-Sr isotope studies in the Archean Hopedale block and the adjacent Proterozoic Makkovik Subprovince, Labrador: report 5; *in* Current Research, Part B, Geological Survey of Canada, Paper 83-1B, p. 127-134.

Jocelyn, J. and Pidgeon, R.T.
1974: Examples of twinning and parallel growth in zircons from some Precambrian granites and gneisses; Mineralogical Magazine, v. 39, p. 587-594.

Korstgård, J.A. and Ermanovics, I.F.
1984: Archean and early Proterozoic tectonics of the Hopedale Block, Labrador, Canada; *in* Precambrian Tectonics Illustrated; E. Schweizerbart'sche Verlagbuchshandlung (Nagele u. Obermiller), Stuttgart, Germany, p. 295-318.

Parrish, R.R., Roddick, J.C., Loveridge, W.D., and Sullivan, R.W.
1987: Uranium-lead analytical techniques at the Geochronology Laboratory, Geological Survey of Canada; *in* Radiogenic Age and Isotopic Studies: Report 1; Geological Survey of Canada, Paper 87-2.

Pidgeon, R.T. and Aftalion, M.
1972: The geochronological significance of discordant U-Pb ages of oval-shaped zircons from a Lewisian gneiss from Harris, Outer Hebrides; Earth and Planetary Science Letters, v. 17, p. 269-274.

Ryan, A.B., Kay, A., and Ermanovics, I.F.
1983: Notes to accompany maps 83-38 to 83-41 showing the geology of the Makkovik Subprovince between Kaipokok Bay and Bay of Islands, Labrador; Department of Mines and Energy, Newfoundland.

Silver, L.T.
1969: A geochronologic investigation of the Adirondack Complex, Adirondack Mountains, New York; *in* Origin of Anorthosite and Related Rocks, ed. Y.W. Isachsen; Memoir 18, New York State Museum and Science Service.

Sullivan, R.W. and Loveridge, W.D.
1980: Uranium-lead age determinations on zircon at the Geological Survey of Canada: current procedures in concentrate preparation and analysis; *in* Loveridge, W.D., Rubidium-strontium and uranium-lead isotopic age studies, Report 3; *in* Current Research, Part C, Geological Survey of Canada, Paper 80-1C, p. 161-246.

van Breemen, O., Davidson, A., Loveridge, W.D., and Sullivan, R.W.
1986: U-Pb zircon geochronology of Grenville tectonites, granulites and igneous precursors, Parry Sound, Ontario; *in* The Grenville Province, ed. J.M. Moore, A. Davidson and A.J. Baer; Geological Association of Canada Special Paper 31, p. 191-207.

Voner, F.R.
1985: Crustal evolution of the Hopedale Block, Labrador, Canada; unpublished Ph.D. dissertation, Miami University, Oxford, Ohio.

APPENDIX

Zircon descriptions and sample locations

Maggo Gneiss, EE78-189   13N/8E Coast of Labrador

Location: Lat. 55°25′30″N
Long. 60°13′00″W

Four kilometres SSE of Hopedale and one kilometre northeastward along shore from a point marked Uivak Point on mapsheet 13N/8E. Same location as the sample marked "UP" on the map of Grant et al., (1983).

Zircon grain shapes are euhedral with rounded outline to completely rounded equidimensional; length to breadth ratios range from 1:1 to 4:1. Many grains exhibit irregular shapes with embayments and necking. Side to side and end to end parallel growth are also noted. Colour varies from light to brown-purple through hyacinth. Rod, bubble and black speck inclusions are prevalent. Grains in the less magnetic splits are clear, more magnetic grains are less clear.

Examination of etched polished sections indicated that most crystals are zoned throughout except for a thin veneer of unzoned zircon on the surface. Embayments and necking disrupt the pattern of zonation. Some broken grains show a thin veneer of clear zircon on the broken surface indicating that the break occured in situ.

Kanairiktok Plutonic Suite, EE78-178   13N/8E
Coast of Labrador

Locations: Lat. 55°17′30″N
Long. 60°05′55″W

An unnamed island 8 x 3.5 km, immediately SE of Kikkertavak Island. For purposes of the present study this island is named Kijuksuatalik Island derived from Inuit records. On the NW side, this island has a large L-shaped bay that has a NNW-, and then a NE- trending reach. The sample was taken 0.5 km south of the contact with migmatite, on a point on the left side of the NNW-trending part of the bay. Same location as the sample marked "KT" on the map of Grant et al. (1983).

Zircon grain shapes are euhedral with rounded tips to completely rounded, with terminations showing indistinct crystal faces. Etched, polished sections reveal a central, euhedrally zoned, relatively high U, euhedral zircon surrounded by a thick outer coating of featureless or faintly zoned, low U zircon. The thickness of the external zircon may approximate the radius of the central zircon, being thinner on the sides and thicker on the ends. Under a binocular microscope the external coating is visible as clear, hyacinth to colourless sides or tips to the otherwise translucent brown zircon grains; one fraction was specifically picked to include grains with clear ends.

Elongations range from 2:1 to 4:1, rod and bubble shaped inclusions are visible in many grains. The external zircon coating of most grains is permeated with a network of fractures.

Island Harbour Plutonic Suite, EE78-composite   13O/5W
Coast of Labrador

Locations: Lat. 55°17′00″N
Long. 59°46′30″W

One kilogram samples of similar pink granites were collected from seven localities along shore of a bay with the above location. On the map of Grant et al. (1983) the sample location is on the coast, opposite Striped Island.

Zircon crystals are euhedral; etched polished grain mounts reveal euhedral zoning throughout. Crystals are colourless but many are strongly stained red orange; some are clear, most are translucent. About 80 % are broken with the freshness of the broken surfaces indicating breakage during rock crushing and grinding. Most grains show a network of surficial cracks. Length to breadth ratios in whole grains range from 2:1 to 5:1.

A U-Pb age on zircon from a granite pluton, Kamilukuak Lake area, District of Keewatin, establishes a lower limit for the age of the Christopher Island Formation, Dubawnt Group

W.D. Loveridge[1], K.E. Eade[1], and J.C. Roddick[1]

Loveridge, W.D., Eade, K.E., and Roddick, J.C., A U-Pb age on zircon from a granite pluton, Kamilukuak Lake area, District of Keewatin, establishes a lower limit for the age of the Christopher Island Formation, Dubawnt Group; in Radiogenic Age and Isotopic Studies: Report 1, Geological Survey of Canada, Paper 87-2, p. 67-71, 1987.

Abstract

A U-Pb study of zircon from a young granite pluton 22 km north of Carruthers Lake in the Kamilukuak Lake area yielded a concordia intercept age of 1753 +3/−2 Ma. Since this pluton intrudes alkaline mafic trachytes and volcanoclastic sediments of the Christopher Island Formation, Dubawnt Group, this result establishes a minimum age for the Christopher Island Formation. This age agrees with published whole-rock Rb-Sr isochron ages on the very similar Nueltin Lake granite 220 km to the south.

Résumé

La datation par la méthode U-Pb appliquée au zircon, d'un pluton granitique récent situé à 22 km au nord du lac Carruthers, dans la région du lac Kamilukuak, a indiqué un âge de concordance de 1753 +3/−2 Ma au point d'intersection avec la courbe de concordance. Étant donné que ce pluton traverse les trachytes mafiques de nature alcaline et les sédiments volcanolastiques de la formation de Christopher Island dans le groupe de Dubawnt, cet âge représente l'âge minimum de la formation de Christopher Island. De plus, il concorde avec les datations publiées de l'isochrone de la roche totale, calculées à l'aide de la méthode Rb-Sr, sur le granite très similaire du lac Nueltin, situé à 220 km au sud.

GEOLOGICAL SETTING

The simplified geological sketch map, Figure 1, adapted from Eade (1985) and Tella and Eade (1985), shows the general distribution of rock types in the area.

Aphebian plutonic rocks are confined to the northern half of the map area. They consist of two different suites, one pre-Dubawnt Group and the other, post-Dubawnt Group. The former are commonly associated areally with faults, either east- or northeast-trending. For the most part they consist of granite to quartz monzonite and syenite, but some small bodies of diorite or quartz diorite are also present.

Dubawnt Group rocks occur in three distinct areas. The easternmost area, the Yathkyed Lake Basin, is separated from the central area, the Angikuni Lake Basin by a major northeast-trending fault that is a subsidiary of the Tulemalu Fault. The latter fault forms the western margin of the Angikuni Lake Basin. West of the Tulemalu Fault, the Dubawnt Group rocks are part of the major Baker Lake Basin extending to the northeast, but within this area the basin is dismembered by younger intrusive plutonic rocks. Mafic trachytes, flows and pyroclastics, and derived volcaniclastic sediments of the Christopher Island Formation comprise the majority of the Dubawnt Group rocks in this area, with minor occurrence of sandstone and conglomerate of the South Channel Formation sporadically exposed at margins of the basins, and a small area of siltstone and sandstone of the Angikuni formation at the southwest end of the Angikuni Lake Basin. Small lamprophyre dykes, assumed to be genetically related to the mafic trachytes of the Christopher Island Formation, are common in the older rocks adjacent to the Dubawnt Group basins.

The post-Dubawnt Group plutonic rocks present in the northwest part of the area are mostly granite, commonly a distinctive coarse grained porphyritic rock with rapakivi texture that is fluorite bearing. The granite intrudes trachytes of the Christopher Island Formation.

A zircon concentrate was obtained from a sample of a young granite pluton, WN-615-79, collected approximately 22 km north of Carruthers Lake. The granite is a coarse grained, porphyritic, rapakivi textured rock that is fluorite-bearing. Compositionally it is an alkali granite, hornblende-bearing, and is part of a widespread post-tectonic intrusive

[1] Geological Survey of Canada, 601 Booth Street, Ottawa, Ontario K1A 0E8

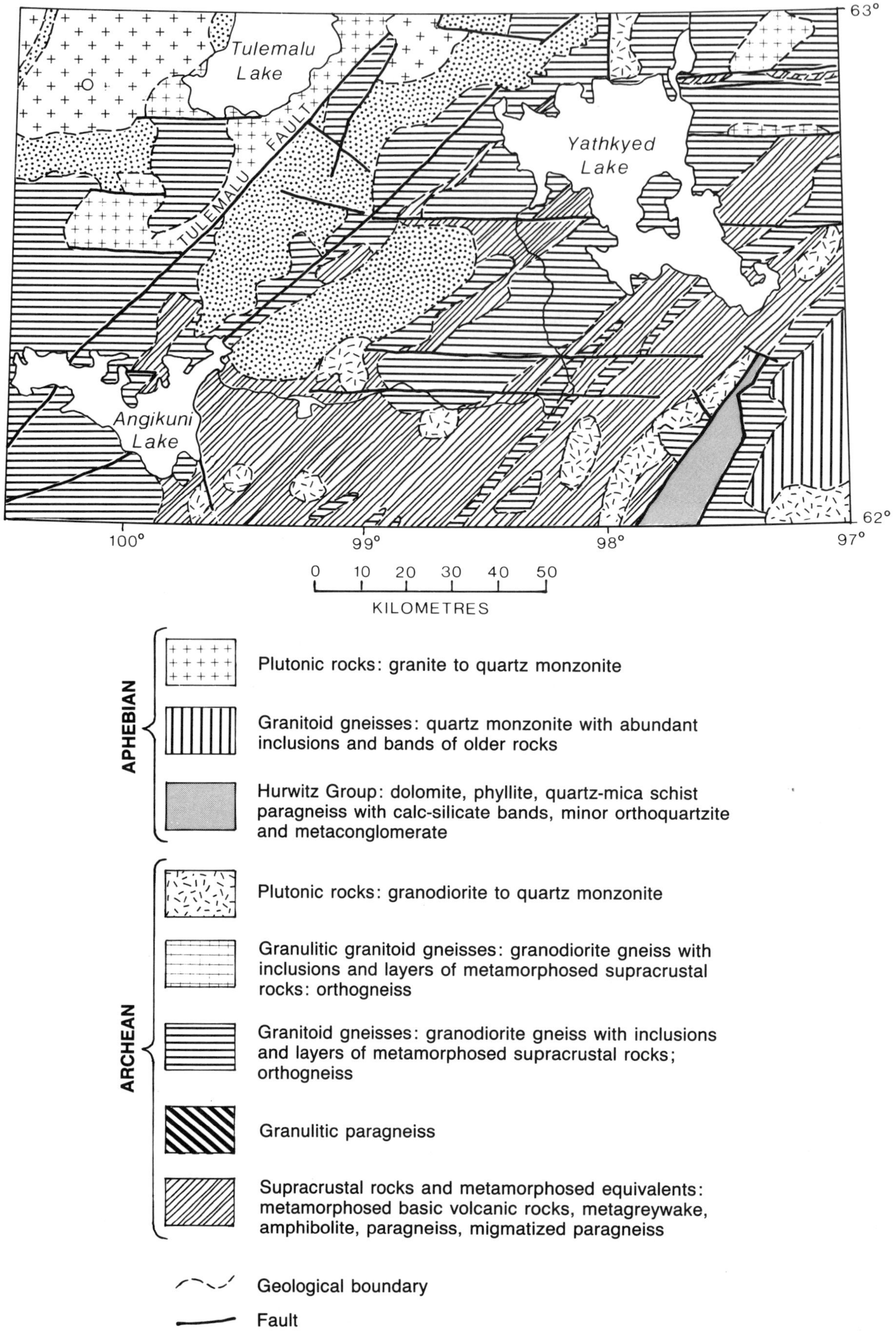

Figure 1. Geological sketch map, Kamilukuak Lake area, showing sample location of WN-615-79 (circle).

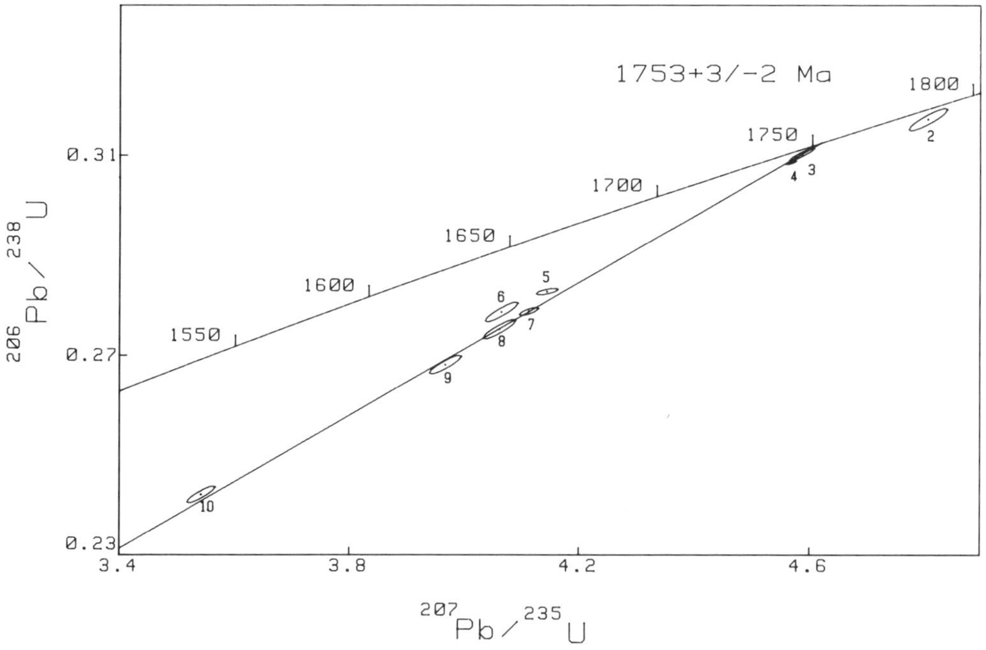

Figure 2. Concordia diagram showing the results of U-Pb analyses of zircon fractions from sample WN-615-79. The data point for fraction 1 falls outside the limits of the diagram.

suite. It is unmetamorphosed and the only structural dislocation is some late faulting, for the most part at the margins of the pluton. A brief description of the granite is included in Tella and Eade (1980).

ANALYTICAL PROCEDURES AND RESULTS

Zircon fractions 2, 6, 8, 9, and 10 were analyzed in 1980 using procedures described by Sullivan and Loveridge (1980). The near concordant result obtained on fraction 2 was thought at that time to have established the geological age, and this age is published on the geological map (Tella and Eade, 1985). However, results from 5 additional fractions analyzed in 1986 and 1987 demonstrated that the zircon systematics were more complex than originally envisioned and provided a revised age estimate. Analytical procedures in use in 1986-87 (Roddick et al., 1987) allowed analysis of quantities of zircon two orders of magnitude smaller than in 1980.

Results are presented in Table 1 and displayed on a concordia diagram (Fig. 2). Analytical uncertainties are estimated for the 1980 measurements but are precisely calculated for the 1986-87 results (Roddick, 1987). A general description of zircon morphology is presented in the Appendix; descriptions of the individual fractions are presented in Table 2. Linear regression of U-Pb data and estimation of uncertainty in age result are based on Davis (1982). The sample location is shown in Figure 1; co-ordinates are given in the Appendix.

Three of the 10 zircon fractions analyzed, fractions 3, 7 and 8, were specifically hand picked to include the clearest, most colourless, most inclusion-free grains. Fractions 3 and 7 (1986) were picked to duplicate fraction 8 (1980) resulting in adjacent error cnvelopes for the data points of fractions 7 and 8. Fraction 3, however, was moderately abraded (25 minutes) producing a data point collinear with those of 7 and 8 but much more concordant. The uranium contents of these fractions (102 to 129 ppm) are considerably lower than those of the other fractions analyzed.

A regression line with 74 % probability of fit, the primary chord, may be fitted to these 3 points yielding concordia intercept ages of 1753 +3/-2 and 58 +57/-58 Ma. The interpretation is straightforward: low uranium zircons crystallized at 1753 +3/-2 Ma and suffered approximately 12 % surficial Pb loss at or near 58 Ma. The discordant surficial zircon was mostly removed from fraction 3 by abrasion.

Table 1. Analytical data, zircon fractions from sample WN-615-79

Fraction no.	Weight (mg)	U (ppm)	Pb (ppm)	Measured $^{206}Pb/^{204}Pb$	Isotopic abundance, ^{206}Pb = 100 ^{204}Pb	^{207}Pb	^{208}Pb	$^{206}Pb/^{238}U$	$^{207}Pb/^{235}U$	$^{207}Pb/^{206}Pb$ age (Ma)
1	0.0733	5382	3353	945.6	0.1054	13.298	49.696	0.45576	7.4663	1938
2	3.42	760.2	275.1	1231	0.071	11.957	23.813	0.31717	4.8073	1798
3	0.0330	101.6	35.83	974.3	0.0354	11.206	22.506	0.31017	4.5857	1753
4	0.0370	2060	669.3	3324	0.0232	11.054	12.481	0.30850	4.5672	1755
5	0.0561	306.6	96.03	618.8	0.1477	12.644	22.877	0.28271	4.1438	1737
6	4.85	1636	516.7	634.0	0.1536	12.675	26.012	0.27865	4.0647	1728
7	0.0533	129.1	40.40	739.1	0.0994	12.057	22.933	0.27868	4.1121	1749
8	4.00	128.1	38.89	708.7	0.1168	12.292	21.274	0.27522	4.0610	1749
9	2.94	363.8	105.8	703.0	0.1193	12.356	19.229	0.26812	3.9672	1754
10	1.93	430.5	113.2	464.5	0.1703	12.935	21.161	0.24212	3.5429	1734

A second group of collinear data points is also included in these results. Fractions 1, 2, 5 and 6 form a secondary chord (Table 3) but in this case the systematics are entirely different.

The data point for fraction 1 shows reverse discordance in that it falls above the concordia curve in the U loss/Pb gain region. Pb gain is unlikely due to the highly radiogenic (95.6 %) nature of the Pb in the zircon, therefore we are led to assume relative U loss. The zircons of this fraction, which we denote type 1, are clear yellow, subhedral, relatively nonmagnetic and have a high measured U content of 5382 ppm. At this U level (compare with 102-129 ppm, fractions 3, 7, 8) radiation damage to the crystal structure would rapidly accumulate, predisposing the zircon to open system behaviour with regard to U-Pb due to minor environmental influences.

The high $^{207}Pb/^{206}Pb$ age of fraction 1, 1938 Ma, is not interpreted as reflecting the age of crystallization of the zircon in that fraction. U loss systematics are such that loss of relatively more U than Pb at 933 Ma, from a zircon that crystallized at 1753 Ma, could produce the measured Pb/U results. In this particular case, where the U content is very high, a more probable history would involve more than one U and/or Pb loss event.

Collinear with fraction 1 are two other high U data points, fractions 2 and 6, with 760 and 1636 ppm U respectively. These fractions sampled all morphologies in the less and more magnetic splits of the + 140 μm grain size component. We interpret this secondary chord as a mixing line consisting of type 1 (fraction 1) zircon as one end number and bulk zircon showing some lead loss as the other. The position of the data

Table 2. Characteristics of zircon fractions.

Fraction no.	Grain size (μm) magnetic split	Comment
1	– 149 +105 N3°, M1°	Light yellow, clear, subhedral, 2 populations, clear yellow subhedral grading to clear colourless euhedral.
2	+ 149 N2°	Colourless to red orange, clear to cloudy white, subhedral with clear grains approaching euhedral.
3	– 149 + 105 N2°	Clearest colourless grains, most inclusion free, abraded 25 minutes.
4	– 149 + 105 N3°, M1°	A subfraction of fraction 1. The best clearest (yellow) grains of fraction 1 from the remaining 50% after analysis.
5	– 149 + 105 N3°, M1°	Similar to fraction 1.
6	+ 149 M2°	Clear euhedral, 2 populations, clear yellow subhedral grading to clear colorless euhedral.
7	– 149 + 105 N2°	Similar to fraction 3 but not abraded.
8	– 149 + 105 N – 1°	Clearest colourless grains, euhedral, bubble and rod inclusions present.
9	– 149 + 105 N0°, M – 1°	Clear, light to yellow reddish brown, euhedral, bubble, black speck and reddish-brown inclusions.
10	– 105 + 74	75% clear colourless euhedral, more elongated than in the coarser splits, 25% clear to highly coloured red brown subhedral.

N = nonmagnetic at given angle
M = magnetic at given angle

Table 3. Results of regressions of various combination of data.

Fraction numbers	Probability of fit (%)	Estimates of age (Ma)	
Primary chord			
3, 7, 8	74.3	1753 + 3/- 2	58 + 57/- 58
3, 4, 7, 8,	27.4	1755 + 2/- 2	87 + 46/- 47
3, 7, 8, 9	5.3	1753 + 3/- 2	30 + 76/- 79
3, 7, 8, 9, 10	0.0002	1753 + 3/- 2	62 + 71/- 62
Secondary Chord			
1, 2, 5, 6	95.1	1807 + 4/- 4	757 + 17/- 17

point for fraction 2, as well as its 760 ppm U content, indicates a substantial component of type 1 zircon. Conversely, fraction 6 zircon which is a more magnetic split and therefore, presumably, excludes type 1 zircon, falls to the left of the primary chord. This suggests earlier lead loss for the fraction 6 zircons than for those generating the primary chord, an argument compatible with their higher U content.

A similar argument with regard to early lead loss would also apply to fractions 5 (on the secondary chord, 307 ppm U) and 10 (431 ppm U, greater relative lead loss due to smaller grain size). The data point for fraction 9 falls just to the right of the primary chord, although the $^{207}Pb/^{206}Pb$ age for this fraction is 1754 Ma which agrees with the primary age within analytical uncertainty. The position of this data point, together with the 364 ppmU content, may indicate the presence of a small component of type 1 zircon. Fraction 4 is a subfraction of fraction 1. Half the zircon picked for fraction 1 was used in the original analyses; the very clearest yellow grains in the remainder were later selected for analysis as fraction 4. A very precise analysis was obtained with the error envelope for fraction 4 immediately adjacent to that of fraction 3, falling on the primary chord but not on the secondary chord. We consider the position on the primary chord fortuitous given its association with fraction 1 and its high U content (2060 ppm).

Results of regressions of various combinations of data points are presented in Table 3. Due to the near concordance of data point 3, all combinations of results selected to approximate the primary chord give essentially the same age result. Therefore points 3, 7 and 8 were chosen to define the age as they also allow the most straightforward interpretation.

DISCUSSION

Hornblende from sample WN-615-79 yields a K-Ar age of 1793 ± 53 Ma (GSC 81-149, Stevens et al., 1982), in agreement with the U-Pb zircon age. A fluorite-bearing granite occurring to the east-northeast, WN-608-79, similar to WN-615-79, gives a K-Ar age on biotite of 1747 ± 32 Ma (GSC 81-152, op. cit.) Muscovite from a granite considered to be part of the same intrusive suite occurring west of the area of Figure 1 gives a K-Ar age of 1757 ± 23 Ma (GSC 81-148, op. cit). These ages are all in agreement with the U-Pb age on zircon of 1753 +3/-2 Ma, within analytical uncertainty.

Since granite, WN-615-79 intrudes the alkaline mafic trachytes and volcaniclastic sediments of the Christopher Island Formation of the Dubawnt Group, this age provides a lower limit on the age of the Christopher Island. Also, it is possible that the calc-alkaline igneous suite, of which the granite is a part, includes the felsic volcanic Pitz Formation of the Dubawnt Group. The age of this granite could therefore be very close to the age of extrusion of the Pitz Formation.

The granite is similar to the Nueltin Lake granite occurring approximately 220 km to the south, differing only in that rapakivi texture is more pronounced in the northern rocks. Both granites are characteristically coarse grained, porphyritic, fluorite-bearing, post-tectonic plutons. Two Rb-Sr whole-rock isochron age measurements on the Nueltin Lake granite (Wanless and Loveridge, 1972, p. 33 and 37) resulted in ages of 1760 ± 16 Ma, initial ratio 0.7052 ± 0.0018 and 1775 ± 61 Ma, initial ratio 0.7060 ± 0.0040. Stated ages, recalculated with ^{87}Rb decay constant = $1.42 \times 10^{-11}\ a^{-1}$, are in good agreement with the U-Pb age on zircon for WN-615-79.

REFERENCES

Davis, D.W.

1982: Optimum linear regression and error estimation applied to U-Pb data; Canadian Journal of Earth Sciences, v. 19, no. 11, p. 2141-2149.

Eade, K.E.

1985: Precambrian geology of the Tulemalu Lake-Yathkyed Lake area; Geological Survey of Canada, Paper 84-11.

Roddick, J.C.

1987: Generalized numerical error analysis with application to Geochronology and thermodynamics; Geochimica et Cosmochemica Acta, v. 51, p. 2129-2135.

Roddick, J.C., Loveridge, W.D., and Parrish, R.P.

1987: Precise U/Pb dating of zircon at the sub-nanogram Pb level; Isotope Geoscience, v. 66, p. 111-121.

Stevens, R.D., DeLabio, R.N., and Lachance, G.R.

1982: Age determinations and geological studies, K-Ar isotopic ages, Report 16; Geological Survey of Canada, Paper 82-2.

Sullivan, R.W. and Loveridge, W.D.

1980: Uranium-lead determinations on zircon at the Geological Survey of Canada: current procedures in concentrate preparation and analysis; *in* Loveridge, W.D., Rubidium-strontium and uranium-lead isotopic age studies, Report 3; *in* Current Research, Part C, Geological Survey of Canada, Paper 80-1C, p. 161-246.

Tella, S. and Eade, K.E.

1980: Geology of the Kamilukuak Lake map area, District of Keewatin: a part of the Churchill Structural Province; *in* Current Research, Part B, Geological Survey of Canada, Paper 80-1B, p. 39-45.

1985: Geology, Kamilikuak Lake, District of Keewatin, Northwest Territories; Geological Survey of Canada, Map 1629A, scale 1:250 000.

Wanless, R.K. and Loveridge, W.D.

1972: Rubidium-strontium isochron age studies, Report 1; Geological Survey of Canada, Paper 72-23.

APPENDIX

WN-615-79, Lat. 62°51'05''N, Long. 100°11'25''W.

Zircon from WN-615-79 comprises a heterogeneous assemblage of equant clear to highly coloured grains of subhedral to euhedral habit. 50 to 60 % are broken, probably during crushing of the original rock for mineral separation. Up to 50 % are cloudy and translucent while the remainder are clear and colourless or pale yellow, but both translucent and clear grains are often heavily stained red orange. Length to breadth ratios range from 1 to 2 reaching 3 in the finer fractions. Magnetic separation concentrates the clearer, colourless to pale yellow grains into the less magnetic fraction while favouring the translucent grains in the more magnetic split. The clearer grains in the more magnetic fraction tend to be a light tan to yellow reddish brown. In the less magnetic split, the clear colourless grains are usually euhedral and sharply terminated whereas the clear yellowish grains have more rounded terminations. Faint traces of euhedral zoning may be seen in some crystals but no distinct zircon cores are visible.

Proterozoic geochronology in the Taltson Magmatic Zone, N.W.T.

H.H. Bostock[1], O. van Breemen[1], and W.D. Loveridge[1]

Bostock, H.H., van Breemen, O., and Loveridge, W.D., Proterozoic geochronology in the Taltson Magmatic Zone, N.W.T.; in Radiogenic Age and Isotopic Studies: Report 1, Geological Survey of Canada, Paper 87-2, p. 73-80, 1987.

Abstract

In the Taltson Magmatic Zone of the northwest Canadian Shield, Proterozoic plutons extend west at least as far as the Paleozoic cover. The westernmost Deskenatlata Granodiorite has a U-Pb zircon age of 1986.4 + 2.4/−2.0 Ma. U-Pb monazite analyses yield ages of 1955 ± 2 Ma for the Slave Monzogranite, 1935 ± 3 Ma for the Natael Monzogranite and 1922 ± 2 Ma for the megacrystic Konth Syenogranite. These monazite ages are tentatively interpreted as the age of crystallization which survived later metamorphism. The three younger granites feature deformation fabrics.

Résumé

Dans la zone magmatique de Taltson, située dans le nord-ouest du Bouclier canadien, des plutons protérozoïques s'étendent vers l'ouest au moins jusqu'à la couverture paléozoïque. La granodiorite de Deskenatlata, la plus occidentale, a été datée par la méthode U-Pb appliquée au zircon, à 1986, 4 + 2,4/−2 Ma. Les datations par la méthode U-Pb faites sur la monazite indiquent 1955 ± 2 Ma pour le monzogranite de la province des Esclaves, 1935 ± 3 Ma pour le monzogranite de Natael et 1922 ± 2 Ma pour le syéonogranite mégacristallin de Konth. Ces âges, obtenus sur la monazite, sont interprétés provisoirement comme étant l'âge de la cristallisation, non oblitérée par un métamorphisme ultérieur. Les trois granites les plus jeunes montrent des fabriques de déformation.

INTRODUCTION

A regional north-trending aeromagnetic low marks the western limit of the exposed Churchill Province south of Great Slave Lake. Early mapping by Camsell (1916), Wilson (1941), and Henderson (1939) showed that the region is underlain mainly by granitic rocks which include abundant scattered remnants of high grade paragneiss and metavolcanic rocks. Reconnaissance mapping of the Fort Smith and Taltson Lake areas (Charbonneau, 1980; Bostock, 1981, 1982), showed that the aeromagnetic low is underlain by a megacrystic syenogranite batholith. The syenogranite gives way eastward to a complex zone in which a predominantly foliated granitic terrane passes eastward into a predominantly gneissic terrane. West of the syenogranite a terrane composed mostly of monzogranite and granodiorite is exposed as far as the Paleozoic overlap. The Great Slave Lake Shear Zone (Hanmer and Lucas, 1985; Hanmer and Connelly, 1986) first recognized as a major mylonite zone by Reinhardt (1969), truncates the northerly trending structures of the whole area.

Supracrustal rocks of low metamorphic grade, thought to form parts of at least two groups of predominantly sedimentary rocks of different age, are scattered within the gneisses east of the syenogranite batholith. The presumably older group, tentatively correlated with Tazin Group (Mulligan and Taylor, 1954; Bostock, 1984) and best exposed at Hill Island Lake, has uncertain field relations with the gneisses. The presumably younger Nonacho Group (Aspler, 1985) lies unconformably upon the gneisses but is variably to severely deformed along its eastern margins. A single occurrence of quartzite with local pebble clusters, possibly correlative with Wilson Island Group (Bostock, 1986), is probably downfaulted within granodiorite west of the syenogranite batholith.

In northeastern Alberta, Riley (1960) showed that several granite types occur along strike with the region of low aeromagnetic anomalies. Detailed mapping by Godfrey and Langenberg (1978) and Rb-Sr isochron geochronology by Baadsgaard and Godfrey (1972) suggest that high grade metasedimentary rocks and some of the granitic rocks along the east margin of the syenogranite batholith, which are on strike with and therefore possibly correlative with similar rocks farther north, are Archean. These rocks were shown to have been intruded by the major granite bodies culminating at about 1939 Ma.

This paper reviews the geology of exposed Precambrian rocks of the Little Buffalo River (85A), Ft. Smith (75D),

[1] Geological Survey of Canada, 601 Booth Street, Ottawa, Ontario K1A 0E8

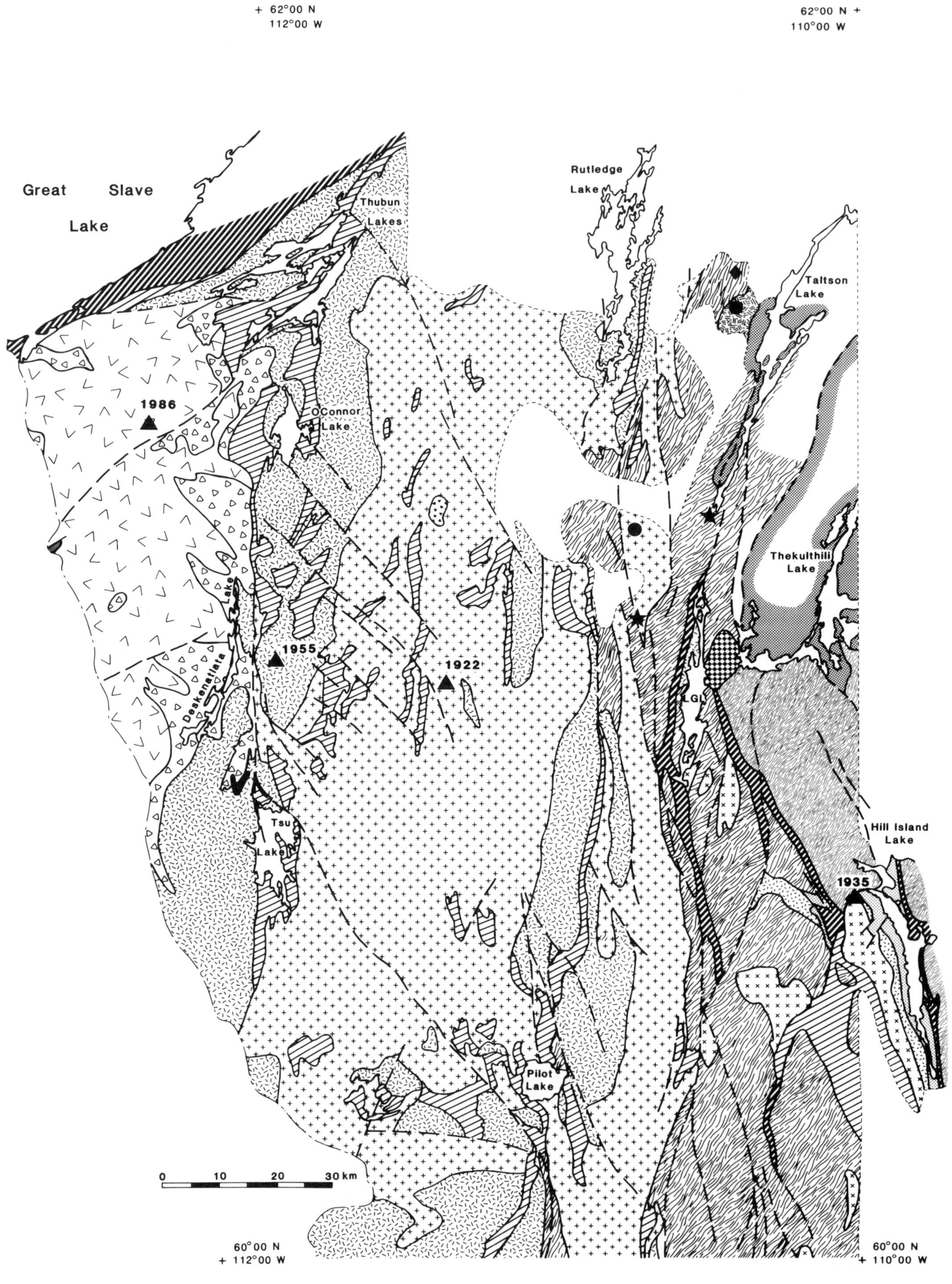
+ 62°00 N
112°00 W
62°00 N +
110°00 W
Great Slave Lake
Thubun Lakes
Rutledge Lake
Taltson Lake
1986
OConnor Lake
Thekulthili Lake
Deskenatlata Lake
1955
1922
LGL
Hill Island Lake
Tsu Lake
1935
Pilot Lake
0 10 20 30 km
60°00 N
+ 112°00 W
60°00 N
+ 110°00 W

Ft. Resolution (85H) and parts of Taltson Lake (75E) and Hill Island Lake (75C) NTS areas. Four new U-Pb ages for granitic plutons in this region are reported and their significance within the regional setting is discussed. The extensive exposure of three of the dated plutons and the grouping of all four ages, suggest that they form part of a single Proterozoic magmatic zone, herein named the Taltson Magmatic Zone.

REGIONAL GEOLOGY

The oldest rocks in the area (unit 1, Fig. 1), established through observation of widespread crosscutting by the major plutons, are mainly pelitic and semipelitic gneisses, with some quartzite and calc-silicate gneiss plus scattered occurrences of amphibolite. Some small bodies of granite in the eastern part of the area, resembling a granite described by Baadsgaard and Godfrey (1972) in Alberta, may also be Archean. Microcline-sillimanite-cordierite assemblages are widespread in the pelitic rocks, and mafic bands containing hypersthene- plagioclase-quartz-biotite suggest granulite facies metamorphism. Complex variations of metamorphism in both time and space, however, are suggested by irregular distribution of hypersthene, sillimanite and andalusite, by local evidence that sillimanite succeeded andalusite, and by extensive alteration of higher grade minerals to chlorite-biotite in eastern parts of the area. Kyanite has been reported in an inclusion of paragneiss within strongly foliated

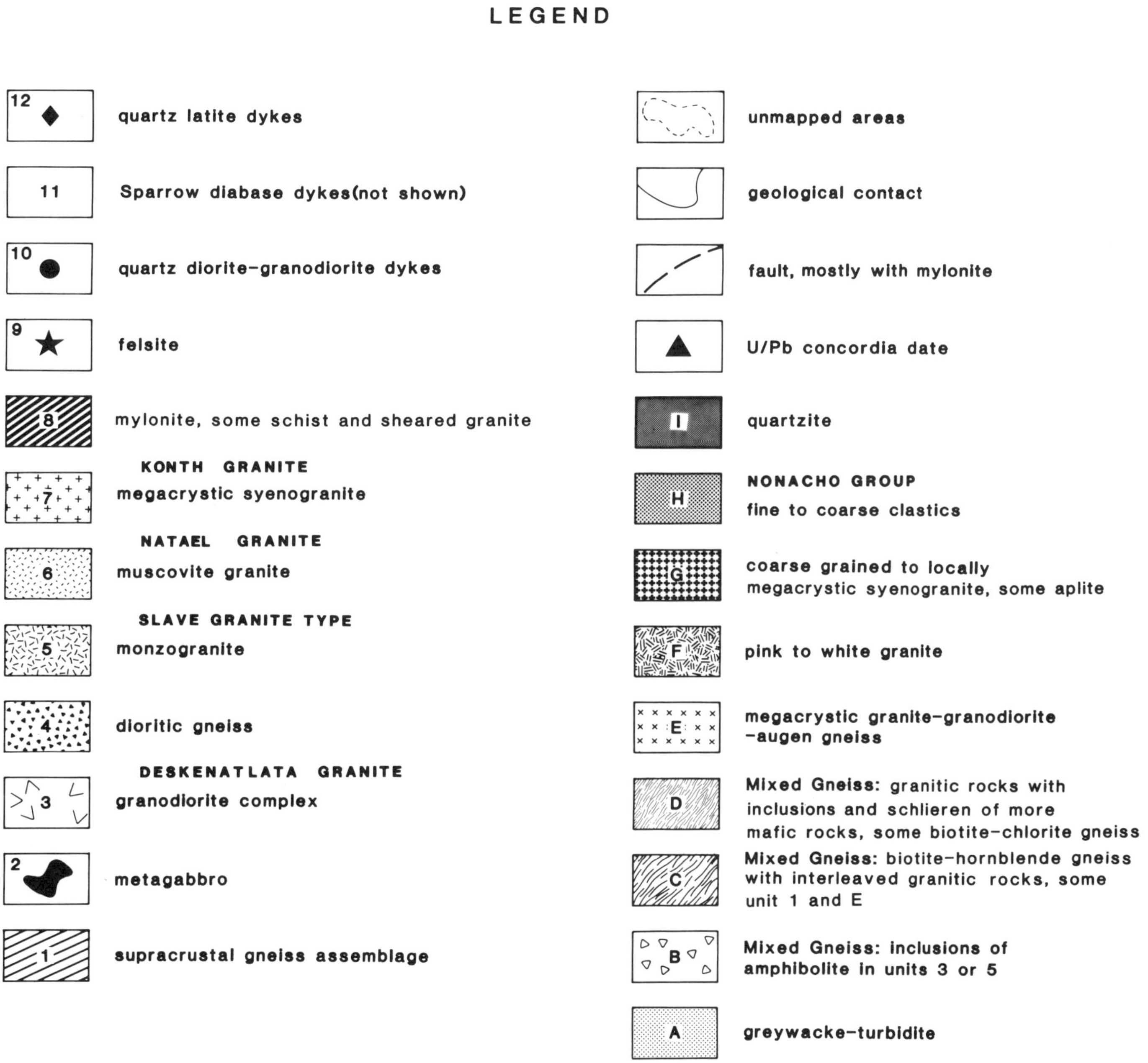

Figure 1. Geological map showing sample locations and ages.

syenogranite at Pilot Lake (Burwash and Cape, 1981), but no other occurrence is known to the writers. Graded bedding is preserved in the paragneiss and mafic rocks may be volcanic origin. Metagabbro and mafic gneiss, engulfed by later granites and dismembered, or attenuated by ductile shear, may be of about the same age (unit 2). Similar metasediments in Alberta have been assigned an Archean age on the basis of crosscutting pegmatites dated by Rb-Sr whole-rock isochron (Baadsgaard and Godfrey, 1972).

The oldest recognized Proterozoic plutonic rocks of the region are part of the Deskenatlata Granodiorite (unit 3) which is exposed westward from O'Connor and Deskenatlata lakes to the Paleozoic overlap. This unit comprises grey to pink, massive, medium-to coarse-grained, quartz diorite to granite. The more mafic members have colour indices in excess of 20. Biotite, the most abundant mafic mineral, commonly occurs in poikiloblasts 5-10 mm in diameter, but hornblende or chlorite predominates locally. The rocks contain variable proportions of pale pink subhedral microcline megacrysts and accessory sphene. The unit is unusual within this region because it lacks foliation except along its northern and eastern margins. Its eastern margin contains variable proportions of massive to banded amphibolite inclusions (mixed gneiss, unit B) derived from the preceeding supracrustal sequence and is cut, particularly in the northeast, by pink granite veins and dykes. Dioritic gneiss (unit 4), engulfed within later granites, may be of similar age or older than the Deskenatlata Granodiorite. A zircon concordia for granodiorite from the central part of the unit of 1986 $\pm$ 5 Ma (this report) is considered the age of emplacement.

Monzogranite (unit 5) in the southern part of the area is continuous with Slave Granite described by Godfrey and Langenberg (1978) in Alberta. Discontinuous remnants of similar granite, albeit of widely different degrees of deformation and metamorphic overprint, extend along the margins of, and lie within, the younger Konth syenogranite batholith. These remnants are herein referred to as "Slave Granite type" because the inference that all remnants are related is less clear at the current stage of investigation as compared to similar inferences regarding the parts of Deskenatlata Granodiorite complex, and of the Konth syenogranite batholith.

The Slave Granite type is a grey, white, buff, or pink, medium grained, equigranular to slightly megacrystic, biotite monzogranite varying locally to granodiorite or syenogranite. It lies to the east of the Deskenatlata Granodiorite and is mostly separated from it by a narrow belt of metasediments but locally the two are in deformed contact. Remnants of high grade paragneiss of varying size and abundance belonging to unit 1 are widely distributed. Potassium feldspar in Slave Granite type is typically mesoperthitic to submesoperthitic in contrast to that in Deskenatlata Granodiorite which is microcline perthite. Garnet, cordierite and accessory green spinel are common in the monzogranite but are absent through extensive areas west of Tsu Lake and east of Pilot Lake. Sillimanite, and more locally hypersthene or corundum armoured by spinel, andalusite or biotite, are present in some specimens. Although locally massive, the monzogranite is widely foliated and in its eastern exposures commonly severely deformed. Near Deskenatlata Lake where mixtures of monzogranite, granodiorite, metasedimentary and metamorphosed mafic igneous rocks occur, the rocks are mapped as part of the western mixed gneiss (unit B). The age of emplacement is 1955 $\pm$ 2 Ma (zircon-monazite U-Pb concordia, this report).

Pink to buff or grey megacrystic syenogranite of the Konth Granite, (unit 7) intrudes monzogranite of Slave Granite type and abundant remnants of paragneiss. Garnet and cordierite are common and hypersthene, usually partly altered to chlorite, is present locally. Potassium feldspar is generally perthitic to mesoperthitic. Accessory minerals include green spinel, monazite, ilmenite and zircon. The pluton is widely but variably foliated with strongest foliation in its eastern regions. Foliation trends northerly with moderate to steep dips, but southwest of the fault zone between Tsu Lake and west of Pilot Lake dips are mostly shallow. Lineation typically plunges gently northward or southward. A U-Pb concordia zircon- monazite date from the central part of the pluton suggests emplacement at 1922 $\pm$ 2 Ma (this report). A small plug of massive, coarse grained to locally megacrystic syenogranite (unit G) intrudes gneisses and mylonite at Lady Grey Lake. Preliminary U-Pb geochronology and the unfoliated state of this granite suggest that it may be somewhat younger than the chemically similar syenogranite of the main batholith.

Mixed gneisses (unit C) along the margin of the Konth Granite pluton extend eastward to a prominent shear zone (unit B) that extends from the southwest end of Nonacho Basin at Lady Grey Lake southeast through Hill Island Lake. These consist of biotite and hornblende-biotite gneisses interleaved with foliated, equigranular to megacrystic granitic bodies (unit E), and are intruded by mostly small bodies of pink to white granite (unit F). In places the granitic rocks contain numerous inclusions of banded and massive amphibolite and of pelitic gneiss. Conformable to crosscutting bands of amphibolite may be mafic dykes. Bands of pelitic metasediment contain remnants of sillimanite and garnet. These gneisses have undergone northerly trending ductile shear and have been variably retrograded to greenschist facies. East of the Lady Grey Lake-Hill Island Lake shear zone, mixed gneiss (unit D) consists of granitic rocks with widely variable proportions of schlieren, inclusions and larger remnants of amphibolite and hornblende-bearing gneiss with some biotite-chlorite gneiss. In contrast to unit C, no pelitic gneiss and few remnants of quartz-rich gneiss have been found. Metamorphic grade is reduced to greenschist or lower amphibolite facies. Zones of northerly trending brittle deformation which transect these eastern gneisses, contrast with more ductile late deformation in the gneisses (unit C) farther west. These easternmost gneisses comprise a large, fault bounded, southward tapering block extending east of Figure 1 upon which the low grade to unmetamorphosed Nonacho Group lies unconformably. This block is bordered still farther east by mostly foliated granitic plutonic rocks which extend southward into Saskatchewan where a number of zircon U-Pb dates yield ages close to 2600 Ma (Koster and Baadsgaard, 1970; Stevenson, 1985; Van Schmus et al., 1986).

Greywacke (unit A) with minor calc-silicate horizons at Hill Island Lake (Bostock, 1984) has been isoclinally folded. Metamorphic grade varies from sub- biotite zone along the lake to cordierite zone near the Natael Granite (unit 6) contact to the west. Andalusite-bearing schists appear locally along the shear zone which follows the east margin of the lake. West of the north end of Hill Island Lake, unit A is migmatite formed by deformation of muscovite granite in the metasediments. At west Thekulthili Lake small remnants of similar low grade metasediments are cut by muscovite-bearing granite dykes. Foliated muscovite monzogranite (Natael Granite, unit 6) intrusive into the Hill Island Lake greywackes, has yielded a concordant U-Pb monazite date of 1935 $\pm$ 3 Ma (this report). The date provides an age of intrusion for the granite and a minimum age for the metasediments at Hill Island Lake.

Low grade sediments similar to unit A and mafic volcanics have been described at Wauch Lake, Alberta (Watanabe, 1963). Crosscutting relations between these rocks and the surrounding granite were not observed but, like the Hill Island Lake metasediments, they are isoclinally folded and have undergone mostly greenschist facies prograde metamorphism. These two groups together comprise a lithological assemblage similar to the high grade (but commonly retrograded) gneisses farther west within the Taltson Magmatic Zone. If complex faulting, likely at the east margin of the zone, has separated high grade retrograded from low grade prograded metasediments, then it is possible that the two are correlative. A significant confinement of the maximum ages for the low grade metasediments at Hill Island and Wauch lakes is therefore an important project for future consideration.

Nonacho Group (unit H), consisting almost entirely of fine-to coarse-grained terrestrial clastic sediments, occupies the Nonacho Basin north of Thekulthili Lake. It is deformed most severely along its western margin where bedding is locally overturned. A large lens of folded Nonacho conglomerate is enclosed between mylonite bands between Lady Grey and Hill Island lakes. A second small outlier lies unconformably, tilted but unfolded, upon weathered red-stained basement in the northern arm of Hill Island Lake. The group is intruded by Sparrow diabase dykes of about 1700 Ma age, (McGlynn et al., 1974), but the maximum age is not closely confined. Aspler (1985) has interpreted the Nonacho Group as intracontinental molasse.

An isolated hill composed of white to purplish massive quartzite containing a few local quartzite pebble clusters occurs within granodiorite along the Paleozoic overlap northwest of Deskenatlata Lake (unit I). It lies close to the southwestern projection of a mylonite-bearing zone and is therefore likely a downfaulted remnant. This quartzite appears more mature than Nonacho sandstone- conglomerate and is perhaps lithologically more similar to parts of the Wilson Island Group.

Prominent northerly trending fault zones with associated mylonite, schist and sheared granitic rocks (unit 8) occur along the periphery of the syenogranite, and minor, mostly northwesterly trending, mylonites crosscut it. A mylonite belt up to about 1 km wide skirts the western shore of Lady Grey Lake beyond which it contracts but continues southward with several splays to join with Allen Fault Zone (Godfrey and Langenberg, 1978) at the Alberta border. A second major fault with mylonite locally reaching similar width extends southeastward from Lady Grey Lake to northern Hill Island Lake. There it appears to offset en echelon to the east and to continue southward along the lake valley into Saskatchewan. Blocks of mylonite are included in a megacrystic phase of the syenogranite plug at Lady Grey Lake suggesting that at least some major ductile faulting occurred before final syenogranite plutonism. Lesser faults with mylonite belts mostly less than a few tens of metres wide extend northward through Tsu and Deskenatlata lakes. This zone is linked by a southeasterly trending lineament between Tsu Lake and the Alberta border with Warren Fault Zone (Godfrey and Langenberg, 1978).

Mylonite is also present along a northeasterly trending lineament from the Paleozoic overlap to Thubun Lakes. Where it cuts the Deskenatlata Granodiorite this zone is about 100 m wide but at Thubun Lakes exposures on the islands suggest it may be wider.

Common mafic dykes of the Sparrow dyke swarm (unit 11, not shown on Fig. 1) trend northwesterly cutting all the major map units. The dykes are generally less then 1 m thick but some reach about 35 m; they have K-Ar dates of approximately 1700 Ma (McGlynn et al., 1974).

Local dykes of felsite (unit 9), granodiorite (unit 10) and quartz latite (unit 12) intrude the gneisses east of Konth Granite batholith. Further mapping of these minor bodies may contribute to the chronology of events that succeeded emplacement of the major plutons in the Taltson Magmatic Zone.

STRUCTURAL GEOLOGY

Examination of textures of the dated Proterozoic plutons within Taltson Magmatic Zone has shown that northerly trending foliation is a prominent feature in most parts of the younger three plutons. This foliation is strongest within the eastern part of Konth Granite and is continuous into the gneisses still farther east. Sinistral kinematic indicators have been widely recognized within this zone, but late belts of dextral movement are evident locally. Dextral kinematic indicators have also been found within Slave Granite type and older gneisses west of Konth Granite from Thubun Lakes south to Deskenatlata Lake, but the zone within which they occur appears to contract farther south. East of Konth Granite most major mylonites associated with the Allen fault zone and with a fault zone through Hill Island Lake appear to be associated with sinistral movements. West of Konth Granite mylonites with dextral movement are relatively minor in comparison. This implies that displacement has been concentrated along the east margin of the foliated zone. It further suggests that although the net displacement across the foliated zone has been west side south, there has been some lesser overall southward displacement of the zone as a whole.

U-PB ZIRCON AND MONAZITE ISOTOPIC AGES

Analytical techniques have changed significantly during the period of analysis (1978 to 1986) as reflected by the hundredfold reduction in the weight of analyzed material (Table 1). Zircon abrasion and dissolution methods are similar to Krogh (1972, 1982) while mass spectrometric analysis of small samples has been described by Roddick et al. (1986). Recent Pb blanks are in the range 20-40 picograms. Uncertainties (2 sigma) in U/Pb isotope ratios are indicated in the concordia diagram (Fig. 2) and reflect a propagation of all analytical uncertainties (Roddick, in press). Regression calculations have been done according to the methods of York (1969).

Zircons from the Deskenatlata Granodiorite are euhedral to subhedral and prismatic with length to breadth ratios (L/B) of 3:1. The zircons were clear with some inclusions and few cores, which were avoided, as far as possible, in the preparation of four size fractions. U-Pb zircon data points are colinear and are more discordant with decreasing grain size. U concentrations are moderate but show no consistent variation with discordance. The analytical uncertainty for the coarsest fraction is comparatively large and there is a possibility that this fraction contains some older zircon component (Fig. 2). A regression analysis of the three remaining fractions yields an upper intercept age of 1986.4 +2.4/−2.0 Ma and a lower intercept age of 50 ± 155 Ma. The regression line fits the data points as the mean square of weighted deviates (MSWD) is 0.71.

Zircons from the Slave Granite type monzogranite show a great variety of shapes from prisms with L/B of 7 to equidimensional round grains. Even the prismatic crystals have rounded perimeters. Twins and necked crystals are common. Cores are abundant and have been avoided as far as possible in four size and magnetic fractions of prismatic grains. On a concordia diagram, data points corresponding to these four fractions, not plotted, would show considerable scatter which probably reflects variable memory of older cores. U concentrations are high and it is also possible that Pb loss has occurred at different times for different fractions. As this sample contains monazite, two fractions were picked and slightly abraded. These concentrates yield concordant and slightly discordant U-Pb ages with $^{207}Pb/^{206}Pb$ ages averaging at 1955 ± 2 Ma.

Zircons from the Konth Syenogranite are irregular in shape and range from prismatic (L/B=4.1) to equidimensional. Some of the latter were rounded. Cracks are common. Cores are rare and have been avoided in picking three size and magnetic fractions. Two monazite fractions were also picked and slighty abraded. The fractions yield slightly discordant U-Pb ages with $^{207}Pb/^{206}Pb$ ages averaging at 1922 ± 2 Ma. The coarsest zircon fraction yields a significantly older $^{207}Pb/^{206}Pb$ age and may contain a component

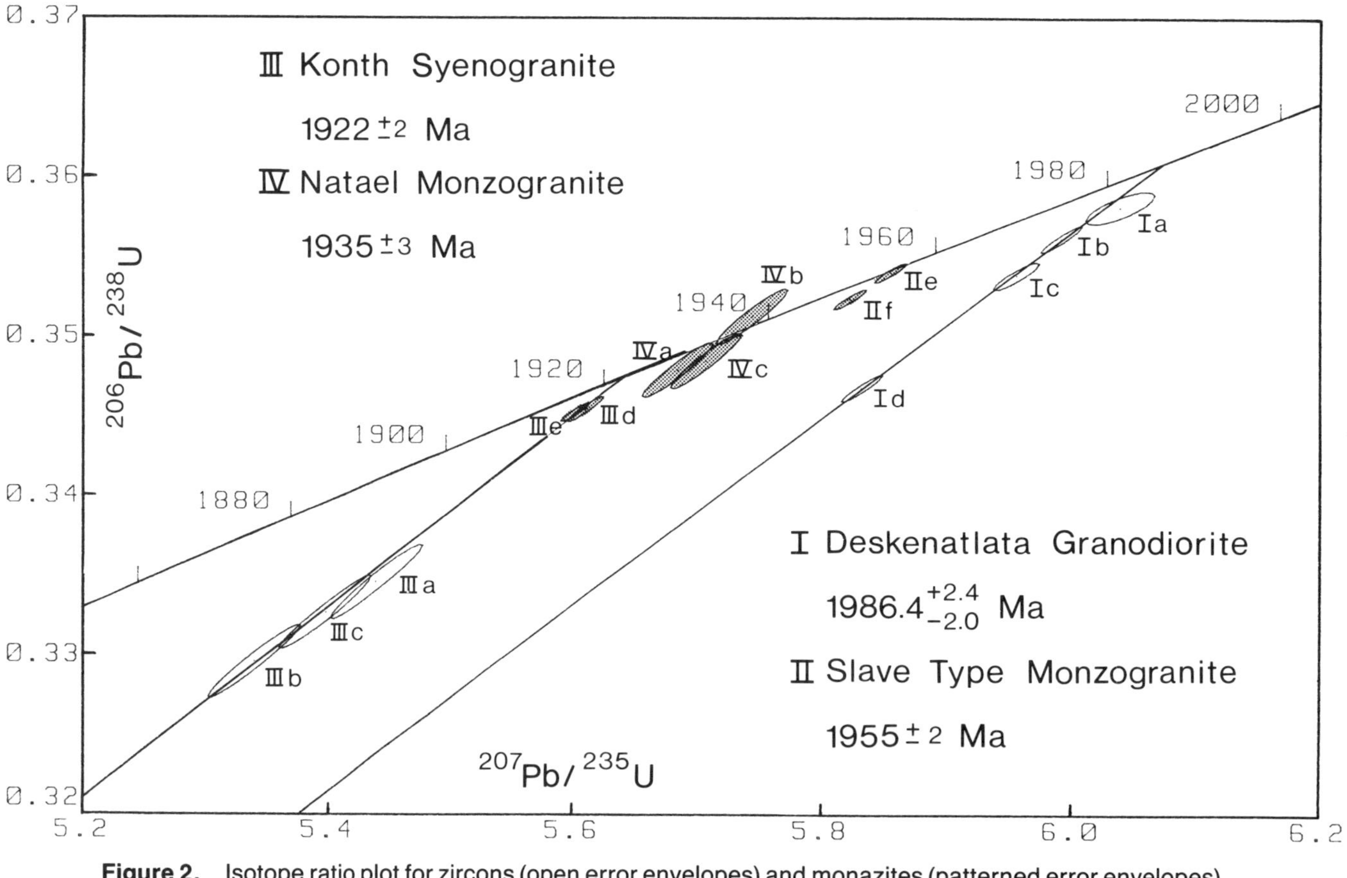

Figure 2. Isotope ratio plot for zircons (open error envelopes) and monazites (patterned error envelopes) from four granitoid units. Numbers and letters with data points correspond to those of Table 1.

of older zircon. If the two monazite data points are regressed with the two fine zircon data points, they yield a best fit line (MSWD=1.83) with an upper intercept age of 1922.4 + 1.7/−1.5 Ma and a lower intercept age of 100 ± 145 Ma.

Zircons from the Natael Monzogranite are of poor quality, translucent with abundant cores. Three monazite fractions were analyzed yielding virtually concordant U-Pb and $^{207}Pb/^{206}Pb$ ages with the latter averaging at 1935 ± 3 Ma.

In view of the resistance of zircon U-Pb isotopic systems to metamorphic resetting (van Breemen et al., 1987a) the upper intercept age for the zircon isotopic data from the Deskenatlata Granodiorite is interpreted in terms of the time of igneous crystallization. The 1922 ± 2 Ma age for the Konth Syenogranite, indicated by both zircon and monazite data, is also interpreted in terms of granite emplacement and crystallization. Monazite U-Pb ages could reflect regional cooling through temperatures corresponding to upper amphibolite facies conditions (Cliff, 1985). However, an age older than 1955 ± 2 Ma is not supported by the U-Pb zircon data from the Slave Granite type monzogranite. These monazite ages have not been reset by the proximate Konth Syenogranite batholith, and by analogy, the 1935 ± 3 Ma U-Pb monazite age for the Natael Monzogranite is also assumed to have survived thermal resetting. We tentatively interpret monazite ages as reflecting crystallization ages.

DISCUSSION

Recent reconnaissance geology of Precambrian rocks south of Great Slave Lake has suggested that an extensive zone of Early Proterozoic plutonism, the Taltson Magmatic Zone, extends westward from a presently only broadly defined north-south boundary near Hill Island and Thekulthili Lakes. Aeromagnetic anomaly patterns suggest that the western limit of this zone lies beneath younger rocks an unknown but apparently large distance west of the Paleozoic overlap.

Comparison of the period of plutonism in Taltson Magmatic Zone with that in Thelon Tectonic Zone (van Breemen et al., 1987b) indicates that the periods are closely similar for both of them. Furthermore, northerly trending deformation, possibly offset along the Great Slave Lake Shear Zone, has affected both zones. On the other hand, when aeromagnetic anomaly patterns west of the Paleozoic overlap are considered, it appears that the width of the Taltson Magmatic Zone is far greater than that of Thelon Tectonic Zone. No suture with rocks comparable to those of Slave Province is known south of the Great Slave Lake Shear Zone. No zones of high pressure metamorphism suggestive of thrusting are evident along the east margin of Taltson Magmatic Zone such as have been found farther north. These contrasts suggest that although the eastern margin of each zone is the Churchill craton, and the west margin of Thelon Tectonic Zone is the Slave craton, the west margin of Taltson Magmatic Zone may have been of a different sort. Furthermore, the later deformation history of the two zones may have differed considerably.

Gibb's indentor concept (Gibb, 1978), modified by P.F. Hoffman (pers. comm., 1987) for evolution of the Slave-Churchill boundary in Early Proterozoic time provides a tectonic framework which is consistent with these similarities and contrasts. Thus resolution of forces expressed along the

Table 1. U-Pb isotopic data on zircon and monazite

Sample size fraction (in microns)	Weight mg	U ppm	Pb ppm	Measured $^{206}Pb/^{204}Pb$	Isotopic abundances ^{206}Pb = 1000 ^{204}Pb	^{207}Pb	^{208}Pb	Isotopic ratios $^{206}Pb/^{238}U$	$^{207}Pb/^{235}U$	Age, Ma $^{207}Pb/^{206}Pb$
I. Deskenatlata Granodiorite										
a, +105, N1°	.034	221.6	92.26	632	1.409	141.1	179.3	.35816	6.0388	1989.8
b, −105+74, N1°	.042	247.8	97.80	4009	.116	123.6	161.5	.35618	5.9910	1985.9
c, −105+74, N1°	.040	184.0	73.50	1588	.442	128.0	174.8	.35386	5.9561	1986.8
d, −74+62, N1°	.050	272.5	106.4	3822	.158	124.1	181.2	.34679	5.8323	1985.3
II. Slave Type Monzogranite										
a, −105+74, N0°	0.051	1211	449.5	2328	.406	123.1	154.0	.33553	5.4444	1921.3
b, +105, N1°	0.76	818.9	290.5	2168	.093	120.5	101.1	.33713	5.5416	1944.4
c, −74+62, N0°	2.21	1316	464.9	1358	.685	126.9	103.0	.33030	5.3614	1921.8
d, −74+62, M0°	1.04	1467	487.2	1236	.705	124.8	114.6	.30792	4.8950	1884.4
e, Monazite	0.008	4055	7735	8145	.051	120.6	5096	.35404	5.8556	1955.6
f, Monazite	0.016	5062	11310	17930	.028	120.2	6197	.35236	5.8228	1954.0
III. Konth Syenogranite										
a, −149+105, N-1°*	2.32	528.3	181.6	5108	.046	118.5	75.03	.33457	5.4395	1924.8
b, −149+105, M-1°*	1.89	474.2	165.3	3011	.145	119.4	106.8	.32957	5.3397	1918.6
c, −105+74, N-1°*	1.87	360.0	123.2	2055	.230	120.8	71.03	.33267	5.3979	1921.2
d, Monazite	.016	2613	10030	11750	.032	118.2	11699	.34544	5.6101	1922.9
e, Monazite	.002	3138	11740	10080	.066	118.6	11368	.34525	5.6018	1921.2
IV. Natael Monzogranite										
a, Monazite	1.10	3492	4248	1300	.734	128.3	2868	.34791	5.6840	1933.6
b, Monazite	1.25	2976	3640	1961	.414	124.1	2875	.35128	5.7439	1935.1
c, Monazite	0.95	3513	4251	3140	.188	121.3	2881	.34848	5.7070	1937.9

Fractions are zircon unless indicated otherwise
N, nonmagnetic at given angle
M, magnetite at given angle

eastern and southern margins of Slave Province as it penetrated Churchill Province along the Great Slave Lake Shear Zone may have modified the development of an earlier tectonic margin. Could thrusting along Thelon Tectonic Zone have accompanied sinistral strike slip along the east margin of Taltson Magmatic Zone? Completion of mapping along the contact between the Great Slave Lake Shear Zone and the Taltson Magmatic Zone, and opportunities it provides for refining the timing between shearing and plutonism, may shed further light on these questions.

ACKNOWLEDGMENTS

This paper has been critically read by S.K. Hanmer, M.B. Lambert and R.R. Parrish.

REFERENCES

Aspler, L.B.
1985: Geology of Nonacho Basin (Early Proterozoic) N.W.T.; unpublished Ph.D thesis, Carleton University, 384 p.

Baadsgaard, H. and Godfrey, J.D.
1972: Geochronology of the Canadian Shield in northeastern Alberta II. Charles-Andrew-Colin Lake area; Canadian Journal of Earth Sciences, v. 9, p. 863-888.

Bostock, H.H.
1981: A granitic diapir of batholithic dimensions at the west margin of the Churchill Province; *in* Current Research, Part B, Geological Survey of Canada, Paper 81-1B, p. 73-82.
1982: Geology of the Fort Smith map-area, District of Mackenzie, Northwest Territories, (NTS 75D); Geological Survey of Canada, Open File 859, 53 p.
1984: Preliminary geological reconnaissance of the Hill Island Lake and Taltson Lake areas, District of Mackenzie; *in* Current Research, Part A, Geological Survey of Canada, Paper 84-1A, p. 165-170.
1986: Reconnaissance geology of Precambrian rocks of the Fort Resolution, Taltson Lake and Fort Smith areas, District of Mackenzie; *in* Current Research, Part A, Geological Survey of Canada, Paper 86-1A, p. 35-42.

Burwash, R.A. and Cape, D.F.
1981: Petrology of the Fort Smith-Great Slave Lake radiometric high near Pilot Lake, N.W.T.; Canadian Journal of Earth Sciences, v. 18, p. 842-851.

Camsell, C.
1916: An exploration of the Tazin and Taltson Rivers, N.W.T.; Geological Survey of Canada, Memoir 84, 124 p.

Charbonneau, B.W.
1980: The Fort Smith Radioactive Belt, Northwest Territories; *in* Current Research, Part C, Geological Survey of Canada, Paper 80-1C, p. 45-57.

Cliff, R.A.
1985: Isotope dating in metamorphic belts; Geological Society of London, Journal, v. 142, p. 97-110.

Gibb, R.A.
1978: Slave-Churchill collision tectonics; Nature, v. 271, p. 50-52.

Godfrey, J.D. and Langenberg, C.W.
1978: Metamorphism in the Canadian Shield of northeastern Alberta *in* Metamorphism in the Canadian Shield, ed. J.A. Fraser and W.W. Heywood; Geological Survey of Canada, Paper 78-10, p. 129-138.

Hanmer, S. and Connelly, J.N.
1986: Mechanical role of the syntectonic Laloche Batholith in the Great Slave Lake Shear Zone, District of Mackenzie, N.W.T.; *in* Current Research, Part B, Geological Survey of Canada, Paper 86-1B, p. 811-826.

Hanmer, S. and Lucas, S.B.
1985: Anatomy of a ductile transcurrent shear; the Great Slave Lake Shear Zone, District of Makenzie, N.W.T.; *in* Current Research, Part B, Geological Survey of Canada, Paper 85-1B, p. 121-131.

Henderson, J.F.
1939: Taltson Lake; Geological Survey of Canada, Map 525A with marginal notes.

Koster, F. and Baadsgaard, H.
1970: On the geology and geochronology of northwestern Saskatchewan I. Tazin Lake region; Canadian Journal of Earth Sciences, v. 7, p. 919-930.

Krogh, T.E.
1973: A low contamination method for hydrothermal dissolution of zircon and extraction of U and Pb for isotope age determination; Geochimica et Cosmochimica Acta, v. 37, p. 485-494.
1982: Improved accuracy of U-Pb zircon ages by the creation of more concordant systems using an air abrasion technique; Geochimica et Cosmochimica Acta, v. 46, p. 637-649.

McGlynn, J.C., Hanson, G.N., Irving, E., and Park, J.K.
1974: Paleomagnetism and age of Nonacho Group sandstones and associated Sparrow Dykes, District of Mackenzie; Canadian Journal of Earth Sciences, v. 11, p. 30-42.

Mulligan, R. and Taylor, F.C.
1954: Hill Island Lake; Geological Survey of Canada, Map 1203A with marginal notes.

Reinhardt, E.W.
1969: Geology of the Precambrian rocks of Thubun Lakes map-area in relationship to the McDonald Fault System, District of Mackenzie; Geological Survey of Canada, Paper 69-21, 29 p.

Riley, G.C.
1960: Fort Fitzgerald; Geological Survey of Canada, Map 12-1960 with marginal notes.

Roddick, J.C., Loveridge, D.W., and Parrish, R.R.
1986: Precise U-Pb dating of sub-miligram quantities of zircon; Terra Cognita, v. 6, p. 152.

Roddick, J.C.
1987: General numerical error analysis with applications to geochronology and thermodynamics; Geochimical et Cosmochimica Acta, v. 51, p. 2129-2135.

Stevenson, R.K.
1985: Implications of amazonite to sulfide-silicate equilibria; unpublished MSc thesis, McGill University, 311 p.

van Breemen, O., Henderson, J.B., Loveridge, W.D., and Thompson, P.H.
1987: U-Pb zircon and monazite geochronology and zircon morphology of granulites and granite from the Thelon Tectonic Zone, Healey Lake and Artillery Lake map areas, N.W.T.; *in* Current Research Part A, Geological Survey of Canada, Paper 87-1A, p. 783-801.

van Breemen, O., Thompson, P.H., Bostock, H.H., and Loveridge, W.D.
1987b: Timing of plutonism in the northern Thelon Tectonic Zone and the Taltson Magmatic Zone; Geological Association of Canada, Saskatoon meeting (Abstract).

Van Schmus, W.R., Persons, S.S., Macdonald, R., and Sibbald, T.I.I.
1986: Preliminary results from U-Pb zircon geochronology of the Uranium City region, northwest Saskatchewan; *in* Summary of Investigations 1986, Saskatchewan Geological Survey; Saskatchewan Energy and Mines, Miscellaneous Report 86-4, p. 108-111.

Watanabe, R.Y.
1963: Geology of the Wauch Lake metasedimentary complex, Northeastern Alberta; Canadian Mining Journal, v. 84, p. 193 (abstract of unpublished M.Sc. thesis, University of Alberta, 1971).

Wilson, J.T.
1941: Fort Smith; Geological Survey of Canada, Map 607A with marginal notes.

York, D.
1969: Least squares fitting of a straight line with correlated errors; Earth and Planetary Science Letters, v. 5, p. 320-324.

U-Pb zircon and monazite geochronology from the northern Thelon Tectonic Zone, District of Mackenzie

O. van Breemen[1], P.H. Thompson[1], P.A. Hunt[1], and N. Culshaw[1].

van Breemen, O., Thompson, P.H., Hunt, P.A., and Culshaw, N., U-Pb zircon and monazite geochronology from the northern Thelon Tectonic Zone, District of Mackenzie; in Radiogenic Age and Isotopic Studies: Report 1, Geological Survey of Canada, Paper 87-2, p. 81-93, 1987.

Abstract

A granitoid migmatite complex 6 km east of the western boundary of the Thelon Tectonic Zone has been dated at ca. 2.6 Ga. 100 km south-southeast and 50 km farther east across the strike of the zone, garnet granite derived by melting of pelitic metasediments has been dated at 1908 ± 2 Ma and granulite facies metamorphism prior to mylonitization has been dated in the 2000-1980 Ma interval. In between these localities, samples from agmatitic granodiorite and ''charnockite'' from the granitoid migmatite complex as well as plutons of megacrystic granodiorite and clinopyroxene granodiorite have yielded Proterozoic ages in the 2020-1910 Ma interval. Internal zircon structure has been used to interprete conditions during both granite magma formation and emplacement.

Résumé

Un complexe de migmatites granitoïdes situé à 6 km à l'est de la limite ouest de la zone tectonique de Thelon a été daté à environ 2,6 Ga. À 100 km au sud-sud-est et 50 km plus à l'est, perpendiculaire à l'orientation de la zone, un granite à grenat dérivé de l'anatexie des roches pélitiques a été daté à 1908 ± 2 Ma et le métamorphisme du faciès des granulites survenu avant la formation de mylonite date d'un intervalle de temps se situant entre 2000 et 1980 Ma. Entre ces endroits, on a établi que l'âge des échantillons provenant de la granodiorite amagtitique et de la « charnockite » provenant du complexe de migmatites ganitoïdes, ainsi que des plutons de granodiorite « à phénocristaux » et de granodiorite à clinopyroxène, correspondait à l'ère protérozoïque, soit à un intervalle de temps se situant entre 2020 et 1920 Ma. La structure interne du zircon a été utilisée pour interpréter les conditions dans lesquelles le magma granitique a été formé et mis en place.

INTRODUCTION

This contribution is part of a continuing study directed toward obtaining absolute ages of rock units, magmatism, metamorphism and deformation across the polyorogenic transition from the Slave Province eastward into the Thelon Tectonic Zone. The Tinney Hills (NTS76J) and Overby Lake (76I, west half) map areas (Fig. 1) include part of the western boundary of the northern segment of the Thelon Tectonic Zone. Data presented here are from nine of the samples collected during the last field season (1985) of the 1:250 000 scale mapping project.

GENERAL GEOLOGY

The westernmost part of the map area (Fig. 1) is underlain by weakly metamorphosed clastic and carbonate sedimentary rocks of the Lower Proterozoic Goulburn Supergroup (Campbell and Cecile, 1981; Grotzinger and Gall, 1986; Grotzinger et al., 1987) which overlie with angular unconformity metamorphosed Archean supracrustal and plutonic rocks. East of the unconformity the area can be divided into two lithological associations or terranes: 1) a metasedimentary migmatite terrane comprising mainly migmatitic metasedimentary rocks of the Late Archean Yellowknife Supergroup (unpatterned unit, Fig. 1) and subordinate plutonic rocks; 2) a granitoid migmatite terrane in the east half and the northernmost part of the west half of the map area that is made up mainly of variably deformed and migmatized, granitic to tonalitic plutonic rocks of the granitoid migmatite complex (irregular dashed pattern, Fig. 1). A heterogeneous package of mafic, pelitic, carbonate-bearing and quartzofeldspathic rocks, for the most part of supracrustal origin, is prominent along the contact between the metasedimentary migmatite terrane and the extensive eastern part of the granitoid migmatite terrane, and as belts and lenses within the latter. The western boundary of the Thelon Tectonic Zone is located along the contact between the metasedimentary migmatite terrane and the eastern granitoid migmatite terrane. The eastern part of the granitoid migmatite terrane is subdivided by a somewhat arbitrary line (the ''megacrystic

[1] Geological Survey of Canada, 601 Booth Street, Ottawa, Ontario K1A 0E8

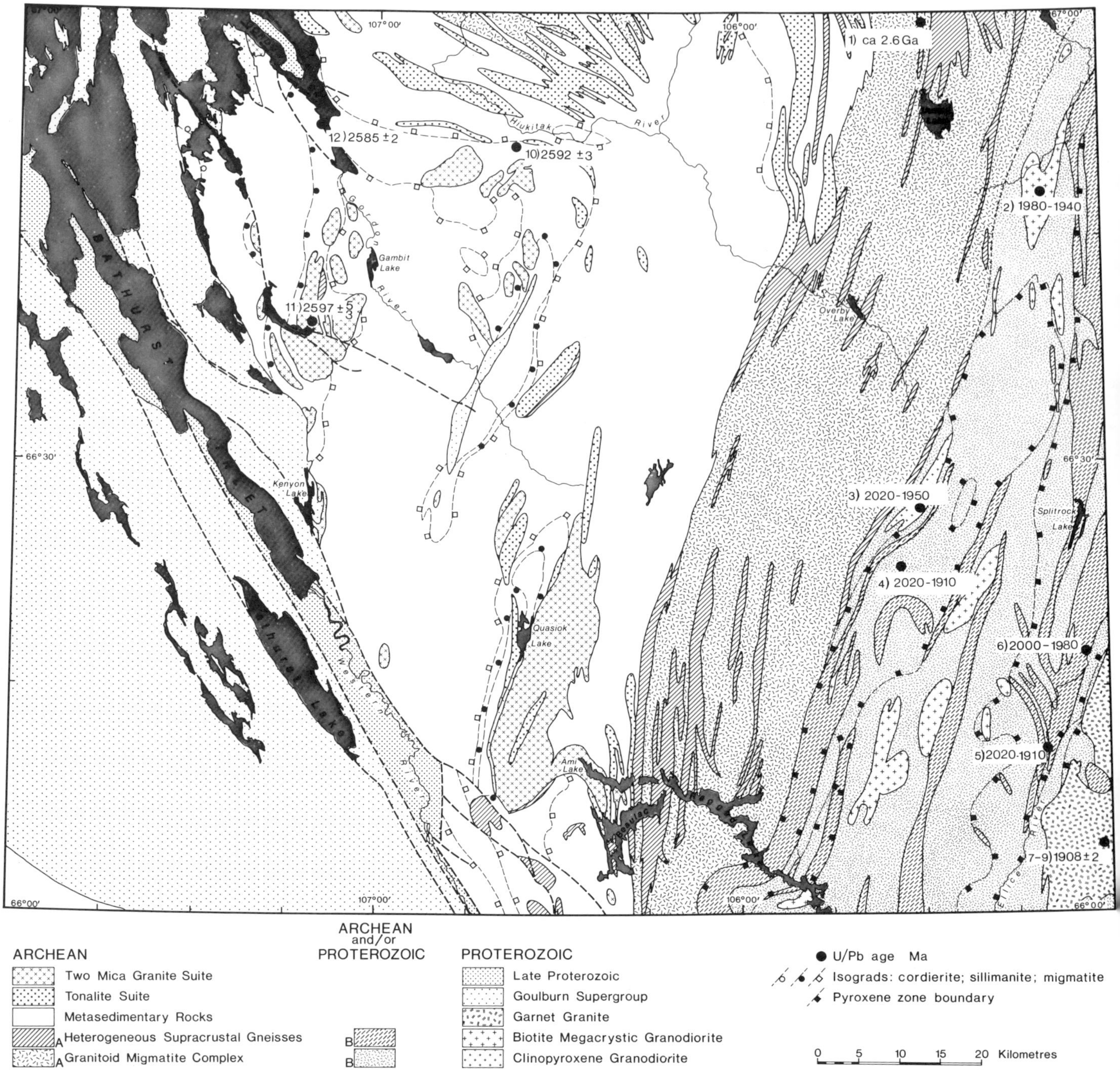

Figure 1. Geology of the Tinney Hills (75J) and Overby Lake (76I west half) map areas simplified from Thompson (1986) with sample locations. Pyroxene zone boundary is either an isograd, a lithological contact, or a tectonic contact. Crystalline rocks are divided into a metasedimentary migmatite terrane comprising mainly metasedimentary rocks and subordinate plutonic rocks, and a granitoid migmatite terrane (northernmost west half and east half of the map area) made up mainly of variably migmatized and deformed granitoid rocks. The subdivision of the eastern part of the granitoid migmatite terrane into Archean rocks variably reworked in the Proterozoic (A) and rocks that may be Archean, Proterozoic or both (B) occurs along the dashed line and/or pyroxene zone boundary that extends from east of Uumati Lake to Ragged Lake. This corresponds to the "megacrystic granodiorite line" of Thompson (1986).

granodiorite line"; Thompson, (1986); separating coarse irregular and fine irregular dashed pattern on Fig. 1) marking the western limit of Proterozoic megacrystic K-feldspar biotite granodiorite and/or extensive granulite facies rocks. Geochronological evidence to date supports geological evidence of Archean plutonism, metamorphism and deformation west of the line whereas to the east, three plutonic units, two high grade, high strain supracrustal rocks and two samples from the migmatitic granitoid complex yield Proterozoic ages (*see* below). Given these data and the possibility that the similar lithologies formed at different times in the map area, east of the "megacrystic granodiorite line", migmatitic granitoid complex and the heterogeneous supracrustal gneiss map units have been subdivided (Fig. 1) into: A) Archean rocks affected to a varied degree by Proterozoic events; B) rocks that may be Archean, Proterozoic or both.

The predominant structural and metamorphic patterns west of the Thelon Tectonic Zone have been attributed to Late Archean tectonism that, in its later stages produced a northerly trending straight belt along the eastern border of the Slave Province (Thompson et al., 1986; Thompson, 1986). A Proterozoic event variably deformed and metamorphosed Archean rocks, post-Archean basic dykes and Proterozoic cover, with the degree of basement involvement in the Proterozoic deformation increasing eastward as the metamorphic grade increases from anchizone in the west to lower amphibolite facies (Thompson and Frey, 1984; Thompson et al., 1986) along the eastern border of the metasedimentary migmatite terrane. At the boundary and to the east in the granitoid migmatite terrane, straight and anastomosing greenschist to amphibolite facies high strain zones deform amphibolite to granulite facies rocks and syn- to post-metamorphic Proterozoic plutons.

GEOCHRONOLOGY

Sample localities are indicated on Figure 1. A biotite granodiorite (sample 1) from the granitoid migmatite complex was collected just east of the boundary of the Thelon Tectonic Zone. The megacrystic granodiorite (sample 2) and clinopyroxene granodiorite (sample 3) occur along the line subdividing the eastern granitoid migmatite terrane. "Charnockite" (sample 4) is interlayered with granulite facies gneisses in the granitoid migmatite complex 4 km south of the clinopyroxene granodiorite. Along the Ellice River, samples were collected from a migmatitic, inclusion-rich granodiorite in the granitoid migmatite complex (sample 5) and a mylonitic, garnet-rich quartzofeldspathic layer in a granulite facies outcrop of heterogeneous supracrustal gneiss. In the southeastern corner of the map area, garnet granite (sample 7), a large inclusion of garnet sillimanite gneiss (sample 8) and a lineated mylonitic garnet granite tectonite (sample 9) were collected. Samples from Archean metatonalite, two-mica granite and pegmatite (10,11,12) from the northwestern corner of the area will be described elsewhere.

Images of polished and etched sections of zircons from five of the samples are presented in Figure 2. Techniques of zircon abrasion and dissolution are similar to Krogh (1982, 1973). Isotope dilution and mass spectrometry have been described by Roddick et al. (1987) and Parrish et al. (1987). U-Pb isotopic data are presented in Table 1 and have been graphically presented in Figures 3, 4, 5 and 6. Uncertainties indicated in the figures are at the 2 sigma level. Regression analyses are after York (1969).

Foliated biotite granodiorite-granitoid migmatite complex
(sample 1)

At this locality the unit consists of pale grey weathering, fine- to medium-grained, foliated biotite granodiorite containing inclusions of hornblendite, migmatitic quartzofeldspathic gneiss, and fine grained granodiorite. Contacts with supracrustal rocks to the east and west are severely tectonized. To the west, staurolite-grade metamorphism has overprinted adjacent Archean migmatites. In thin section, the metamorphic mortar texture indicates grain size reduction and high strain was accompanied by medium-to high-grade metamorphism and subsequent recrystallization. Rounded to subhedral grains (1-4 mm) of plagioclase, some partially broken, are surrounded by a finer grained matrix (0.1-0.8 mm) of quartz, plagioclase and microcline that appears to flow around the porphyroclasts. Irregular blebs of quartz (1-4 mm) are also present. Fine grained, well recrystallized biotite (8-10%) occurs as aggregates rimming plagioclase or in the matrix. Apatite and magnetite are prominent accessories. Scapolite, possibly pseudomorphing calcic plagioclase, and very fine grained aggregates of amphibole (?) after an unknown mineral are scattered through the rock.

Zircons are generally colourless to clear, euhedral to subhedral with length: breadth (L:B) ratios of 3:2 to 2:1. Some cores are apparent in transmitted light. However polished and etched sections (van Breemen et al., 1986) show a two-stage growth history for the majority of the crystals (Fig. 2A). Euhedral to subhedral igneous (Poldervaart, 1956) zoning characterizes most inner cores. This zoning, inferred from the strong leaching by the etching acid, becomes irregular and discordant near the outer margin of the core and is followed by a mantle of high U zircon. The inner mantle is also characterized by irregular, but igneous zoning. The outer mantle consists of lower U euhedral to subhedral igneous zircon. A number of twins are found. Some indicate synneusis during the inner mantle growth stage. Others are growth twins.

Four fractions were analyzed, two of each of two size fractions which were lightly (1b and 1d; 30 min.) and more strongly (1a and 1c; 140 min.) abraded. Isotopic ratios are strongly clustered, slightly discordant (1.3%) and yield $^{207}Pb/^{206}Pb$ model ages of 2608 Ma and 2601 Ma for the strongly and weakly abraded fractions respectively (data points not graphically displayed). A regression line through the four data points yields an upper intercept age of $2656 \pm 55/24$ Ma, a lower intercept of 1665 ± 370 Ma with a MSWD of 1.7.

In view of the problematical nature of cores and rims which were generally not apparent in transmitted light and

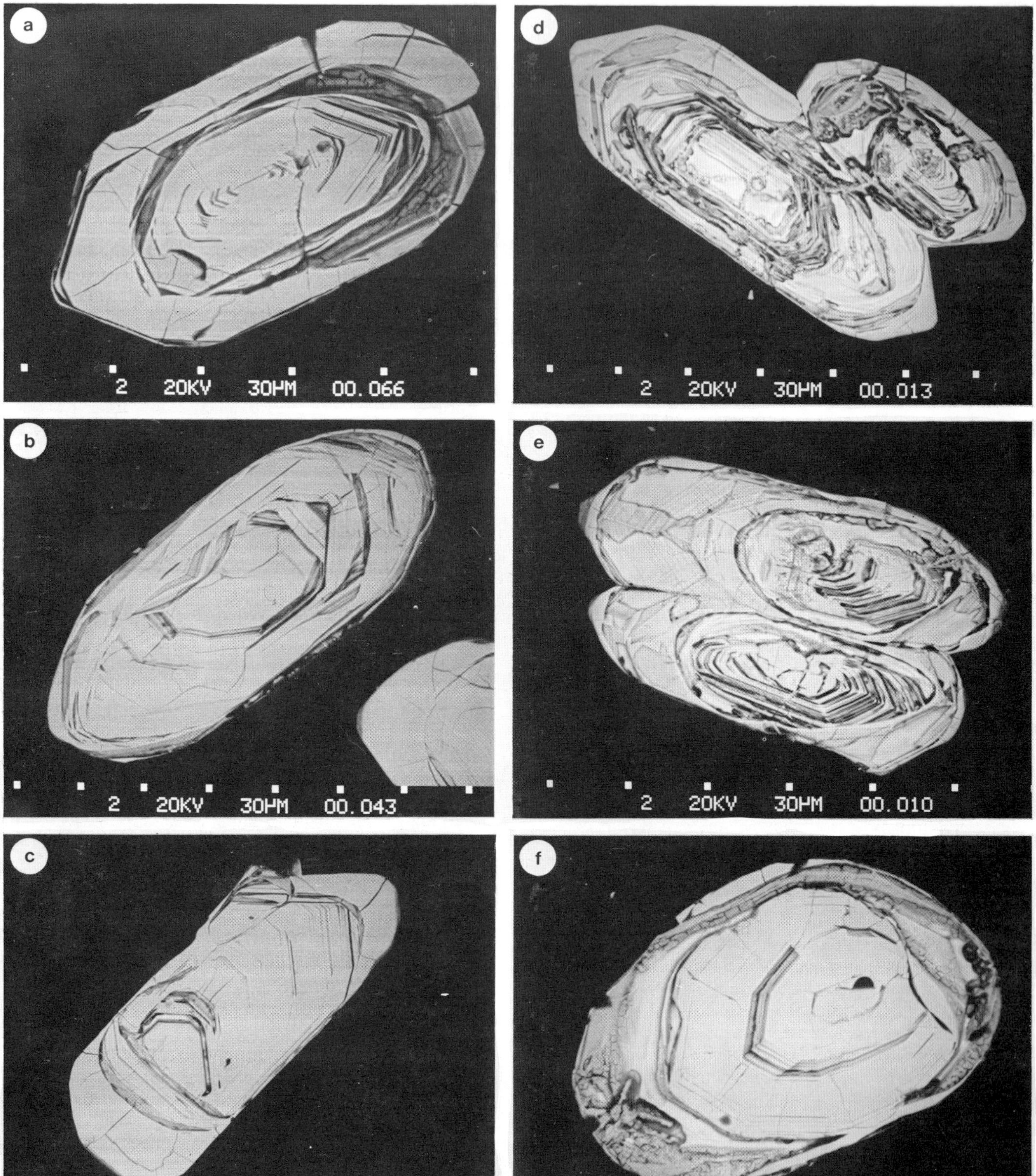
a
2 20KV 30µM 00.066
d
2 20KV 30µM 00.013
b
2 20KV 30µM 00.043
e
2 20KV 30µM 00.010
c
2 20KV 30µM 00.044
f
2 20KV 30µM 00.019

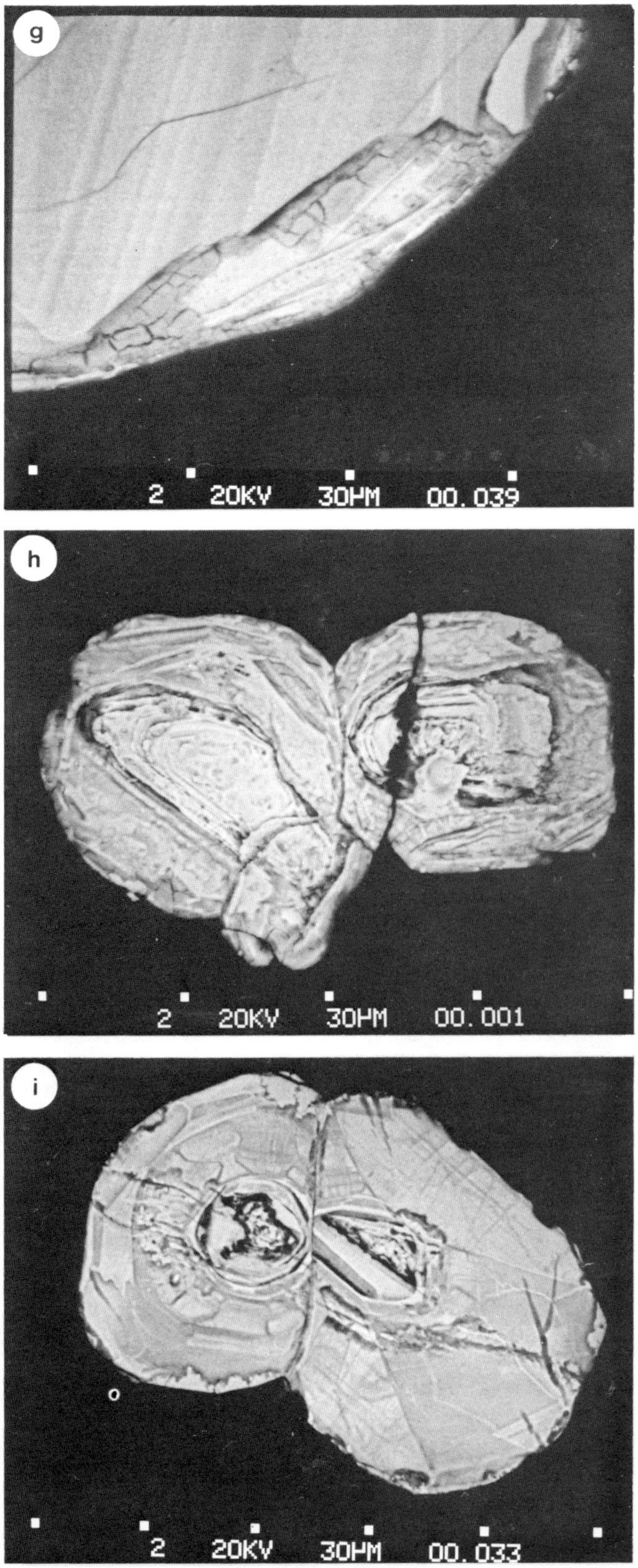

Figure 2. Scanning electron microscope images of polished and etched (HF) sections of zircon. All images are taken in the backscattered electron mode.

A. Zircon from biotite granodiorite-granitoid migmatite complex (sample 1) showing two stage history. An inner core shows regular euhedral (igneous) zoning which becomes more irregular and discordant approaching an inner mantle of apparently high U zircon which has been leached out by the acid. The inner mantle is surrounded by an outer mantle of regularly zoned igneous zircon.

B. Prismatic zircon from clinopyroxene granodiorite (sample 3) featuring irregular and discordant zircon. Note apparent rounding resulting from multiple facets along outer perimeter and possible thin rim of metamorphic zircon.

C. Prismatic zircon from clinopyroxene granodiorite in which early euhedral igneous zircon has been broken and overgrown by low U zircon, which is generally characterized by apparently igneous zonation.

D. Synneusis twin from "charnockite" - granitoid migmatite complex (sample 4). Note how subhedral prismatic centres are surrounded by more irregularly zoned igneous zircon and that synneusis occurred at a late stage of zircon growth.

E. Synneusis twin from "charnockite". Again docking has occurred at a late stage of igneous growth. Note possible thin rim of metamorphic zircon. Also note apparent "core" rounded by magmatic corrosion.

F. Ovoid zircon from mylonitic garnet-quartzofeldspathic gneiss (sample 6), from heterogeneous supracrustal gneiss unit, featuring regular multifaceted zoning followed by irregular zoning along the outer perimeter. The outer perimeter has been more strongly etched and presumably was higher in U.

G. Close-up of outer perimeter of zircon from mylonitic garnet gneiss features strongly etched discordant zoning.

H. Synneusis twin of three ovoid zircons from garnet granite (sample 7), synneusis occurred late in zircon growth. Zoning is irregular and discordant.

I. Synneusis twin of ovoid zircons from garnet granite. Multifaceted zoning is only slightly irregular. Parts of both zircons have been removed during docking. The planar join suggests fracture during synneusis.

the paucity of material, further work was not attempted. The slight shift to older ages with strong abrasion is interpreted in terms of progressive removal of younger rims. However in view of the high proportion of outer mantle, an Archaean-Proterozoic mix is considered unlikely. In view of the clustering of the data points, it is more likely that the U-Pb isotope systematics and zircon morphology reflect a multiple zircon growth history near 2.6 Ga. Inherited cores may be a reflection of the xenoliths in this rock.

Megacrystic K-feldspar granodiorite (sample 2)

Pinkish grey to pink and black weathering, this unit is characterized by anhedral to euhedral phenocrysts of potassium feldspar that now form microperthitic microcline augen (2-10 cm) surrounded by a fine grained (0.1-1 mm) matrix of quartz, plagioclase, biotite and epidote. Sphene and plagioclase porphyroclasts (1-2 mm) are also present. Accessory apatite and magnetite are prominent. Augen and matrix define a weakly to well developed foliation and, where strain is high, a distinctive pink (augen) and black (matrix) rock has formed. Megacrystic granodiorite is associated with migmatitic granitoid rocks and high grade supracrustal gneisses that locally attain the granulite facies, but most contacts are covered or tectonic. There is, however, no evidence indicating the pluton was metamorphosed to such a high grade. In thin section the sample dated exhibits a prominent protomylonitic fabric that was imposed on an originally coarser grained rock under metamorphic conditions compatible with recrystallization of biotite, quartz and feldspar and pervasive formation of epidote in plagioclase augen and matrix biotite and as rims on magnetite. Minor chlorite is present.

Zircons are euhedral to subhedral with L:B ratios of 2:1 to 3:1 and have mostly regular igneous zoning. There are some inclusions and some apparent cores which may not have been entirely avoided even during hand-picking of repeated (2c and 2f) small fractions. The presence of some cores appears to be indicated by the scatter of data points (Fig. 3). A regression analysis of data points 2c, 2e and 2g yields a concordant line with upper and lower intercept ages of 1963 +4/−3 Ma and 507 ± 125 Ma respectively. The lower intercept age is higher than those found for less recrystallized plutons of similar age in the Taltson Magmatic Zone and attributed to recent Pb loss (Bostock et al., 1987).

Most grains show rounded outer perimeters and some feature very thin discordant rims which may have formed during subsequent protomylonitization. Thus slight lead loss not long after granite emplacement cannot be ruled out. These rims are, however, likely to have been largely removed by abrasion, especially in the strongly abraded fractions 2c and 2f. Given the greater likelihood of inheritance the age of granite intrusion is placed in the interval 1980 -1940 Ma.

Clinopyroxene granodiorite (sample 3)

This massive to weakly foliated, medium grained, lenticular pluton ranges in composition from tonalite to ferrogranodiorite (Newman, 1986). Augite rimmed by hornblende and a prominent positive aeromagnetic anomaly distinguish this unit from other plutonic rocks in the area. Inclusions of mafic gneiss and, at one locality, a relatively fine grained contact zone indicate the pluton is intrusive into the adjacent migmatitic granitoid and upper amphibolite to granulite facies gneisses. The hornblende coronas have been attibuted to late magmatic auto-alteration (Newman, 1986) and a regional metamorphic overprint (Thompson et al., 1986). Newman (1986) suggested, on the basis of major and trace element geochemistry, that this pluton was a product of a subduction-related continental margin tectonic environment. Younger low grade shear zones have developed locally in the pluton.

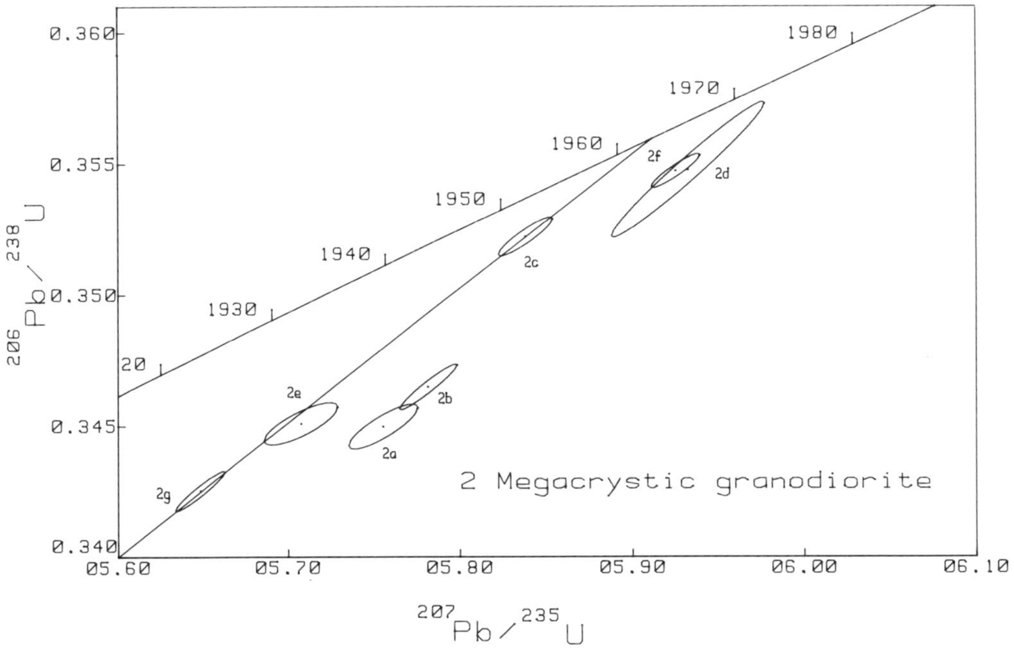

Figure 3. Isotope ratio plot for zircons from sample 2, megacrystic K-feldspar granodiorite. Numbers and letters with data points correspond to those of Table 1. Regression line through data points 2c, 2e, and 2g passes through upper and lower concordia intercepts of 1963 Ma and 507 Ma respectively.

Table 1. U-Pb isotopic data on zircon and monazite

Sample and Size fraction (in microns)	Weight (mg)	U (ppm)	Pb* (ppm)	Measured $^{206}Pb/^{204}Pb$	Isotopic abundances (^{206}Pb=1000)			Isotopic ratios		Age, Ma
					^{204}Pb	^{207}Pb	^{208}Pb	$^{206}Pb/^{238}U$	$^{207}Pb/^{235}U$	$^{207}Pb/^{206}Pb$
1. Biotite granodiorite - granitoid migmatite complex										
a, −105 + 88, N0	.072	267.9	140.1	6456	.111	176.7	78.37	.48629	11.7529	2608.8
b, −105 + 88, N0	.072	246.6	128.3	7493	.065	175.3	79.88	.48317	11.6222	2600.9
c, −88 + 74, N0	.072	261.3	137.2	4462	.144	176.9	83.13	.48700	11.7608	2607.5
d, −88 + 74, N0	.054	287.3	149.3	5128	.117	176.0	79.21	.48331	11.6278	2601.2
2. Megacrystic K−feldspar granodiorite										
a, +149, N-1	.064	235.2	87.05	1013	.882	132.8	154.9	.34498	5.7557	1971.1
b, −149 + 105, N-1	.052	189.5	70.06	4165	.078	122.1	121.8	.34649	5.7820	1971.4
c, −105 + 74, N-3	.024	238.7	88.46	1514	.338	124.7	114.3	.35222	5.8383	1959.5
d, −105 + 74, N-1	.021	164.1	63.04	1127	.506	128.1	153.6	.35477	5.9326	1975.2
e, −105 + 74, N-1	.227	240.5	93.63	853	1.102	134.7	153.0	.34509	5.7076	1955.5
f, −74, N−4	.038	178.7	69.21	2142	.195	123.8	141.4	.35473	5.9255	1973.3
g, −74, N-1	.217	223.1	80.41	6523	.070	120.5	105.7	.34252	5.6484	1950.3
3. Clinopyroxene granodiorite										
a, + 149, N-1	.040	141.4	55.05	1464	.439	126.0	183.7	.35078	5.8107	1958.3
b, −149 + 105, N-1	.040	197.5	74.95	3519	.105	121.1	153.4	.34739	5.7346	1952.1
c, −105 + 74, N-5	.021	228.5	89.48	2068	.165	123.4	164.3	.35548	5.9408	1974.0
d, −105 + 74, N-1	.043	324.7	124.6	2478	.266	123.7	150.5	.34775	5.7593	1957.9
e, −105 + 74, N-1	.049	212.8	80.93	4934	.066	120.4	165.0	.34458	5.6771	1948.6
f, −74 + 62, N-1	.082	255.2	94.87	7461	.056	120.2	152.1	.33920	5.5885	1948.6
4. "Charnockite" − granitoid migmatite complex										
a, −149 + 105, N1	.059	587.1	214.9	6552	.102	118.1	167.6	.33210	5.3432	1906.1
b, −149 + 74, N1, cores	.069	661.3	237.7	4959	.164	119.3	128.3	.33699	5.4400	1912.1
c, −105 + 74, N0	.026	717.5	258.3	6018	.056	118.4	115.9	.33955	5.5325	1928.8
d, −105 + 74, N1	.042	740.0	265.2	4352	.165	120.8	115.3	.33922	5.5478	1935.4
e, −105 + 74, N1	.079	512.3	184.3	9827	.059	118.1	128.2	.33632	5.4387	1915.2
f, −74 + 62, N1	.050	280.9	97.48	2422	.307	120.9	125.7	.32731	5.2709	1907.8
5. Agmatitic granodiorite − granitoid migmatite complex										
a, + 149, N0	0.083	289.6	124.8	2221	.364	124.7	329.5	.34770	5.7451	1953.8
b, −149 + 130, N0	0.067	291.8	120.7	2797	.248	121.3	309.3	.33827	5.5039	1926.3
c, −130 + 105, N0	0.065	393.2	163.0	4297	.164	120.9	299.5	.34066	5.5733	1936.1
d, −105, N0	0.061	316.7	129.5	5534	.090	120.8	266.0	.34350	5.6636	1950.0
e, All sizes, N3	0.059	226.0	87.22	3537	.156	126.8	145.1	.35469	6.0985	2024.5
6. Mylonitic garnet quartzofeldspathic gneiss										
a, + 149, N0	.072	520.8	204.4	6819	.092	123.2	158.8	.35701	6.0015	1984.5
b, −149 + 105, N0	.029	398.4	158.9	3803	.086	123.2	180.5	.35663	6.0011	1986.3
c, −105 + 74, N0, prismatic	.036	424.9	172.9	4655	.079	123.2	216.5	.35385	5.9581	1987.5
d, −105 + 74, N0, ovoid	.031	472.8	189.0	4674	.076	122.6	187.5	.35545	5.9584	1979.5
e, monazite	.023	3165	7176	10580	.066	123.7	6101	.36213	6.1303	1996.9
f, monazite	.023	5391	7913	10840	.0673	123.6	3577	.36073	6.1036	1996.1
g, monazite	.021	2876	7023	12910	.0264	122.2	6744	.35828	6.0174	1982.9
h, monazite	.031	2067	6402	19210	.018	121.7	8876	.35745	5.9843	1977.3
7. Garnet granite										
a, −130 + 74,	.056	2537	780.2	903	1.092	127.9	109.5	.30156	4.7052	1850.8
b, −74 + 44, prismatic	.060	2794	859.1	736	1.346	132.1	69.17	.31463	4.9423	1862.9
c, −74 + 44, ovoid	.034	3027	921.6	900	1.089	128.2	59.25	.31201	4.8843	1856.7
d, monazite	.039	11266	13820	52090	.014	117.0	3005.9	.34382	5.5382	1908.2
e, monazite	.027	15095	18080	44210	.017	117.0	2917.7	.34303	5.5223	1907.2
8. Garnet, sillimanite gneiss										
a, + 62, N0, ovoid	0.014	2947	952.5	1027	.931	125.9	707.6	.30931	4.8362	1854.6
b, −62, N0, prismatic	0.016	2647	786.1	584	1.668	135.1	92.60	.30085	4.6693	1841.2
c, −62, N0, ovoid	0.011	1411	436.2	424	2.238	142.5	147.7	.30352	4.6987	1836.6
9. Garnet granite tectonite										
a, + 74, N1, ovoid	0.021	4399	1408	1028	.944	128.3	51.27	.32837	5.2325	1888.
b, −74, N1, ovoid	0.006	5230	1700	1091	.839	127.1	48.96	.33289	5.3146	1892.2

Fractions are zircon unless indicated otherwise
* radiogenic lead
N, nonmagnetic at given angle
M, magnetic at given angle

Zircons are prismatic with L:B ratios of 2:1 to 4:1 and are euhedral to slightly rounded. Fractures are common. In polished and etched sections, most grains feature regular igneous zoning near the centre, which towards the outer margin becomes more irregular and discordant (Fig. 2B). Some grains appear to have been broken at a late stage and then overgrown by discordantly zoned zircon (Fig. 2C). The pattern of zoning is consistent with early unimpeded growth in a liquid, followed by dynamic interference of zircon growth in a crystal mush within a tectonically active environment (van Breemen and Hanmer, 1986). It is essentially impossible during hand picking, to distinguish magmatic zircons with this type of zoning from zircons containing older cores. Outer rims are consistently low in U and are grey in transmitted light in contrast to brown centres. Faint zoning in rims appears to be of igneous origin in that it appears to be gradational with inner discordant zoning. However, a metamorphic origin for part of the low U rims cannot be ruled out.

The distribution of U-Pb zircon data points for six fractions corresponds to a curved array where the more concordant fractions have older $^{207}Pb/^{206}Pb$ ratios (Figure 4). Reference Pb loss lines on Figure 4 have been drawn with lower intercepts of 0 Ma and 500 Ma. An older lower intercept age cannot, however, be ruled out. A line drawn through data points 3a and 3c would intersect the concordia at 2001 Ma and 1526 Ma. An emplacement age greater than 2000 Ma for the pyroxene granodiorite requires a significant metamorphic zircon component possibly related to Aphebian hornblende corona development (Thompson et al., 1986). It is not likely that a significant proportion of Pb was lost by solid diffusion in the absence of time cumulative lattice damage, even during granulite facies conditions (cf. van Breemen et al. 1986, 1987a). In view of the rather tentative textural evidence concerning the nature of the low U outer rims, the upper igneous age limit is placed at 2020 Ma. The younger age limit is placed at 1950 Ma; corresponding to the apparent 0 - 500 Ma Pb loss upper intercept ages of fractions 3b, 3e, and 3f.

"Charnockite"-granitoid migmatite complex (sample 4)

The greasy green fresh surface on samples of this orange-grey, weathering, medium grained, massive to foliated granite led to use of the term charnockite (orthopyroxene granite) as a field term. Subsequent petrography revealed that orthopyroxene was rare or absent and that biotite, hornblende and augite are the main mafic minerals. The unit is a layer or lense several hundred metres across within a heterogeneous package of granulite facies supracrustal gneisses, migmatites and pyroxene-bearing granitoids that is included in the granitoid migmatite complex (unit B on Fig. 1). Contact relations are not definitive but inclusions of leucocratic quartzofeldspathic gneiss demonstrate its intrusive nature. The overall texture is mainly equigranular granoblastic with mafic minerals forming interstitial aggregates between coarser grains of quartz, oligoclase-andesine, and microperthitic K-feldspar, locally transformed to microcline. Exsolution features in feldspar suggest relatively high initial crystallization temperatures. Both pyroxenes are variably altered to hornblende and quartz or very fine grained aggregates of an unknown mineral.

Mineralogically, except for the orthopyroxene, this rock is similar to the clinopyroxene granodiorite, but whereas the latter has well preserved textural evidence of an igneous origin, in thin section the "charnockite" is a metamorphic rock.

Zircons from the "charnockite" are similar to those of the clinopyroxene granodiorite except that a) irregular zoning started early in the growth history and b) there is an abundance of synneusis twins (Fig. 2D and 2E). Both these differences are indicative of greater crystal interaction in a tectonic environment (van Breemen and Hanmer, 1986). There are also a number of growth twins. Metamorphic zircon rims are relatively insignificant; in a number of cases the irregular igneous zoning can be shown to extend to the outer margin. Also, late synneusis must have occurred in a partially

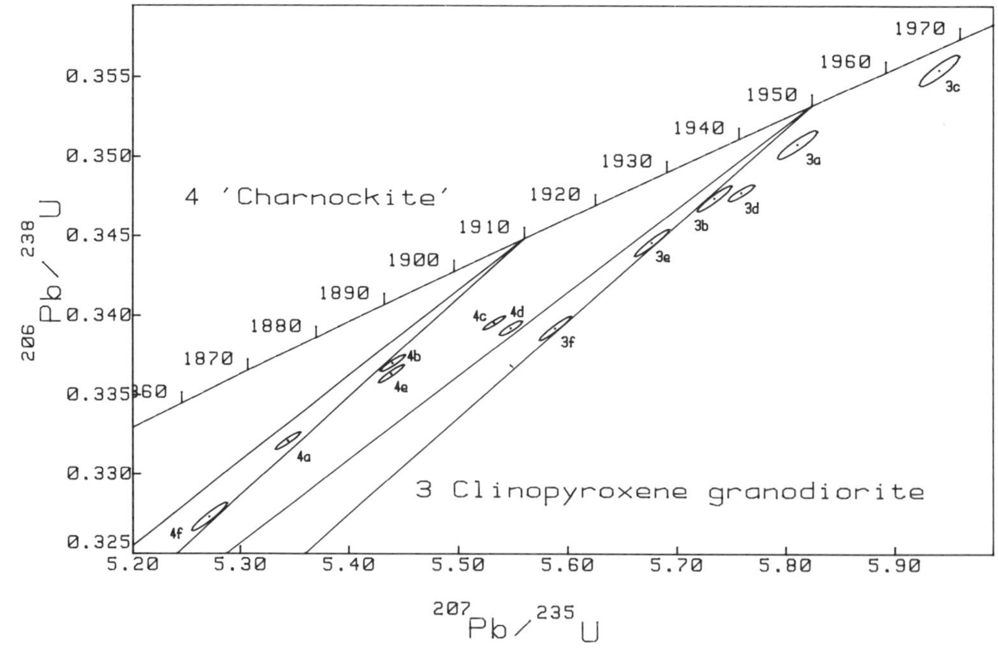

Figure 4. Isotope ratio plot for zircons from sample 3, clinopyroxene granodiorite and sample 4, "charnockite". Numbers and letters with data points correspond to those of Table 1. Hypothetical Pb loss lines have been drawn through lower intercepts of concordia of 0 Ma and 500 Ma.

liquid intrusive stage as this metamorphic rock is weakly foliated and shows no evidence for migmatization.

The distribution of data points for the "charnockite" is similar to that of the clinopyroxene granodiorite except that they are more discordant, younger and less regularly distributed (Fig. 4). Although a significant component of inherited zircon cannot be ruled out, fraction 4b picked specifically for apparent cores (in transmitted light) has the youngest $^{207}Pb/^{206}Pb$ age of the four most concordant points 4b, 4c, 4d and 4e (Table 1; Fig. 4). It is likely therefore that apparent cores are zircon centres with euhedral zoning rounded by magmatic corrosion (Fig. 2E) during a complex magmatic history.

Metamorphic recrystallization of zircon is not indicated by the igneous zoning while thin rims of possible metamorphic origin are likely to have been removed during abrasion. In view of the comparatively high U concentrations (Table 1) some Pb loss through lattice damage and intense subsequent Aphebian metamorphism cannot be ruled out. The more concordant data points appear to follow the same trend as the data points of the pyroxene granodiorite (Fig. 4). A similar, 2020 Ma, upper age limit is inferred while the younger age limit is placed at 1910 Ma.

Agmatitic granodiorite-granitoid migmatite complex (sample 5)

This pinkish grey weathering, moderately to highly strained agmatitic granodiorite is characterized by a heterogeneous association (10-30%) of rounded blebs, blocks and lenses of mafic gneiss and plutonic rocks in a pink fine-to medium-grained granitoid matrix or leucosome that contains clinopyroxene rimmed by hornblende (*see* Fig. 68-3 of Thompson et al., 1985). In thin section 1-4 mm grains of plagioclase, microcline, hornblende and clinopyroxene form porphyroclasts in a (mylonitic) matrix that must have formed during mylonitization under medium to high grade conditions. High strain also occurred as the rocks passed into a greenschist facies regime.

This heterogeneous rock contains rather poor zircons with fractures and many cores. Crystals are prismatic (L:B = 2:1) slightly rounded but feature regular igneous zoning. A minor population of ovoid zircons, also present, were not picked for analysis. Four size fractions (5a-d) were picked from one sample. From a separate and adjacent rock sample only one composite fraction (5e) could be prepared.

Data points corresponding to the four size fractions from the first sample form an array subparallel to the concordia (Fig. 5), 2.3% discordant, between 1954 Ma and 1926 Ma ($^{207}Pb/^{206}Pb$ ages). A regression line through data points 5a, 5b and 5c yields upper and lower intercept ages of 1992 Ma and 1306 Ma. Analysis of the composite zircon fraction from the second rock sample yields a data point (5e, not plotted) 4% discordant with a $^{207}Pb/^{206}Pb$ age of 2024.5 Ma. Assuming the likelihood of an inherited zircon component in fractions 5a-5d the age of this igneous unit is also placed between 2020 Ma and 1910 Ma.

Mylonitic garnet-quartzofeldspathic gneiss (sample 6)

Interlayered with isoclinally folded, mafic granulite layers, this beige weathering garnet-biotite quartzofeldspathic straight gneiss is the fine grained mylonitic equivalent of migmatitic metasedimentary rocks. The extremely high strain began under high grade conditions sufficiently dry for orthopyroxene to remain stable during grain size reduction in associated rocks and continued or recurred while physical conditions decreased to those of the greenschist facies. Inclusions of hercynitic spinel inside porphyroclasts of garnet confirm the high temperatures attained prior to mylonitization.

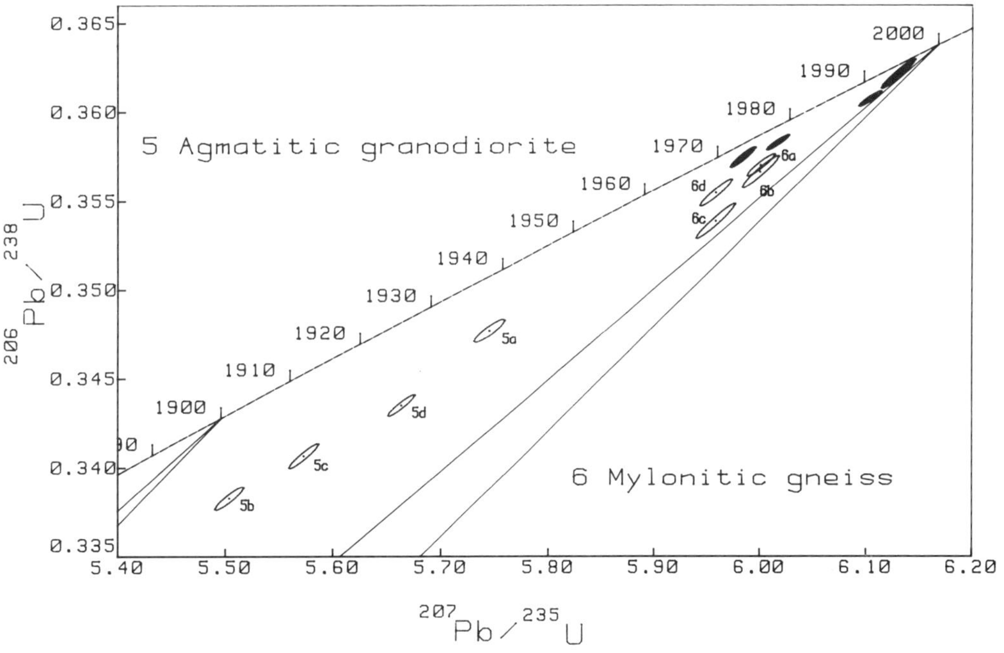

Figure 5. Isotope ratio plot for zircons (open error ellipses) and monazites (closed error ellipses) from sample 5, (agmatitic granodiorite-granitoid migmatite complex) and sample 6, mylonitic garnet gneiss. Hypothetical Pb loss lines have been drawn through lower intercepts of concordia at 0 Ma and 500 Ma.

Zircons are clear, equidimensional and rounded at the edges. Polished and etched sections show that the internal zoning conforms to the equidimensional shapes. This zoning ranges from simple to multifaceted growth, generally regular (Fig. 2F). Only near the margins, which are higher in U, can the zoning be seen to be irregular and discordant (Fig. 2F, 2G). Using the criteria of van Breemen et al. (1987a), the zircon grew mainly during conditions of static metamorphism and was overtaken by mylonitization late in the crystal growth history. However, in view of the relatively simple internal zoning of many of the crystals, an early growth history in a melt cannot be discounted on the basis of morphology alone.

Data points corresponding to four fractions are slightly discordant (0.1%) and clustered with $^{207}Pb/^{206}Pb$ model ages ranging from 1979.5 Ma to 1987.5 Ma (Fig. 5). The oldest $^{207}Pb/^{206}Pb$ age corresponds to the only prismatic fraction picked, with L:B ratios of 2:1 to 3:1. The youngest $^{207}Pb/^{206}Pb$ age corresponds to the small ovoid zircons of fraction 6d. The range in apparent ages could be the result of inherited older zircons or an extended period of growth. Evidence supporting the latter idea comes from the analyses of four monazite fractions, all nearly concordant, whose $^{207}Pb/^{206}Pb$ model ages range from 1977.3 Ma to 1996.9 Ma. It is possible that the oldest monazite ages (6e and 6f averaging at 1996.5 ± 2 Ma) are older than the youngest zircon age. With an average Pb loss age of 500 Ma, the upper intercept age for this zircon fraction (6d) would be 1982 Ma. A line through zircon point 6d and the oldest monazite data point 6e yields upper and lower intercept ages of 2002 Ma and 1328 Ma. The data suggest but do not prove that the monazite U-Pb system remained closed during conditions permitting growth of zircon. In view of the pattern of internal zircon zoning, the zircon is not likely to have grown after the ductile shearing during which orthopyroxene was stable. The isotopic data are consistent with high grade metamorphism followed by ductile shearing in the time interval 2000-1980 Ma.

Garnet granite, sillimanite-garnet gneiss, garnet granite tectonite (samples 7, 8, 9)

Greyish pink weathering, the weakly to well foliated and lineated, medium grained to megacrystic garnet granite is associated with a marked aeromagnetic low in the southeast corner of the map area. Garnet is usually rimmed by biotite where it was in contact with K-feldspar. Accessory sillimanite, spinel and cordierite are present. Inclusions and screens of mylonitic sillimanite-garnet gneiss (sample 8) with similar mineralogy but a higher proportion of aluminous minerals occur in the granite. Rims of cordierite on sillimanite or garnet, biotite on garnet, and cordierite and sillimanite on spinel indicate an evolution of P-T conditions with temperature constant or decreasing as pressure decreased. Locally, adjacent to the mylonitic gneiss, the granite is a lineated tectonite.

Zircons from the garnet granite, sillimanite-garnet gneiss and the lineated tectonite are similar, with the majority in each sample consisting of equidimensional rounded grains. Rounded prismatic grains (L:B = 2:1) constitute a minority. Euhedral igneous cores appear to be least common in the zircons from the garnet granite. Most crystals have a wavy zoning which in places becomes discordant. Synneusis twins are also common, with points of docking corresponding to locations of discordant zoning which are generally late in the growth history (Fig. 2H and 2I).

It would be difficult to describe with confidence the environment of growth of the zircons from the garnet granite. Many grains feature the equidimensional multifaceted shapes characteristic of metamorphic zircons, while others have more euhedral shapes and zoning characteristic of growth in a liquid. Irregular and discordant zoning is not constrained to a particular stage of growth. Impingement with other crystals in a tectonic environment may have contributed to the rounding of many grains. The similarity of the zircons in the garnet granite and in the garnet sillimanite schist suggest

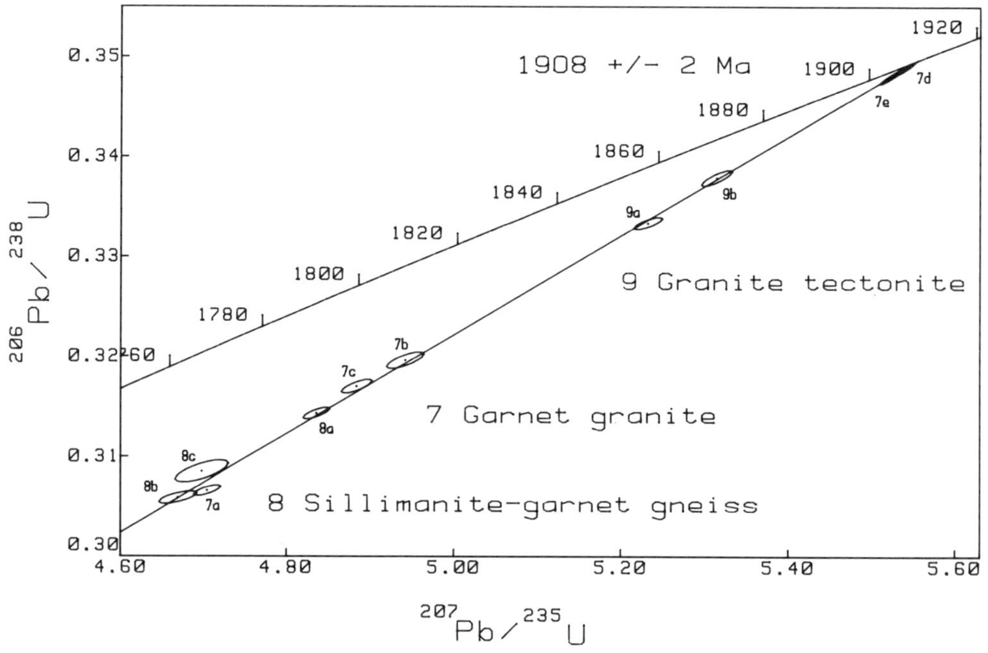

Figure 6. Isotope ratio plot for zircons (open error ellipses) and monazites (closed error ellipses) from sample 7, garnet granite sample 8, sillimanite-garnet gneiss, and sample 9, lineated granite tectonite. Regression line passes through upper and lower concordia intercepts of 1909 Ma and 755 Ma.

that many of the zircons of the former are also of metamorphic origin. Others may have grown in a partially molten, moving crystal mush, probably with a viscous melt phase.

A common origin between the zircons from the garnet granite and the sillimanite gneiss is supported by their isotope systematics. Two monazite fractions from the garnet granite, one concordant and one slightly discordant, define an age of 1908 ± 2 Ma which in view of the monazite isotope systematics in the garnet quartzofeldspathic gneiss (above) is interpreted in terms of granite emplacement. U concentrations from the garnet granite and sillimanite gneiss are both high, from 2500 ppm to 3000 ppm, while zircons from the lineated granite tectonite are even higher at 5000 ppm. Two prismatic and one ovoid zircon fractions from the garnet granite and two ovoid and one prismatic fractions from the sillimanite gneiss are strongly discordant and linearly aligned (Fig. 6). Two ovoid fractions from the lineated granite are less discordant but also aligned with the zircon and monazites from the garnet granite. A regression line through all 10 data points yields upper and lower intercepts of 1909 ± 2.5 Ma and 755 ± 45 Ma with a MSWD of 3.2. A lower intercept age greater than 500 Ma may be attributed to the high U contents of the zircons permitting relatively early cumulative lattice damage. There is no evidence for U-Pb zircon disturbance after metamorphic-anatectic crystallization as indicated by both the linear alignment of the zircons and agreement of zircon and monazite ages.

The slight scatter of data points about the regression line could be interpreted in terms of variable times of lead loss. However, the scatter could also be the result of variable amounts of inheritance. Such a view is supported by the fact that the coarsest prismatic fraction of the garnet granite plots slightly to the right of the regression line. In all three rocks, it is likely that the small component of inherited zircon Pb was swamped by Pb generated by the high U contents of the new zircon. These high U contents further support the morphological evidence that zircons in the sillimanite gneiss and garnet granite were formed by the same process and that ultrametamorphism of sediments, melting and granite crystallization events were closely related. Preservation of metamorphic zircon and monazite through the melting stage of an S-type granite is supported by the experimental data of Watson and Harrison (1983) and Rapp and Watson (1986) respectively.

Thus in view of the high proportion of new zircon of metamorphic aspect, the minor zircon Pb inheritance, agreement of zircon and monazite ages, the discordant internal zoning and synneusis twins, and the S-type nature of the garnet granite with sillimanite-garnet gneiss inclusions, this granite is likely to have formed during syntectonic heating, melting and mobilization of a sedimentary pile of pelitic or semi-pelitic composition at or close to 1908 ± 2 Ma.

DISCUSSION

Conditions of zircon and monazite growth and isotopic disturbance

Granite emplacement appears to have been syntectonic within an extended orogenic interval. The megacrystic granodiorite is a protomylonite indicating intense deformation after igneous emplacement which may be reflected in the rounding of otherwise igneous zircons. The clinopyroxene granodiorite of igneous texture and weak foliation has zircons featuring some irregular zoning indicative of movement when the rock was a mixture of liquid and crystals (van Breemen and Hanmer, 1986). The ''charnockite'' again in the abscence of textures indicative of strong deformation has zircons with irregular zoning and abundant synneusis at a late stage of crystallization. Significant strain in these unmigmatized weakly foliated rocks is believed to have occurred when they were in a semi-molten state during magmatic emplacement (cf. Hollister and Crawford, 1986).

Age ranges for these igneous units excluding the garnet granite are both broader and older than those presented in van Breemen et al., (1987c). It should be stressed that both groups of ages are model related. The former age assessment assumed a maximum recent lead loss event of 500 Ma. The broader and older revised ages are based on the fact that the isotope systematics do not support a large amount of inheritance implying another cause for data dispersion.

Zircon lattice anneals at ca. 350°C (Gebauer and Grunenfelder, 1976). Thus even though Aphebian orogenic events occurred over a span of at least 90 Ma it is likely that damaged zircon lattice was continuously annealed. The high proportion of Aphebian lead loss required to explain the isotope systematics (Fig. 4) cannot easily be explained by solid diffusion (Cliff, 1985). Yet the zircon data from the ''charnockite'' (4c, 4b) follow the same trend as the data points for the pyroxene granodiorite (3c, 3a). It is possible, therefore, that the greater discordance and younger ages of the ''charnockite'' correlate with both its strong metamorphic fabric and the high U content of the zircons. U concentration may also correlate with a younger Proterozoic lead loss event following more extensive lattice damage without annealing. Finally with possible xenocrysts blending into complex magmatic growth zonation, better igneous age resolution may be problematic.

Internal structure of 1990-1980 Ma zircons from the granulite facies mylonite (sample 6) show that mylonitization occurred late in the zircon growth history. Much of the zircon growth appears to have been metamorphic as two almost concordant monazite data points yield $^{207}Pb/^{206}Pb$ ages of 1997 Ma. The data, however, raise the possibility that the U-Pb monazite system can survive conditions of mylonitization during which hypersthene is a stable phase. Sawka et al. (1986) have suggested that the closing temperature of the U-Pb system in monazite exceeds that of anatexis in the production of S-type granite: ages some 13 to 22 Ma older than plutonic emplacement (Williams, 1977) are interpreted in terms of prograde metamorphism with dehydration of rhabdophane to monazite at ca. 400°C (Vlasov, 1966). It is possible therefore that in view of the ''metamorphic '' internal zonation of most zircons of the younger garnet granite (sample 7), both zircon and monazite ages predate plutonic emplacement by 10 to 20 Ma (Watson and Harrison, 1983; Rapp and Watson, 1986). However, considering the consistent zircon (three independant data sets) and monazite systematics it is more likely that the garnet granite formed and was emplaced in a comparatively short interval close to 1908 ± 2 Ma.

Regional implications

Thompson and Henderson (1983) suggested that the Thelon Tectonic Zone formed in the Archean and subsequently became a locus for Proterozoic orogenic activity. A suture, if present, was thought to occur within the zone or along its eastern boundary. A subsequent intraplate ensialic model (Thompson et al., 1987) provides an alternative to the interplate model of Gibb and Halliday (1974) and Gibb and Thomas (1977) who proposed that the Thelon Front, the western boundary of the Thelon Tectonic Zone represented a suture produced by a Proterozoic continent/continent collision. Hoffman et al. (1986) elaborated on the Gibb and Thomas model by suggesting oblique convergence and eastward subduction associated with arc magmatism.

The ca. 2.6 Ga zircon age from the granitoid migmatite complex 6 km east of the western boundary of the Thelon Tectonic Zone near the north edge of the map area (sample 1) is characteristic of ages of plutonism within the Slave Province (van Breemen et al., 1987b; Henderson et al., 1987). East of the ''megacrystic K-feldspar granodiorite line'' (Thompson, 1986; Fig. 1) two samples from the granitoid migmatite complex at granulite facies grade have yielded Proterozoic ages. The agmatitic inclusion-rich granodiorite (sample 5) and the ''charnockite'' (sample 4) appear to have been emplaced in the interval 2020-1910 Ma. Data for the megacrystic granodiorite (sample 2) and clinopyroxene granodiorite (sample 3) are interpreted as dating emplacement in the intervals 1980-1940 Ma and 2020-1950 Ma respectively.

A belt of plutonic magmatism in the interval 2.02-1.91 Ga has now been documented within the Thelon Tectonic Zone. Granulite facies metamorphism accompanied by mylonitization has been dated in the 2000-1950 Ma interval both north of the Bathurst Fault (this study) and between the Bathurst and Macdonald faults (van Breemen et al., 1987a) 200 km to the south. Given the metamorphic texture of the ''charnockite'' its discordantly zoned zircons and the tectonites within the 1908 $\pm$ 2 Ma old garnet granite, a Proterozoic orogenic history of 100 Ma or more is indicated. Remaining geochronological objectives in the Thelon Tectonic Zone are: 1) to determine the distribution of Archean ''basement'' in or east of the zone; 2) to determine the relationship between the sedimentological record of Proterozoic events in the sediments of the Kilohigok (foreland) Basin (Campbell and Cecile, 1981; Grotzinger et al., 1987), metamorphism and deformation of these rocks (Thompson and Frey, 1984; Tirrul, 1985; Thompson et al., 1985) and the events documented in the Thelon Tectonic zone; and 3) age and tectonic significance of supracrustal rocks in the Thelon Tectonic Zone.

ACKNOWLEDGMENTS

Scanning electron photographs were taken by S.V. Thompson. We thank colleagues in the Geochronology Section of the GSC for technical help and John B. Henderson for comments on an earlier draft of the manuscript, which also benefitted from critical reading by R.R. Parrish. Discussions with L.M. Heaman, T.E. Krogh and U. Scharer have also been helpful.

REFERENCES

Bostock, H.H., van Breemen, O. and Loveridge, W.D.
1987: Proterozoic geochronology in the Talston Magmatic Zone; *in* Radiogenic Age and Isotopic Studies: Report 1; Geological Survey of Canada, Paper, 87-2.

Campbell, F.H.A. and Cecile, M.P.
1981: Evolution of the Early Proterozoic Kilohigok Basin, Bathurst Inlet-Victoria Island, Northwest Territories; *in* Proterozoic Basins of Canada, Ed. F.H.A. Campbell; Geological Survey of Canada, Paper 81-10, p. 103-131.

Cliff, R.A.
1985: Isotope dating in metamorphic belts; Geological Society of London, Journal, v. 142, p.97-110.

Gebauer, D. and Grunenfelder, M.
1976: U-Pb zircon and Rb-Sr whole rock dating of low grade metasediments: Montagne Noire; Contributions to Mineralogy and Petrology, v. 59, p. 13-32.

Gibb, R.A. and Halliday, D.W.
1974: Gravity measurements in southern District of Keewatin and southeastern District of Mackenzie, N.W.T., Energy Mines and Resources Canada, Gravity Map Series.

Gibb, R.A. and Thomas, M.D.
1977: The Thelon Front: a cryptic suture in the Canadian Shield?; Tectonophysics, v. 38, p. 211-222.

Grotzinger, J.P. and Gall, Q.
1986: Preliminary investigations of early Proterozoic Western River and Burnside formations: evidence for foredeep origin of Kilohigok Basin, N.W.T., Canada; *in* Current Research, Part A, Geological Survey of Canada, Paper 86-1A, p. 95-106.

Grotzinger, J.P., McCormick, D.S. and Pelechaty, S.M.
1987: Progress report on the stratigrapy, sedimentology and significance of the Kimerot and Bear Creek groups, Kilohigok Basin, District of Mackenzie; *in* Current Research, Part A, Geological Survey of Canada, Paper 87-1A, p. 219-238.

Henderson, J.B., van Breemen, O. and Loveridge, W.D.
1987: Some U-Pb zircon ages from Archean basement, supracrustal and intrusive rocks, Yellowknife-Hearne Lake area, District of Mackenzie, N.W.T., *in* Radiogenic Age and Isotopic Studies: Report 1; Geological Survey of Canada, Paper 87-2.

Hoffman, P.F., Culshaw, N.G., Hammer, S.K., LeCheminant, A.N., McGrath, P.H., Tirrul, R., and van Breemen, O.
1986: Is the Thelon Front (NWT) a suture?; Geological Association of Canada, Program with Abstracts, v. 11, p. 82 (abstract).

Hollister, L.S. and Crawford, M.L.
1986: Melt-enhanced deformation: a major tectonic process; Geology, v. 14, p. 558-561.

Krogh, T.E.
1973: A low contamination method for hydrothermal dissolution of zircon and extraction of U and Pb for isotopic age determination; Geochimica et Cosmochimica Acta, v. 37, p. 485-494.
1982: Improved accuracy of U-Pb zircon ages by the creation of more concordant systems using an air abrasion technique; Geochimica et Cosochimica Acta, v. 46, p. 637-649.

Newman, R.M.
1986: Petrology, geochemistry, and geochronology of a pluton in the Thelon Tectonic Zone, northwest territories; unpublished BSc thesis, McMaster University, 82 p.

Parrish, R.R., Roddick, J.C., Loveridge, W.D. and Sullivan, R.W.
1987: Uranium-lead analytical techniques at the geochronology laboratory, Geological Survey of Canada; *in* Radiogenic Age and Isotopic Studies: Report 1, Geological Survey of Canada, Paper 87-2, p. 3-7.

Poldervaart, A.
1956: Zircons in rocks. 2. Igneous rocks; American Journal of Science, v. 254, p. 521-554.

Rapp, R.P. and Watson, E.B.
1986: Monazite solubility and dissolution kinetics: implications for the thorium and light rare earth chemistry of felsic magmas; Contributions to Mineralogy and Petrology, v. 94, p.304-316.

Roddick, J.C. Loveridge, W.D. and Parrish, R.R.
1987: Precise U-Pb dating of zircon at the sub-nanogram Pb level; Isotope Geoscience, v. 66, p. 111-121.

Sawka, W.N., Banfield, J.F. and Chappell, B.W.
1986: A weathering-related origin of widespread monazite in S-type granites; Geochimica et Cosmochimica Acta, v. 50, p. 171-175.

Thompson, P.H.
1986: Geology of the Tinney Hills (76J) and Overby Lake (76I west 1/2) map areas; Geological Survey of Canada, Open File 1316 (1:125 000 map and marginal notes).

Thompson, P.H. and Henderson, J.B.
1983: Polymetamorphism in the Healey Lake map area - implications for the Thelon Tectonic Zone; *in* Current Activities Forum, 1983, Program with Abstract, Geological Survey of Canada, Paper 83-8, p. 2 (abstract).

Thompson, P.H. and Frey, M.
1984: Illite "crystallinity" in the Western River Formation and its significance regarding the regional metamorphism of the early Proterozoic Goulburn Group, District of Mackenzie; *in* Current Research, Part A, Geological Survey of Canada, Paper 84-1A, p. 409-414.

Thompson, P.H., Culshaw, N., Thompson, D.L. and Buchanan, B.R.
1985: Geology across the western boundary of the Thelon Tectonic Zone in the Tinney Hills-Overby Lake (west half) map area, District of Mackenzie; *in* Current Research, Part A, Geological Survey of Canada, Paper 85-1A, p. 555-572.

Thompson, P.H., Culshaw, N., Buchanan J.R. and Manojlovic, P.,
1986: Geology of the Slave Province and Thelon Tectonic Zone in the Tinney Hills-Overby Lake (west half) map area, District of Mackenzie; *in* Current Research, Part A, Geological Survey of Canada, Paper 86-1A, p. 275-289.

Thompson, P.H., Henderson, J.B. and Frith R.A.
1987: The Thelon Front — the transition between the Slave Province and the Thelon Tectonic Zone, northwestern Canadian Shield; Geological Association of Canada, Program with Abstracts, v. 12, p. 96 (abstract).

Tirrul, R.
1985: Nappes in the Kilohigok Basin and their relation to the Thelon Tectonic Zone, District of Mackenzie; *in* Current Research, Part A, Geological Survey of Canada, Paper 85-1A, p. 407-420.

van Breemen O. and Hanmer, S.
1986: Zircon morphology and U-Pb geochronology in active shear zones: studies on syntectonic intrusions along the northwest boundary of the Central Metasedimentary Belt, Grenville Province, Ontario; *in* Current Research, Part B, Geological Survey of Canada, Paper 86-1B, p. 775-784.

van Breemen, O., Davidson, A., Loveridge, W.D. and Sullivan, R.W.
1986: U-Pb zircon geochronology of Grenville tectonites, granulites and igneous precursors, Parry Sound, Ontario; *in* New Perspectives on the Grenville Problem, ed. J., Moore, A. Davidson and A.J. Baer; Geological Association of Canada, Special Paper 31, p. 191-207.

van Breemen, O., Henderson, J.B., Loveridge, W.D. and Thompson, P.H.
1987a: U-Pb zircon and monazite geochronology and zircon morphology of granulites and granite from the Thelon Tectonic Zone, Healey Lake and Artillery Lake map areas, N.W.T.; *in* Current Research, Part A, Geological Survey of Canada, Paper 87-1A, p. 783-801.

van Breemen, O., Henderson, J.B., Sullivan, R.W. and Thompson, P.H.
1987b: U-Pb zircon and monazite ages from the eastern Slave Province, Healey Lake area, N.W.T.; *in* Radiogenic Age and Isotopic Studies: Report 1, Geological Survey of Canada, Paper 87-2, p. 783-801.

van Breemen, O., Thompson, P.H., Bostock, H.H. and Loveridge, W.D.
1987c: Timing of plutonism in the northern Thelon Tectonic Zone and the Taltson Magmatic Zone; Geological Association of Canada, Program with Abstracts, v. 12, p. 98 (abstract).

Vlasov, K.A.
1966: Geochemistry and mineralogy of rare elements and genetic types of their deposits; *in* Mineralogy of Rare Elements, v. 2; Israel Program for Scientific Translations Ltd., Jerusalem, p. 293-297.

Watson, E.B. and Harrison, T.M.
1983: Zircon saturation revisited: temperature and composition effects in a variety of crustal magma types; Earth and Planetary Science Letters, v. 64, p. 295-304.

Williams, I.
1977: The Berridale Batholith: a lead and strontium isotope study of its age and origin; unpublished Ph.D. thesis, Australian National University, Canberra.

York, D.
1969: Least squares fitting of a straight line with correlated errors; Earth and Planetary Science Letters, v. 5, p. 320-324.

U-Pb zircon age of an alkaline granite body in the Booth River Intrusive Suite, N.W.T.

S.M. Roscoe[1], M.N. Henderson[1], P.A. Hunt[1] and O. van Breemen[1].

Roscoe, S.M, Henderson, M.N, Hunt, P., and van Breemen, O. U-Pb zircon age of an alkaline granite body in the Booth River Intrusive Suite, N.W.T.; in Radiogenic Age and Isotopic Studies: Report 1; Geological Survey of Canada, Paper 87-2, p. 95-100, 1987.

Abstract

Alkali-feldspar granite in the Booth River mafic-ultramafic- felsic intrusive complex near Bathurst Inlet yielded a U-Pb zircon date of 2023 +4/−2 Ma. This is considered to represent the time of intrusion of the felsic component of the anorogenic intrusive suite and a maximum date for the beginning of deposition of the Goulburn Group. The Booth River Intrusive Suite is similar to the Blachford Intrusive Suite at Great Slave Lake although it is about 150 m.y. younger.

Résumé

On a évalué l'âge du granite alcalin du complexe intrusif de nature mafique, ultramafique et felsique de Booth River, près de Bathurst Inlet, par la méthode U-Pb appliquée au zircon, à 2023 + 4/−2 Ma. Cet âge représenterait l'époque de l'intrusion de la composante felsique de la série intrusive anorogénique et la date la plus ancienne associée au début de la mise en place du groupe de Goulburn. La série intrusive de Booth River ressemble à celle de Blachford dans la région du Grand lac des Esclaves, bien qu'elle soit d'environ 150 millions d'années plus récente.

INTRODUCTION

The anorogenic Booth River Intrusive Suite (Roscoe, 1985) underlies a structural basin of Goulburn Supergroup strata along the Burnside River Synclinorium 50 km west of Bathurst Inlet in NTS 76K as indicated by aeromagnetic anomalies and outcrop along the flanks of the synclinorium (Fig. 1). The suite is characterized by coarse grained gabbronorite ranging from leuconorite to melanonorite with cumulus plagioclase and abundant intercumulus ilmenite, but includes thin basal layers of dunite, troctolite, websterite and oxides. Monzonite and quartz monzonite are present at the southern margin of Goulburn strata. Alkali-feldspar granite bodies outcrop on both limbs of the syncline. Their contacts with noritic rocks are hidden beneath Goulburn cover, but aeromagnetic anomaly and gravity anomaly patterns, textural and mineralogical similarities with monzonitic phases and shared alkaline character indicate that they belong to the Booth River Intrusive Suite, although they may occur in separate but contiguous bodies. Thin dykes of aplite, muscovite granite pegmatite and alkali granite cut monzonitic to noritic rocks along the south flank as well as perthite granite at the north flank.

The mafic and felsic Booth River rocks were intruded into Archean turbiditic metasedimentary rocks, minor iron-formation and granitic rocks. The Archean rocks are extensively brecciated, hornfelsed, and in places intensely altered and invaded by sulphides along their contacts with basal mafic-ultramafic rocks of the Booth River Complex. Contacts are evidently displaced distances as great as several hundred metres along faults in several places. No contacts of alkaline granite with Archean rocks or with overlying strata of the Goulburn Group have been found. Contacts of mafic Booth River rocks with overlying siltstones of the Western River Formation at the base of the Goulburn Group are obscured either by a few metres of drift or, along the northern exposures, by an intrusive sheet of post-Goulburn gabbro. The Goulburn strata of sub-greenschist regional metamorphic grade, however, show no effects attributable to the Booth River rocks. They are not intruded by late granitic dykes that cut both mafic and felsic intrusive rocks of the Booth River suite. They are not disrupted by faults interpreted to displace the latter rocks. The contact overlaps different component units of the Booth River complex such as magnetic, sulphidic, ilmenite-rich, anorthositic, and basal zones.

It is reasonably certain, in the light of relationships outlined above (*see* Roscoe, 1985), that the igneous rocks of the Booth River Intrusive Suite are unconformably overlain by the Goulburn Group. The original roof under which the intrusive suite was emplaced must have been eroded, and the Booth River intrusions themselves eroded prior to deposition of the overlying strata. This roof, like the floor of the intrusive suite, may have consisted of Archean metamorphic

[1] Geological Survey of Canada, 601 Booth Street, Ottawa, Ontario K1A 0E8

and igneous rocks. Anorogenic igneous complexes, however, are commonly emplaced at unconformities so it is possible, indeed perhaps likely, that the missing roof rocks of the Booth River complex were relatively little deformed early Proterozoic (pre-Goulburn) or very late Archean strata.

A date for the intrusive event should provide a maximum age for the deposition of the immediately overlying Goulburn strata and a minimum age for the missing roof rocks of the intrusion. The precise age of the Booth River Intrusive Suite is of course also of interest in relationship to ages of 2175 ± 7 Ma for the similar anorogenic Blachford Intrusive Suite (Davidson, 1978; Bowring et al., 1984 remac; Fig. 1), 2188 +16/−10 Ma for the anorogenic Big Spruce Complex (Cavell and Baadsgaard, 1986) and 2588 ± 8 Ma for the Uist tonalite-leucogranodiorite pluton of the mostly late- to post-orogenic (or anorogenic) Regan Intrusive Suite (Frith and Loveridge, 1982) in the Slave Structural Province.

SAMPLE DESCRIPTION

Sample 85RF185, weighing approximately 30 kg, was collected at 66°37.0′N, 109°40.3′W in the northwestern granitic lobe of the Booth River Intrusive Suite (Fig. 1 and 2). Previously collected 83RF series samples near the 85RF185 sample locality had been examined petrographically and analysed chemically (Table 1). All of the 83RF samples and sample 85RF185 from this rather homogeneous intrusion are brownish pink, coarse grained, biotite-alkaline amphibole-perthite granites containing zircon. They are strikingly similar texturally and mineralogically, megascopically and microscopically, to peralkaline granites described by others (Whalen and Currie, 1983; Davidson, 1982). Sample 85RF185 contains about 50 % microperthitic orthoclase in subhedral grains up to 12 mm long, 35 % quartz, and 5 % albite in small grains and rims on orthoclase, and 10 % amphibole as small interstitial grains. Albite is very pure,

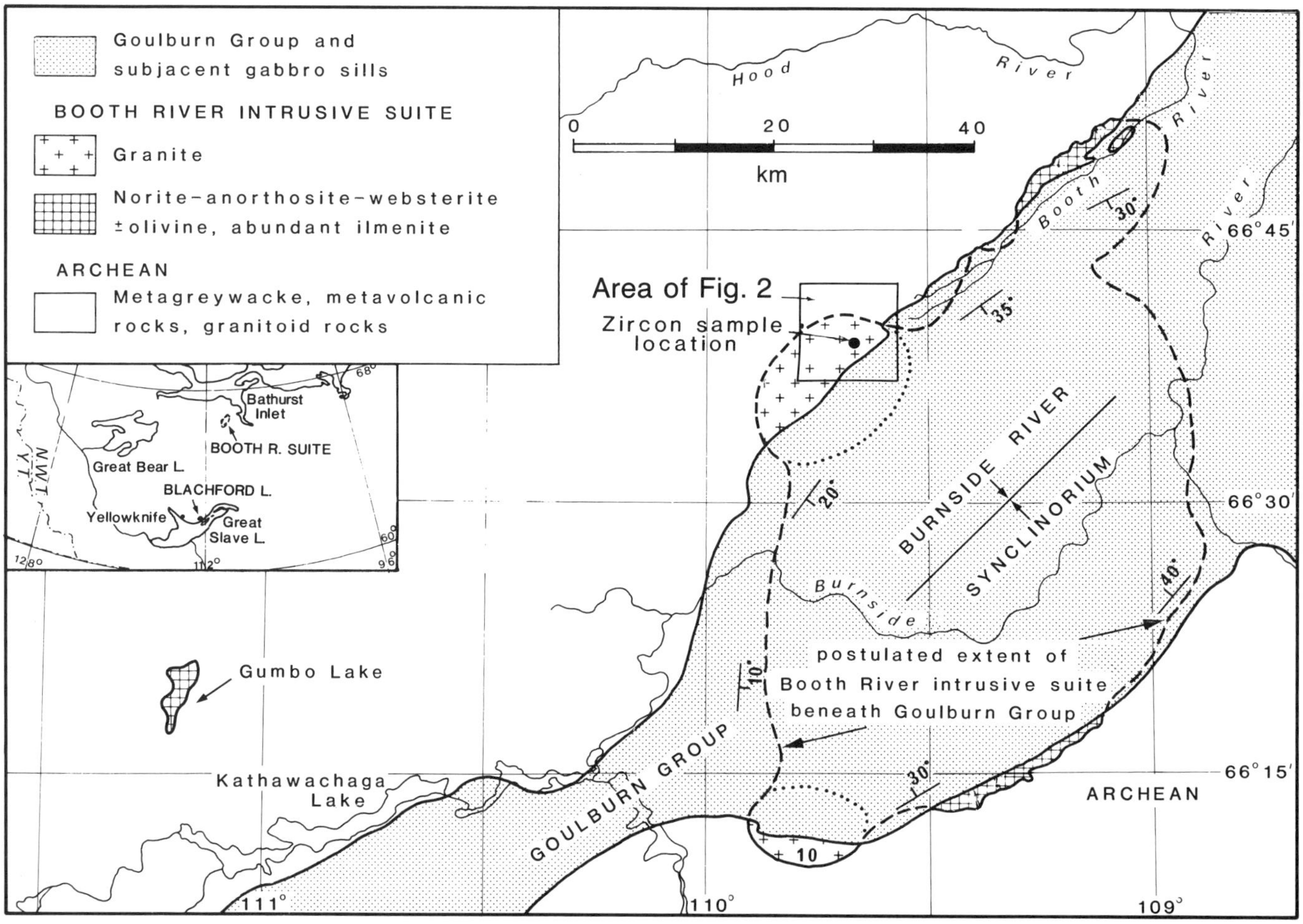

Figure 1. Index map and geological setting of felsic and mafic intrusions of the Booth River Intrusive Suite along the contact between Archean rocks and overlying Aphebian strata of the Goulburn Group in the Burnside River Synclinorium, N.W.T. BR — Booth River, BL — Blachford Lake, BI — Bathurst Inlet, Y — Yellowknife.

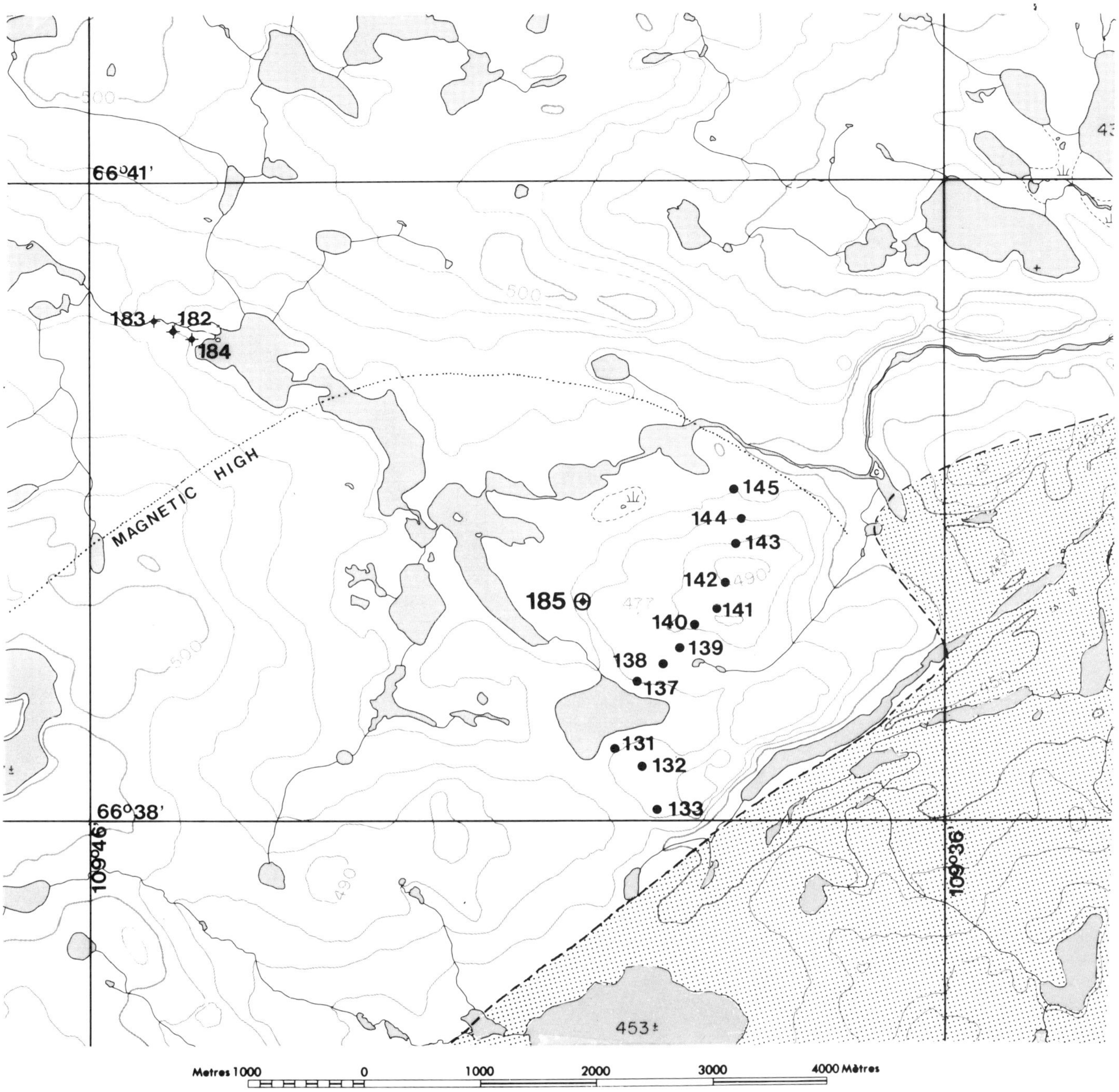

• Sample location 1983
✦ Sample location 1985
⊕ Zircon sample location

Goulburn Group

Contours in metres

Figure 2. Location of geochronological sample 85RF185, zircon-bearing alkaline granite and other samples.

Table 1. Analyses of alkali-feldspar granited of Booth River Intrusive Suite

SAMPLE		83RF132	83RF133	83RF138	83RF139	83RF141	83RF142	83RF143
SiO_2	%	78.8	77.7	77.4	78.3	76.5	76.9	76.2
TiO_2	%	0.20	0.21	0.15	0.20	0.23	0.18	0.16
Al_2O_3	%	10.5	10.2	10.8	10.5	11.1	10.9	10.6
Fe_2O_3(T)	%	2.27	2.56	2.07	2.29	2.04	1.93	1.99
Fe_2O_3	%	1.2	1.3	0.7	1.0	0.9	1.0	1.1
FeO	%	1.0	1.1	1.2	1.2	1.0	0.8	0.8
MnO	%	0.03	0.03	0.04	0.03	0.03	0.04	0.04
MgO	%	0.14	0.09	0.08	0.09	0.21	0.10	0.08
CaO	%	0.52	0.16	0.31	0.42	0.28	0.14	0.15
Na_2O	%	2.80	2.99	3.08	2.88	2.97	2.88	3.23
K_2O	%	4.44	4.75	4.49	4.39	4.79	4.74	4.36
H_2O(T)	%	0.5	0.4	0.3	0.4	0.4	0.5	0.5
CO_2(T)	%	0.5	0.0	0.1	0.2	0.2	0.2	0.0
P_2O_5	%	0.06	0.02	0.02	0.04	0.03	0.04	0.04
S	%	0.02	0.02	0.00	0.00	0.00	0.01	0.01
TOTALS	%	100.7	99.0	98.7	99.7	98.6	98.4	97.1
Ba	ppm	70	40	40	50	50	60	50
Zn	ppm	71	150	180	140	110	200	120
Pb	ppm	8	13	16	32	9	10	0
Ni	ppm	18	19	19	20	19	19	18
V	ppm	12	13	12	13	12	12	12
Sr	ppm	5	2	4	6	1	0	3
Rb	ppm	160	200	200	200	200	190	190
Be	ppm	4.0	4.1	7.6	11	3.7	9.3	4.3
Nb	ppm	120	110	78	140	110	84	64
La	ppm	99	97	100	120	170	130	86
Y	ppm	77	65	82	140	120	92	53
Yb	ppm	5.8	7.3	7.4	15	7.9	9.9	5.4
Th	ppm	26	22	28	49	28	24	23
Zr	ppm	590	810	630	1200	800	720	530

Analyses, analytical laboratories, Geological Survey of Canada; Rb, Sr, Nb, Th, Y, Zr by X-ray fluorescence; FeO, H_2O(T), CO_2(T) by chemical methods; remainder by ICP.

containing only 0.27 mol% Ȧn and 0.51 mol% Or (electron microprobe analyses by G. LeCheminant). Sphene, as well as zircon, is present as an accessory mineral. Small inclusions of fluorite are present in amphibole grains.

The amphibole has pleochroism (α — brownish green, β — light brown, γ — dark blue) characteristic of sodic amphibole. It corresponds chemically (Table 2) to ferrorichterite, which has been reported by Davidson (1981) in peralkaline rocks in the Blachford Intrusive Suite. It is comparable also with "igneous amphiboles" in the Topsails peralkaline granite in western Newfoundland (Strong and Taylor, 1984).

The chemistry of the Booth River granites given in Table 1, including high SiO_2, $Na_2O + K_2O$, Fe/Mg, Zn, Nb, Y, REE, and Zr, and low CaO and Sr, indicates that they are A-type (alkaline, anorogenic) granites (Whalen et al., 1987). The high contents of Rb, Be, Nb, La, Y, Yb, Th, and Zr, particularly in sample 83RF139, are noteworthy as these elements are concentrated in anorogenic peralkaline granites and syenites in the Blachford Intrusive Suite at Great Slave Lake (Davidson, 1981). Altered zones in the core of the Blachford complex contain concentrations of major economic importance of Be and REE and very large volumes of Ta-rich potential Nb-Ta ore (Trueman et al., 1984; Highwood Resources Ltd., 1985). In addition to Be, REE and Nb-Ta, U, Li, Sn, and fluorite are concentrated in some of these zones and might also be enriched in the Booth River granitic rocks.

Table 2. Composition of sodic amphibole (ferrorichterite) from electron microprobe analyses by G. LeCheminant, Geological Survey of Canada.

SiO_2	Al_2O_3	TiO_2	Cr_2O_3	Fe_2O_3	FeO	MnO	MgO	CaO	Na_2O	K_2O
47.95	1.80	1.47	0.13	2.82	33.33	0.55	0.02	4.35	6.43	0.98

ZIRCONS

Zircons are light brown and mostly broken. Unbroken crystals are simple bipyramidal prisms with length : breadth ratios of 2:1 to 3:1 featuring {100} and {101} faces and commonly additional {301} faces. According to Pupin (1980), these forms are characteristic of growth in high temperature alkaline and hyperalkaline syenitic and granitic melts. Polished and etched sections show two stages of growth with the inner crystals having a higher U content. Although there are some irregularities in the zoning of the inner crystals, growth appears to have been continuous and there is no evidence of older cores.

Table 3. U-Pb zircon isotopic data, Booth River granite sample 85RF185

Sample and zircon franction[1]	Weight (mg)	U (ppm)	Pb[2] (ppm)	Measured $^{206/204}Pb$	Isotopic abundances (^{206}Pb-100)			Isotopic ratios		Age, Ma
					^{204}Pb	^{207}Pb	^{208}Pb	$^{206}Pb/^{238}U$	$^{207}Pb/^{235}U$	$^{207}Pb/^{206}Pb$
a+149 M1,A	0.034	290.1	114.8	2122	0.0279	12.82	14.92	0.36378	6.2430	2021.2
b−149+105,N1,A	0.0284	288.8	113.8	3361	0.0102	13.31	13.31	0.36529	6.2679	2020.9
c−105+74,N1,A	0.0293	190.0	74.6	1827	0.0186	12.70	12.47	0.36713	6.3055	2022.5
d−105+74,N0,A	0.0326	355.9	134.8	3115	0.0148	12.65	15.45	0.36584	6.2814	2022.1

[1] A — abraided, N — nonmagnetic at given angle, M — magnetic at given angle
[2] Radiogenic lead

ISOTOPIC ANALYSES

U-Pb analytical techniques have been described by Roddick et al., (in press) and the isotopic data are presented in Table 3. Four abraded size and magnetic fractions yield collinear data points (Fig. 3). A regression analysis according to the methods of York (1969) yields an upper intercept age of 2023+4/−2 Ma, a lower intercept date of ca. 320 Ma and a MSWD of 0.

The collinear data set is consistent with the inner zircon morphological evidence for continuous growth, and the upper intercept age is interpreted in terms of an early stage of crystallization of the granite. There is no isotopic or morphological evidence for an older crustal component in the zircon.

DISCUSSION

The U-Pb zircon date of 2023 +4/−2 Ma on sample 85RF185 establishes that the granitic rocks associated with mafic-ultramafic rocks of the Booth River complex are about 570 Ma younger than the least deformed, youngest Archean granitic plutons — the ca. 2590 Ma Regan intrusions (Frith and Loveridge, 1982). The Booth River Intrusive Suite is evidently about 150 Ma and 165 Ma younger than the similar Blachford Intrusive Suite and the Big Spruce Complex, respectively. The original roof rocks of the intrusive suite were older than 2023 Ma and the presently suprajacent siltstones of the Goulburn Supergroup are younger than 2023 Ma.

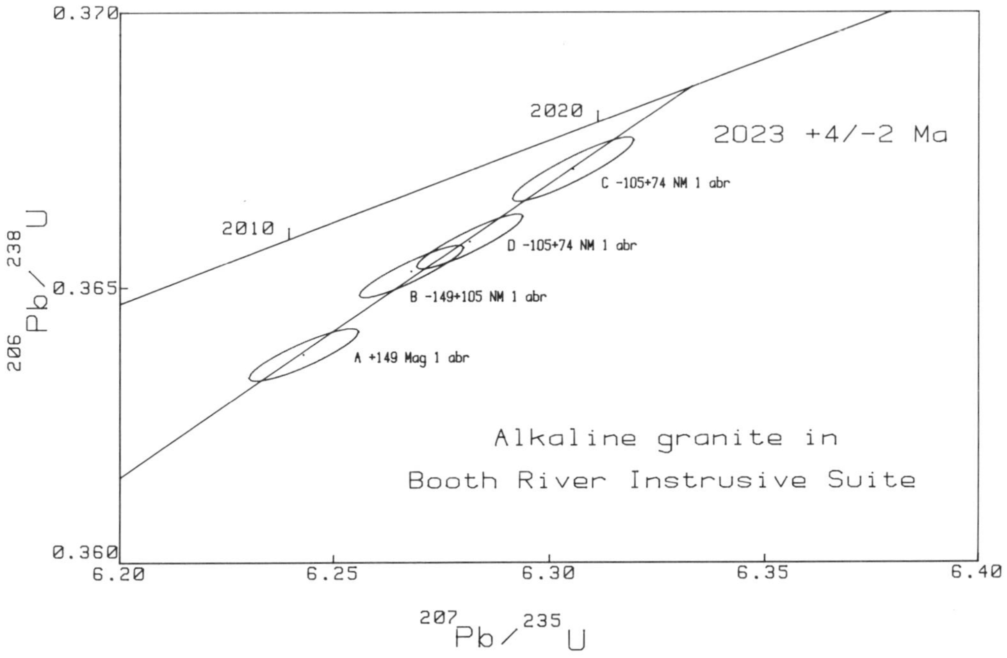

Figure 3. U-Pb concordia plot for zircons from alkaline granite of the Booth River Intrusive Suite.

Grotzinger and Gall (1986) have reasoned that the bulk of the Goulburn Supergroup was deposited in a foredeep developed in response to crustal loading to the southeast in the Thelon Tectonic Zone that was produced by the continental collision that was first postulated by Gibb (1975). The date of 2023 +4/−2 Ma for intrusion of the Booth River alkaline granite would be a ceiling for the time of such a continental collision. The Booth River Intrusive Suite and the Burnside Synclinorium are along a southwesterly-trending gravity high, and it seems likely that the igneous rocks were intruded in an extensional crustal environment, that was subsequently, after uplift, erosion and sedimentation, the site of downfolding of Goulburn strata. The extensional event need not necessarily have had any relationship to tectonic events that led to deposition of the Goulburn Supergroup, but it could have immediately preceded a continental breakup that was followed by deposition of early Goulburn strata on a passive margin appreciably earlier (*see* Evans and Hoye, 1981; Campbell and Cecile, 1976; Hoffman, 1981; Bowring et al., 1984, for discussions of data and correlations relevant to the age of Goulburn strata) than deposition of the bulk of the Goulburn Supergroup in a foredeep developed during continent-continent collision as postulated by Grotzinger and Gall (1986).

The geochronological results together with geochemical data presented in this paper support previous suggestions (Roscoe, 1985) that the Booth River Intrusive Suite is comparable to the Blachford Intrusive Suite and that felsic as well as mafic units should be considered to be favourable for prospecting for types of mineral deposits that are associated with anorogenic intrusions.

ACKNOWLEDGMEMTS

Recognition of the Booth River Intrusive Suite as igneous rocks distinct from a gabbro sill intruded along the base of the Goulburn Supergroup and from Archean granitic rocks was consequent upon A. Davidson's recognition and studies of the Blachford Complex at Great Slave Lake. Courtesies extended and comforts provided during fieldwork by Glen and Trish Warner of Bathurst Inlet Lodge are much appreciated. This paper was reviewed by J.C. Roddick, R.I. Thorpe, and J.B. Whalen, resulting in some significant improvements.

REFERENCES

Bowring, S.A., Van Schmus, W.R., and Hoffman, P.F.
1984: U-Pb zircon ages from Athapuscow aulacogen, East Arm of Great Slave Lake, N.W.T., Canada; Canadian Journal of Earth Sciences, v. 21, p. 1315-1324.

Campbell, F.H.A. and Cecile, M.P.
1976: Tectono-depositional relationships between the Aphebian Kilohigok Basin and the Coronation Geosyncline, N.W.T.; Geological Association of Canada, Program with Abstracts, v. 1, p. 63 (abstract).

Cavell, P.A. and Baadsgaard, H.
1986: Geochronology of the Big Spruce Lake alkaline intrusion; Canadian Journal Earth Sciences v. 23, p. 1-10.

Davidson, A.
1978: The Blachford Lake Intrusive Suite: an Aphebian alkaline, plutonic complex in the Slave Province, Northwest Territories; *in* Current Research, Part A, Geological Survey of Canada, Paper 78-1A, p. 119-127.
1981: Petrochemistry of the Blachford Lake Complex, District of Mackenzie; Geological Survey of Canada, Open File 764 (2 maps, 5 figures, 1 table (11 p.), 5 p. data on 74 samples).
1982: Petrochemistry of the Blachford Lake complex near Yellowknife, Northwest Territories; *in* Uranium in Granites, Ed. Y.T. Maurice; Geological Survey of Canada, Paper 81-23, p. 71-79.

Evans, M.E. and Hoye, G.S.
1981: Paleomagnetic results, Great Slave Lake and Bathurst Inlet areas, N.W.T.; *in* Proterozoic Basins of the Canadian Shield, Ed. F.M. Campbell; Geological Survey of Canada, Paper 81-10, p. 191-202.

Frith, R.A. and Loveridge, W.D.
1982: Ages of Yellowknife Supergroup volcanic rocks, granitoid intrusive rocks and regional metamorphism in the northeastern Slave Structural Province; *in* Current Research, Part A, Geological Survey of Canada, Paper 82-1A, p. 225-235.

Gibb, R.A.
1975: Slave-Churchill collision tectonics; Nature, v. 271, p. 50-52.

Grotzinger, J.P. and Gall, Q.
1986: Preliminary investigations of early Proterozoic Western River and Burnside River Formations: evidence for foredeep origin of Kilohigok Basin; *in* Current Research, Part A, Geological Survey of Canada, Paper 86-1A, p. 95-106.

Highwood Resources Ltd.
1985: Thor Lake rare metals deposit, N.W.T.; *in* Highwood Resources Annual Report, 1985, p. 4-9.

Hoffman, P.F.
1981: Revision of stratigraphic nomenclature, Foreland Thrust-Fold Belt of Wopmay Orogen, District of Mackenzie; *in* Current Research, Part A, Geological Survey of Canada, Paper 81-1A, p. 247-250.

Pupin, J.P.
1980: Zircon and granite petrology; Contributions to Mineralogy and Petrology, v. 73, p. 207-220.

Roddick, J.C., Loveridge, W.D. and Parrish, R.R.
1987: Precise U/Pb dating of zircon at the sub-nanogram Pb level; Isotope Geoscience, v. 66, p. 111-121.

Roscoe, S.M.
1985: The Booth River Intrusive Suite, District of Mackenzie; *in* Current Research, Part A, Geological Survey of Canada, Paper 85-1A, p. 141-144.

Strong, D.F. and Taylor, R.P.
1984: Magmatic subsolidus and oxidation trends in composition of amphiboles from silica-saturated peralkaline igneous rocks; Tschermaks Mineralogische, Petrographische und Mitteilurgen. v. 32, p. 211-222.

Trueman, D.L., Pederson, J.C., and de St. Jorre, L.
1984: Geology of the Thor Lake beryllium deposit: an update; Contributions to the geology of the NWT., v. 1, Department of Indian and Northern Affairs, Economic Geology Series 1984-6, p. 115-120.

York, D.
1969: Least squares fitting of a straight line with correllated errors; Earth and Planetary Science Letters, v. 5, p. 320-324.

Whalen, J.B. and Currie, K.L.
1983: The Topsails igneous terrane of western Newfoundland; *in* Current Research, Part A, Geological Survey of Canada, Paper 83-1A, p. 15-23.

Whalen, J.B., Currie, K.L., and Chappell, B.W.
1987: A-type granites: geochemical characteristics, discrimination, and petrogenesis; Contributions to Mineralogy and Petrology, v. 95, p. 407-419.

U-Pb, zircon and monazite ages from the eastern Slave Province, Healey Lake area, N.W.T.

O. van Breemen[1], J.B. Henderson[1], R.W. Sullivan[1], and P.H. Thompson[1]

van Breemen, O., Henderson, J.B., Sullivan, R.W., and Thompson, P.H., U-Pb, zircon and monazite ages from the eastern Slave Province, Healey Lake area, N.W.T.; in Radiogenic Age and Isotopic Studies: Report 1, Geological Survey of Canada, Paper 87-2, p. 101-110, 1987.

Abstract

A Yellowknife Supergroup felsic volcanic dome yields an age of 2692 ± 2 Ma whereas a sample of meta-quartzdiorite from a heterogeneous granitoid migmatite unit in part presumed to be basement to Yellowknife volcanics yields an age of 2679 ± 3 Ma. Plutonic rocks have been emplaced at 2616 +7/−6 Ma (Tarantula plutonic suite), 2589 +13/−10 Ma (Sinclair granodiorite), 2627 +46/−33 Ma (Taylor plutonic suite) and 2595 ± 10 Ma (megacrystic granodiorite), some 60-100 Ma after volcanism.

Résumé

On a évalué l'âge d'un dôme volcanique à caractère felsique, dans le supergroupe de Yellowknife, à 2692 ± 2 Ma, et celui d'un échantillon de métadiorite quartzique prélevé dans une migmatite granitoïde hétérogène, qui en partie constituerait le socle de terrain volcanique de Yellowknife, à 2679 ± 3 Ma. La mise en place des roches plutoniques a eu lieu aux époques suivantes: série plutonique de Tarantula, 2616 +7/−6 Ma, granodiorite de Sinclair, 2589 +13/−10 Ma, série plutonique de Taylor, 2627 +46/−33 Ma, granodiorite mégacristalline, 2595 ± 10 Ma; c'est-à-dire entre 60 et 100 Ma environ après le volcanisme.

INTRODUCTION

The Healey Lake area contains part of the boundary between the Archean Slave Structural Province and the Thelon Tectonic Zone of the western Churchill Province. The major part of the area is typical of Slave Province geology in that it consists of metasediments and metavolcanics of the Archean Yellowknife Supergroup ranging in metamorphic grade from greenschist to migmatite in the lower pressure facies series, has curvilinear structural trends, and has been intruded by a varied suite of generally massive plutonic rocks. The contrasting dominantly Proterozoic Thelon Tectonic Zone rocks in the southeastern part of the area consist of mainly granulite grade granitoid and supracrustal gneisses and foliated plutons, all with pronounced linear structural patterns (Henderson et al., 1982; Henderson and Thompson, 1982). U-Pb zircon and monazite geochronological data from the Thelon Tectonic Zone are reported in van Breemen et al. (1987).

Nine U-Pb zircon and monazite geochronological studies are presented (Fig. 1). These include 3 samples from the Healey complex, a granitoid and granitoid migmatite and gneiss unit thought to be in part older than the Yellowknife supracrustals (Henderson and Thompson, 1982), one sample from a rhyolite dome in the Back River volcanic complex, and one sample each from a granodiorite pluton of the Tarantula plutonic suite, the Sinclair plutonic suite, the Taylor plutonic suite, a major megacrystic granodiorite in the southern part of the area, and a two-mica granite from the southwestern part of the area.

ANALYTICAL METHODS

General techniques of zircon concentration, chemical dissolution and isotopic analysis (Finnigan MAT 261 mass spectrometer) are similar to those used by van Breemen et al. (1986). Results are presented in Table 1 and Figures 2 and 4. Two sigma uncertainties in the isotopic ratios are 0.5% for $^{206}Pb/^{238}U$ and $^{207}Pb/^{235}U$ and 0.1% or 0.15% for $^{207}Pb/^{206}Pb$. Two samples (2 and 3) as well as the two most concordant fractions of sample 1 and the most concordant fraction of sample 4 were analyzed more recently by improved techniques which are reflected by the smaller amounts of material analyzed (Roddick et al., 1987). Uncertainties for each data point are indicated on the concordia diagrams. Regression calculations for linear alignments of data points were done according to the methods of York (1969).

HEALEY COMPLEX

The Healey complex is a heterogeneous assemblage of granitoids and migmatitic granitoids and gneisses (Henderson et al., 1982; Henderson and Thompson, 1982). The main block

[1] Geological Survey of Canada, 601 Booth Street, Ottawa, Ontario K1A 0E8.

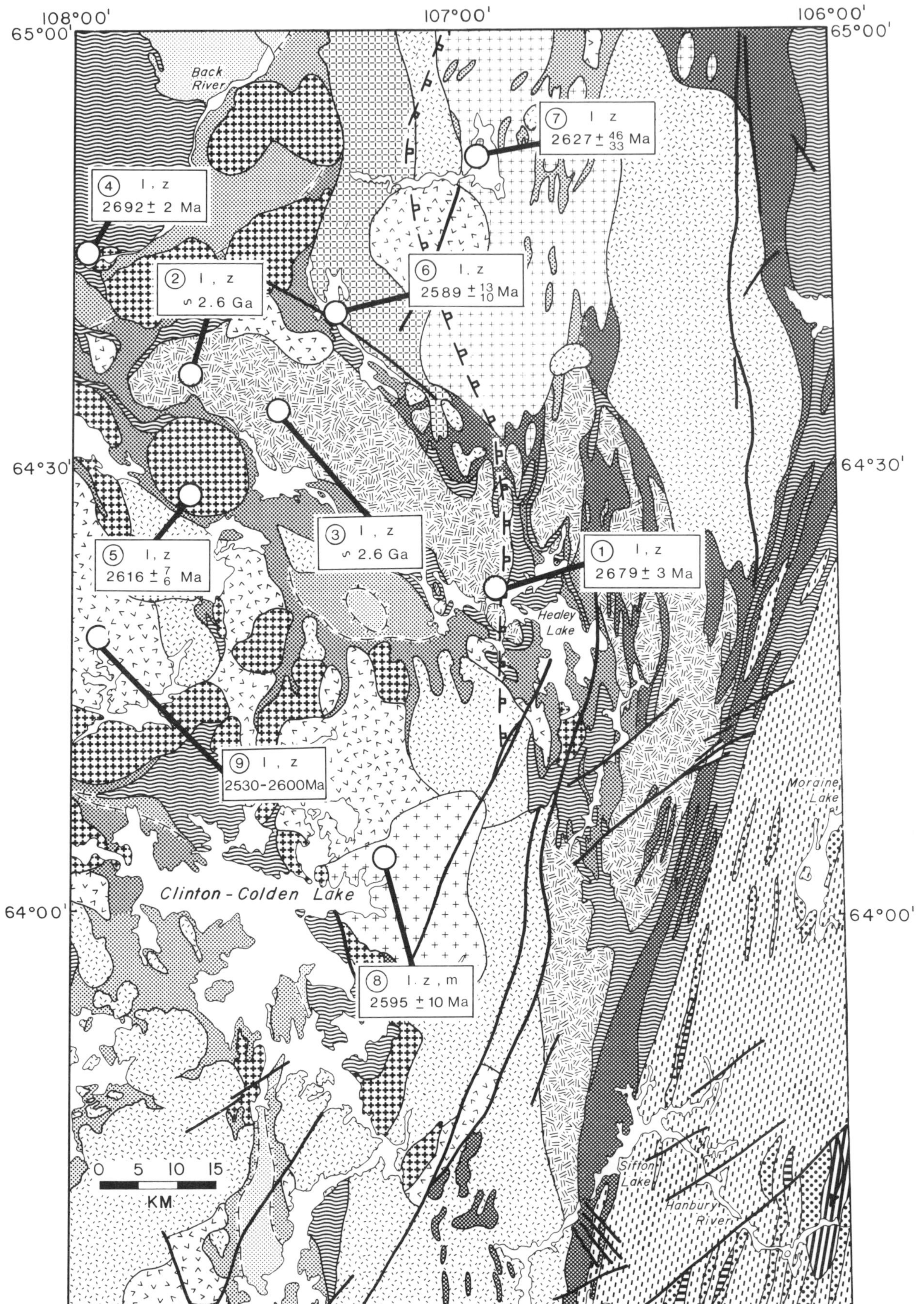
108°00'
107°00'
106°00'
65°00'
65°00'
Back River
7 I z 2627 $\pm^{46}_{33}$ Ma
4 I, z 2692 ± 2 Ma
2 I, z ∽2.6 Ga
6 I, z 2589 $\pm^{13}_{10}$ Ma
64°30'
64°30'
3 I, z ∽2.6 Ga
1 I, z 2679 ± 3 Ma
5 I, z 2616 $\pm^{7}_{6}$ Ma
Healey Lake
9 I, z 2530-2600Ma
Moraine Lake
Clinton - Colden Lake
64°00'
64°00'
8 I, z, m 2595 ± 10 Ma
0 5 10 15
KM
Sifton Lake
Hanbury River
Ptarmigan Lake
Smart Lake
63°30'
63°30'
108°00'
107°00'
106°00'

WEST OF STRAIGHT ZONE

Granitoid rocks, undivided

Granite, granodiorite, muscovite-biotite, massive to strongly foliated, also pegmatite

Granite, megacrystic biotite (muscovite), massive to weakly foliated

Taylor plutonic suite

Granite, granodiorite, tonalite undivided, biotite, minor to rare muscovite, foliated, metasedimentary inclusions locally abundant

Sinclair plutonic suite

Quartz tonalite, biotite, massive to weakly foliated, coarse quartz masses

Tarantula plutonic suite

Tonalite, granodiorite, quartz diorite, biotite-hornblende massive

Yellowknife Supergroup

a b c d — Metagreywacke pelite and high grade equivalents, (a) biotite zone, (b) cordierite zone, (c) sillimanite zone (not defined south of 64°00') (d) migmatite zone

Metavolcanic rocks, mafic to felsic, greenschist to amphibolite grade

Healey complex

Granitoid gneisses, heterogeneous assemblage of complexly deformed granite to diorite gneisses to plutonic units to to granitoid migmatite

EAST OF STRAIGHT ZONE

Granite, granodiorite, biotite, strongly foliated

Undivided foliated granitoid rocks, granitoid gneisses in part at granulite grade

High Grade Supracrustal Rocks

Migmatitic metasediments, with younger granitic sheets and intrusions

Migmatitic amphibolite, with younger granitic sheets and intrusions

Geological contact, defined or assumed

Fault or shear zone

Archean metamorphic isograd

Proterozoic metamorphic isograd (amphibolitization of diabase dykes)

Geochronological data

I Igneous age

M Metamorphic age

z Zircon

m Monazite

Figure 1. Geology of the southeastern Slave Structural Province in the Healey Lake (north of 64° N) and the northern Artillery Lake map areas, modified after Henderson et al. (1982) and Henderson and Thompson, (1982), showing the location of the samples analyzed and summarizing the geochronological data for each sample.

of the unit (50 by 15 km) northwest of Healey Lake, which follows the regional northwesterly trend in the eastern Slave Province, is surrounded by Yellowknife supracrustal rocks and metamorphosed to sillimanite and migmatite grade and is, for the most part, mantled by mafic to felsic metavolcanics, commonly only a few tens of metres thick. At the eastern end of the block is a major, complexly folded, dominantly mafic, volcanic unit. The Healey complex is 'homogeneous in its heterogeneity' typically consisting of foliated and massive granites to diorites, gneisses and migmatitic granitoid rocks of several generations. The complex has not been subdivided at the scale of mapping to date (1:250 000). Few, if any units of supracrustal origin have been recognized within the main part of the complex. However, at the margin of the complex, minor granitoid dykes occur in places within the volcanic carapace and inclusions and screens of Yellowknife supracrustal rocks occur within and near small plutons emplaced at the margin of the complex. Because of the long history implied by the compositional, textural and structural diversity of the complex, and the relatively lower grade and structural simplicity of the surrounding Yellowknife supracrustal rocks, it has been proposed that the complex is at least in part basement to the Yellowknife Supergroup (Henderson and Thompson, 1982). Although no unconformities have been recognized to date, the complex is almost completely mantled by thin volcanic units. It resembles in many respects the Sleepy Dragon complex, basement block 70 km northeast of Yellowknife in the southern Slave Province (Baragar, 1966; Henderson, 1985; Henderson et al., 1987).

Three samples from the Healey complex have been analyzed; a meta quartz diorite (location 1, Fig. 1) and two massive to weakly foliated granites (location 2 and 3) from near the west end of the complex.

Healey complex meta quartz diorite (sample 1)

Sample 1 is a poorly layered, pinkish-grey weathering, medium-to fine-grained, weakly and diffusely migmatitic, quartz diorite composed of hornblende, biotite, quartz and plagioclase with accessory sphene, zircon and apatite. The texture is equigranular granoblastic with a faint suggestion of a coarser relict texture expressed by concentrations of plagioclase. Except for a faint foliation there is no evidence of strain in thin section.

Zircons range from prismatic to equidimensional and are somewhat rounded. Although rounding may be partly due to corrosion, polished and etched sections suggest it also results from growth on facets near the perimeter. Many prismatic grains have spindle shapes. In transmitted light, the grains appear to have cores and rims. However, polished and etched sections indicate that this optical effect can be attributed to one or two narrow zones of high U zircon. Regular igneous zoning continues from the centre to the outer rim.

Six zircon fractions were hand picked for U-Pb analysis: four of prismatic types and two of the round to oval multifaceted variety. All fractions of the prismatic crystals were abraded but only one fraction of the coarse round crystals. U concentrations are close to 200 ppm. Data points show little scatter in spite of the different sizes and morphologies. Ovoid fraction data points 1c and 1g, together with four points corresponding to prismatic fraction data points 1a, 1b, 1d, 1e fit a regression line (MSWD = 2.5, Fig. 2, Table 1) that yields upper and lower intercept ages of 2679 ± 3 Ma and ca. 615 Ma. The lower intercept is consistent with Pb loss occurring after cumulative lattice damage, thus yielding a comparatively young average Pb loss age. These data support the internal morphological evidence that in spite of the high U zones, the zircons are of one generation. Consequently, the upper intercept is interpreted as the age of igneous emplacement of the unit sampled. The apparently older age for fraction 1f is interpreted in terms of a small (one zircon with an older core?) component of inherited zircon in this small fraction. An additional and most discordant fraction was not included in this analysis.

Healey complex, foliated granite (samples 2 and 3)

Two samples were collected from the northwestern end of the Healey complex. Instead of sampling convincing candidates for older basement (i.e. strongly foliated and folded granitoid gneisses or migmatites), two samples were chosen of homogeneous relatively massive granite which are no more intensely foliated than the surrounding plutons which cut adjacent Yellowknife supracrustal rocks. Several helicopter stops were made on strongly heterogeneous and irregularly layered outcrops. Close inspection indicated a variety of relatively melanocratic, foliated to layered crystalline rocks that were pervasively intruded and variably assimilated by granitic material. As the latter represented a relatively constant feature of most outcrops, it was sampled in its purest form at two localities (Fig. 1).

Zircons from sample 2 are strongly magnetic and forms vary from prismatic (length : breadth = 8) to equidimensional. Most grains have subhedral igneous outlines but contain rounded cores, which have significantly lower U contents than the rims (Fig. 3A). Embayments and smooth but irregular shapes of some cores suggest that they were rounded by corrosion. The cores are generally surrounded by discordant zoning, characteristic of impingement of the zircon crystal by a moving liquid and crystal mush (van Breemen and Hanmer, 1986). The irregular/discordant zoning changes to regular/conformable zoning in most grains, thus explaining the outer forms. In three fractions (2a,2b,2d, Table 1) zircons were picked in transmitted light to avoid possible cores. These fractions were strongly abraded, as well as a fraction (2c) that had been picked specifically for cores.

Since the zircons high in U are strongly discordant, they have not been plotted. The least discordant data point (2a) corresponds to the least magnetic fraction (Table 1). Hypothetical discordia drawn through data points 2a and 2b and through data points 2a and 2d yield upper intercept ages of 2570 Ma and 2610 Ma respectively (with lower intercept ages of 750 Ma and 1500 Ma). Isotope ratios for 2b and 2c (cores) are almost identical, in spite of a strong reduction in U content for the abraded sample with visible cores.

The zircon morphological evidence is consistent with the hypothesis of pervasive partial magmatic assimilation or in situ melting of older rocks, although the attempt to analyse the obvious cores (2c) tentatively suggests that the palaeosome was not markedly older than the melt formed at ca.

Table 1. U-Pb isotopic data on zircon and monazite

Sample and zircon fraction	Weight (mg)	U (ppm)	Pb* (ppm)	Measured $^{206}Pb/^{204}Pb$	Isotopic abundances (^{206}Pb = 1000) ^{204}Pb	^{207}Pb	^{208}Pb	Isotopic ratios $^{206}Pb/^{238}U$	$^{207}Pb/^{235}U$	Age,Ma $^{207}Pb/^{206}Pb$
1. Healey complex, meta quartz diorite (64° 21.3′ 106° 52.2′)										
a, +149,N0,A	1.44	188.8	101.3	11124	0.065	193.6	141.5	0.47256	11.739	2654.3
b, −149+105,N−1,A	1.75	160.1	87.4	12431	0.051	181.7	137.7	0.48186	12.033	2663.0
c, −149+105,N−1,R,A	0.45	172.5	95.0	1081	0.837	191.5	151.0	0.49013	12.250	2664.4
d, −105+74.N−1,A	0.80	197.0	109.1	3473	0.247	184.2	152.1	0.48580	12.135	2663.6
e, −105+74,N−3,A	0.038	217.9	125.6	4865	0.116	183.8	143.2	0.50659	12.736	2674.2
f, −105+74,N−3,A	0.031	208.1	119.7	5130	0.079	184.4	121.1	0.51337	12.986	2684.4
g, −74+62,N−1,R	1.78	204.5	107.0	16026	0.011	180.4	95.8	0.47669	11.847	2655.2
2. Healey complex, foliated granite (64° 36.6′ 107° 41.6′)										
a, N3,A	0.012	612.2	319.4	882	0.909	180.9	152.5	0.46887	10.9676	2554.2
b, +74,M3,A	0.007	820.3	331.8	1339	0.385	162.7	167.6	0.35782	7.7876	2432.7
c, +74,M3,A,core	0.038	347.5	140.8	1339	0.586	165.3	174.7	0.35811	7.7967	2433.3
d, −74,M3,A	0.016	755.5	305.6	778	1.123	159.8	145.7	0.37453	7.5149	2294.0
3. Healey complex, foliated granite (64° 34.4′ 107° 28.7′)										
a, +105,A	0.007	785.9	285.7	1571	0.215	151.8	84.78	0.34441	7.0809	2335.8
b, −105+74,A	0.012	973.6	352.8	1993	0.317	150.8	48.81	0.35556	7.1974	2309.1
c, −74+62,A	0.012	1188	336.6	1268	0.597	136.7	50.31	0.28422	5.0551	2084.3
d, −62,A	0.012	1199	372.5	2916	0.176	137.0	33.68	0.31094	5.7742	2159.9
4. Felsic Dome, Back River volcanic complex (64° 44.5′ 107° 58.3′)										
a, +105,N	0.71	1796	946.1	12156	0.038	182.7	132.0	0.46635	11.7149	2672.9
b, −105+74,N2.5	1.45	258.7	136.0	7751	0.106	183.3	139.7	0.46336	11.6263	2671.0
c, −105+74,N2.5,A	1.01	143.0	83.1	1732	0.532	190.5	158.2	0.50938	12.9233	2689.3
d, −105+74,M2.5,A	1.34	317.2	179.7	8353	0.105	184.5	143.5	0.49758	12.5724	2682.5
e, −74+62	0.036	325.4	190.1	9620	0.036	184.4	148.8	0.50962	12.9280	2689.1
5. Tarantula granodiorite (64° 28.6′ 107° 43.7′)										
a, +105,N1	2.66	210.8	105.9	5511	0.158	170.4	165.5	0.43904	10.198	2542.5
b, +105,N1,A	1.02	163.2	94.4	2194	0.228	177.1	220.5	0.48420	11.633	2598.9
c, −105+74,N−1	2.10	154.3	80.5	5439	0.153	172.3	169.1	0.45423	10.671	2561.4
d, −74+62,N−1	0.70	547.9	285.5	7854	0.091	171.3	167.1	0.45382	10.646	2559.0
6. Sinclair Plutonic Suite (granodiorite) (64° 40.5′ 107° 18.1′)										
a, −149+105,N1	1.19	394.6	186.4	3518	0.237	172.5	52.15	0.45261	10.5822	2553.4
b, −105+74,N0	0.151	418.3	194.7	2048	0.392	174.4	48.61	0.44914	10.4987	2553.0
c, −105+74,N0,A	0.208	432.2	205.8	4856	0.130	171.9	42.54	0.45828	10.7595	2560.4
d, −74+62,M0,A	0.192	364.2	175.7	3669	0.199	173.3	48.45	0.46265	10.9002	2566.3
e, −74+62,M0,A	0.160	351.7	171.2	3847	0.170	173.0	46.00	0.46732	11.0101	2566.2
7. Taylor Plutonic Suite (tonalite) (64° 51.0′ 106° 54.5′)										
a, +149,N10	0.328	600.7	263.8	2240	0.410	174.2	48.27	0.42428	9.8911	2548.5
b, −149+105,N10	0.605	642.8	272.3	2895	0.331	172.2	42.90	0.41052	9.5156	2538.9
c, −105+74,N10,	0.284	630.0	280.3	1305	0.741	179.9	66.28	0.42676	10.0416	2564.1
d, −105+74,N10,A	0.379	559.3	259.6	4192	0.211	174.4	44.05	0.44643	10.5767	2575.5
e, −105+74,N10,A	0.208	536.0	234.4	2027	0.443	175.7	53.90	0.42041	9.8655	2559.5
f, −74+62,N10	0.174	501.4	219.6	1693	0.527	175.7	57.64	0.42112	9.8230	2549.5
8. Megacrystic granodiorite (64° 03.4′ 107° 11.0′)										
a, −105+74,N3,A	0.29	751.0	333.5	2317	0.412	163.4	131.5	0.40435	8.8211	2436.7
b, −105+62,N4,A	0.10	1025	429.0	1807	0.503	159.1	144.9	0.37956	7.9911	2376.3
c, −74+62,N3,A	0.60	966.0	414.8	1389	0.609	161.5	165.3	0.38395	8.1390	2388.0
d, monazite	0.017	722.8	4552	5068	0.067	175.3	13543	0.49343	11.8696	2600.9
e, monazite	0.015	838.9	5539	3109	0.194	175.8	14348	0.48971	11.7085	2590.8
9. Two mica granite (64° 18.0′ 107° 55.2′)										
a, coarse,N5,A	0.99	306.6	123.9	1464	0.154	169.1	129.2	0.36510	8.127	2470.8
b, medium,N5	1.45	565.4	154.5	489	1.459	168.0	219.4	0.23917	5.143	2412.2
c, unsized,M5	2.60	1604.2	360.0	558	1.651	158.4	254.9	0.19443	3.673	2189.8

** Sample is zircon unless indicated otherwise
* Radiogenic lead
N,M Respectively non-magnetic or magnetic at given side slope angle on Frantz magnetic separator
R Ovoid shaped
A Abraded

2.6 Ga. Discordant zoning in the zircon rims suggests tectonically active conditions during granite emplacement (van Breemen and Hanmer, 1986.)

Sample 3 comes from a similar geological setting to sample 2. Zircons are scarce and although subhedral and prismatic, are of poorer quality than in sample 2. Ubiquitous cores proved difficult to remove from coarse fraction (3a). All fractions yield strongly discordant isotopic ratios (Table 1). A regression analysis of all four scattered data points yields upper and lower intercept ages of 2590 +230/−150 Ma and ca. 1160 Ma respectively. A regression analysis for the three finer fractions yields an upper intercept age of 2510 +125/−95 Ma (MWSD = 560). The isotope data and abundance of cores are comparable with the data from sample 2.

VOLCANICS

Yellowknife felsic volcanic dome (sample 4)

One of several rhyolite domes in the southern part of the Back River volcanic complex was sampled by M.B. Lambert (Lambert, 1978). The rhyolite is white to pink weathering, pale greenish to bluish grey, sparsely porphyritic with phenocrysts of 2-5 mm plagioclase and quartz in an aphanitic matrix. This unit was previously described and dated at 2667 Ma (zircon $^{207}Pb/^{206}Pb$ age on two size fractions; Lambert and Henderson, 1980).

Zircons from sample 4 have igneous morphology and have long needle-like inclusions characteristic of growth in a liquid (Silver, 1969). The crystals are mainly stubby, with length to breadth ratios of 1 to 3. The poor quality and brown colour generally reflect the high U content (Table 1). Abrasion of three out of five fractions has greatly increased the concordance (Fig. 2).

The four data points fit a regression line (MSWD = 0.97) with an upper intercept age of 2692 ± 2 Ma and a lower intercept of ca. 425 Ma. In view of the linear array of points and apparent absence of cores in the hand-picked fractions, the upper intercept age is interpreted as the time of rhyolite extrusion. This age is significantly older than the ca. 2670 Ma single point $^{207}Pb/^{206}Pb$ and two point discordia ages previously reported from volcanics, synvolcanic intrusions and greywackes of the Yellowknife Supergroup in the Slave Craton (Lambert and Henderson, 1980; Frith and loveridge, 1982).

PLUTONS

Tarantula plutonic suite (sample 5)

The Tarantula plutonic suite includes a series of mainly tonalitic plutons texturally distinct from other intrusions that, locally, are as felsic as granodiorite and as mafic as diorite. They tend to be more mafic towards the south and are typically massive although locally a weak foliation is developed at pluton margins. They have sharp intrusive contacts with the Yellowknife, sediments, project few or no dykes into country rocks, and do not contain inclusions. These dark pinkish grey, medium grained, equigranular rocks contain both hornblende and biotite.

The pluton analyzed is more granodioritic than most bodies in the region. In thin section the rock consists of subhedral laths of plagioclase and varied proportions of anhedral interstitial microcline. Quartz is polygonized. Both blue-green hornblende and olive-green, somewhat chloritized biotite make up the 10 − 15% mafic mineral content of the rock. Sphene is a very abundant accessory mineral along with apatite, allanite and zircon. Abundant, relatively coarse epidote is associated with both the patchy alteration products of the plagioclase and with the mafic minerals.

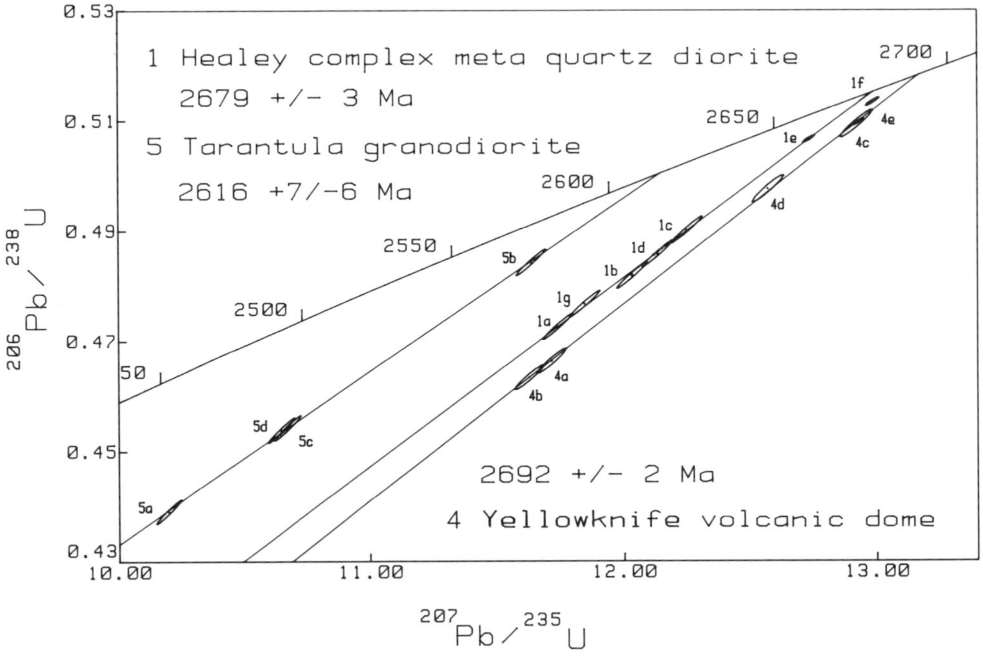

Figure 2. Isotope ratio plot for zircons from 1. Healey complex meta quartz diorite, 4. Felsic dome, Back River Complex and 5. Tarantula plutonic suite. Numbers and letters with data points correspond to those of Table 1.

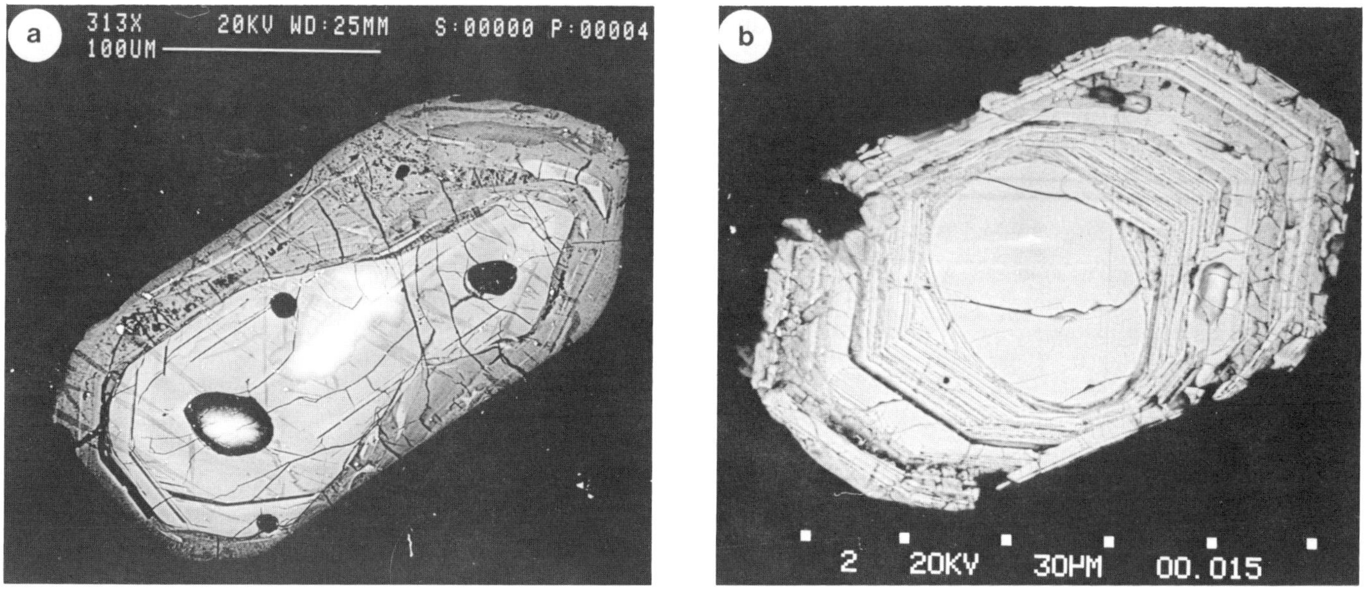

Figure 3. Scanning election microscope images of polished and etched (HF) sections. Images are taken in the backscattered electron mode. **A.** zircon from foliated granite in the Healey complex foliated granite (sample 2). Relatively low U core shows embayment indicative of magmatic corrosion. High U mantle features irregular and discordant zoning, in spite of a nearly euhedral outer perimeter. **B.** Strongly zoned euhedral igneous zircon from megacrystic granodiorite containing oval core devoid of embayments (sample 8).

The anhedral zircons are brownish but clear. They contain clear inclusions but no cores. Length to breadth ratios are generally 2 to 3. Of four zircon fractions analyzed, one (5b) was abraded and shows greatest concordance (Fig. 2). A regression line (MSWD = 1.2) corresponds to an upper intercept age of 2616 +7/−6 Ma and a lower intercept age of ca. 1050 Ma. The upper intercept age is interpreted as the time of emplacement and crystallization of this pluton.

Sinclair plutonic suite (sample 6)

The Sinclair plutonic suite is a northerly trending body of quartz-rich granodiorite to quartz tonalite. It intrudes and contains minor inclusions of, metasediments. Contact relations with the Tarantula plutonic suite to the west and a diorite of the undivided granitoid rocks to the east are unknown but it is intruded by a two-mica granite on its southeast side. The Sinclair plutonic suite is a massive to locally weakly foliated, grey, coarse grained, biotite-rich body that is quite homogeneous throughout most of its outcrop area. It is characterized by abundant coarse masses of irridescent quartz.

The sample consists of equidimensional to slightly elongate, variably altered plagioclase laths, interstitial anhedral microcline, extensively polygonized coarse quartz, and 15% reddish brown, partly chloritized biotite. The rock has a strained aspect with coarsely polygonized quartz within which individual domains have sutured boundaries or are separated by seams of finely polygonized quartz. The coarse plagioclase is also commonly mantled by very fine grained aggregates of quartzofeldspathic material. The biotite is commonly bent.

Zircons are tan, clear and euhedral to subhedral crystals with length:breadth ratios of 2:4. Euhedral zoning is prominent, and a number of crystals have discordant cores which were avoided during hand picking. Six fractions were analyzed, of which three were abraded and show the usual shift to greater concordance on isotope ratio plots. All but the finest unabraded fraction fall in a linear array (Fig. 4) and these five data points fit a regression line (MSWD = 1.3) corresponding to an upper intercept age of 2589 +13/−10 Ma and a lower intercept age of ca. 780 Ma. The upper intercept age is interpreted as the age of emplacement and crystallization of this tonalite.

Taylor plutonic suite (sample 7)

The Taylor plutonic suite consists of an aggregate of undivided granite, granodiorite, tonalite and, locally, diorite bodies. The unit almost everywhere has a strong northerly trending foliation. It contains abundant inclusions to mappable units of Yellowknife Supergroup rocks. Where sampled it is a heterogeneous assemblage of small deformed inclusion-free plutons with intervening, more intensely foliated granodioritic rocks containing abundant metasedimentary inclusions.

This foliated medium grained biotite meta tonalite is composed mainly of plagioclase, quartz and biotite with lesser amounts of potassium feldspar. Plagioclase is anhedral, zoned and prominently twinned. Secondary muscovite after biotite and in cores of plagioclase grains is widespread. Fine grained quartz aggregates are clearly polygonized equivalents of preexisting grains. The mineralogy, grain size, grain boundary relations and spatial relation to the foliation indicate this grain size reduction process reflects strain of the tonalite under moderate conditions of metamorphism.

In thin section the tonalite consists of coarse subhedral to anhedral equant crystals of complexly twinned, moderately to weakly altered, plagioclase and minor microcline with granulated margins wrapped around by a crushed matrix of finely polygonized quartz, shredded brown biotite, feldspar and minor muscovite. The original plutonic rock has been strained, resulting in the reduction of grain size of many of the mineral phases without being annealed.

Zircons are euhedral to subhedral with length to breadth ratios of 2 to 5. Euhedral zoning is prominent and cores are abundant. Many zircons have transverse cracks and some are kinked. Because the sample was not large and rather magnetic, fractions were mainly improved using a needle and cobalt magnet. Using transmitted light, zircons with cores were removed as much as possible.

Of six fractions analyzed, abraded concentrations give the most concordant isotope ratio plots (Fig. 4). The data points show excess scatter beyond the analytical uncertainty. A regression calculation yields a MSWD of 11.6 and the best fit line corresponds to an upper intercept age of 2627 + 46/ −33 Ma and lower intercept age of ca. 800 Ma. Within the broad analytical uncertainty, the upper intercept age is interpreted to represent the time of emplacement of this tonalite. The upper intercept, however, may be biased to an age which is too old. Both the abundance of Yellowknife Supergroup inclusions and cores in the zircons indicate a large crustal component in this plutonic suite.

Megacrystic granodiorite (sample 8)

A large body of megacrystic granodiorite occurs at and northeast of Clinton-Colden Lake. The rock is grey-green to pinkish green, massive, homogeneous, medium grained, equigranular granodiorite with scattered to abundant microcline megacrysts generally 1 to 2 cm in length but locally up to 8 cm. In thin section the granodiorite consists of subhedral to anhedral, moderately altered plagioclase, subhedral microcline, and quartz that is strained and broken but not polygonized. Slightly chloritized reddish brown to brown biotite accounts for 10 – 15% of the rock. Minor muscovite is also present.

Light brown to colourless zircons are euhedral to subhedral with length to breadth ratios of 2 to 5. Zoning is well developed and cores are common. Some cores are well rounded (Fig. 3B), the result of either sedimentary abrasion or magmatic corrosion. Cores were not easily avoided during hand picking. Five zircon fractions were analyzed of which three were abraded and reported in Table 1. The three abraded fractions yielded less discordant data points and only these three were linearly aligned (Fig. 4). Regression analysis yielded an upper intercept age of 2594 + 36/ −31 Ma and a lower intercept age of 1240 Ma with excellent fit (MSWD < 0.1).

In addition to the zircons two fractions of relatively, rare, dark brown, altered-looking, monazite were picked. Near concordant model $^{207}Pb/^{206}Pb$ ages of 2601 Ma and 2591 Ma are in agreement with the zircon age. The time of emplacement of this granite is taken to be 2595 ± 10 Ma.

Two-mica granite (sample 9)

Small to large, commonly irregularly shaped bodies of two mica granite occur throughout the area. They are leucocratic, massive, medium- to coarse-grained, equigranular to, rarely, megacrystic rocks with highly varied amounts and proportions of biotite and muscovite. Associated pegmatites occur locally and in some cases, contain lithium minerals. In thin section, the rock analyzed consists of coarse anhedral poikilitic microperthitic microcline with inclusions of anhedral plagioclase with moderate to weak patchy alteration. Muscovite occurs as coarse, ragged flakes while biotite is a relatively minor and rather chloritized constituent.

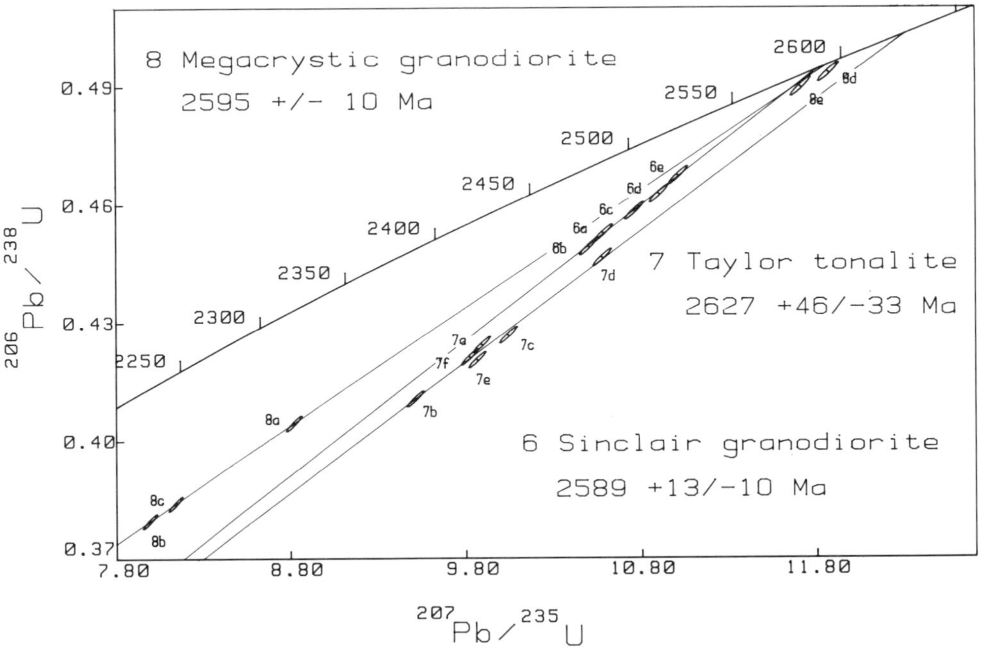

Figure 4. Isotope ratio plot for zircons from 6. Sinclair plutonic suite, 7. Taylor plutonic suite, and 8. megacrystic granodiorite. Numbers and letters with data points correspond to those of Table 1.

Zircons are scarce and altered, probably as a result of metamictization resulting from high U content (Table 1). Cores are ubiquitous and may not have been entirely removed from the hand-picked fractions. In both transmitted light and polished and etched sections, the cores appear to be of varied origin showing features characteristic of igneous, metamorphic and sedimentary types (c.f. van Breemen et al., 1987). Igneous zoning of the rims shows some irregularities suggesting that zircons grew in a crystal mush, (van Breemen et al., 1987).

The three fractions analyzed are strongly discordant especially the two unabraded fractions (Table 1) and have not been plotted. Regression analysis yields an upper intercept age of 2596 + 252/− 179 Ma and a lower intercept of ca. 460 Ma. Coarse muscovite books from a pegmatite phase have, at the same sample locality, yielded Rb-Sr ages ranging from 2560 ± 30 Ma to 2510 ± 30 Ma (unpublished data).

DISCUSSION

At this stage the similarity of ages of the meta quartz diorite (sample 1; 2679 ± 3Ma) on the west shore of Healey Lake and the felsic volcanic dome of the Back River volcanics at 2692 ± 2 Ma does not support the interpretation that the Healey complex is in part basement to the Yellowknife supracrustal rocks (Henderson and Thompson, 1982), although to date only one sample of the older phases of this complex has been dated. If it is assume that the 2679 ± 3Ma date is representative of the older complex as a whole, one of the authors (O.v.B.) suggests the apparent paradox between the age of the complex and the geological criteria used in its original interpretation as basement might be resolved if the supracrustal terrane was tectonic juxtaposed against the Healey complex, although no direct evidence has been recognized to date to support this interpretation. This contradiction between geological and geochronological data is not unique to the Healey complex. To the north, in the Nose Lake-Beechey Lake map area, a ca. 2670 Ma age for the Hackett River gneiss dome granodiorite is now interpreted in terms of a synvolcanic intrusion (Frith and Loveridge, 1982). This gneissic core surrounded by Yellowknife Supergroup rocks had previously been interpreted in terms of an older gneiss dome (Frith and Hill, 1975). East of Yellowknife, the Sleepy Dragon complex, in many respects similar to the Healey complex and long recognized as basement on geological grounds (Baragar, 1966) has phases that have been dated recently as older than the Yellowknife Supergroup (Henderson et al., 1987).

The data from foliated granite samples 2 and 3 are consistent with emplacement or in situ partial melting at ca. 2.6 Ga. Field observations and the nature of zircon cores and overgrowths suggest that at least some of the northwestern part of the Healey complex formed by intrusion of syntectonic granites into more mafic rocks after the volcanics were deposited.

Ages from tonalitic to granodioritic plutons are in the 2630-2580 Ma range. From the Nose Lake-Beechey Lake area, Frith and Loveridge (1982) reported a ca. 2590 Ma U-Pb zircon age for the 'Uist Lake' granodiorite, the core of an annular intrusion of the Regan Intrusive Suite, parts of which are similar to the Tarantula plutonic suite. This time interval compares closely with 2600-2580 Ma period of syn-to-late tectonic plutonism north of the Bathurst Fault (Thompson et al., 1985; van Breemen et al., 1987). There is therefore U-Pb zircon data accumulating from the eastern Slave Province that dates plutonic magmatism which overlaps the main metamorphism and deformation and is between 60 and 100 Ma younger than volcanism.

ACKNOWLEDGMENTS

The manuscript has been critically read by M.B. Lambert and R.R. Parrish.

REFERENCES

Baragar, W.R.A.
1966: Geochemistry of the Yellowknife volcanic rocks; Canadian Journal of Earth Sciences, v.3, p. 9-30.

Frith, R.A. and Hill, J.D.
1975: The geology of the Hackett-Back river greenstone belt preliminary account; *in* Report of Activities, Part C, Geological Survey of Canada, Paper 75-1C, p. 367-370.

Frith, R.A. and Loveridge, D.
1982: Ages of Yellowknife Supergroup volcanic rocks, granitoid intrusive rocks and regional metamorphism in the northeastern Slave Structural Province; *in* Current Research, Part A, Geological Survey of Canada, Paper 82-1A, p. 225-237.

Henderson, J.B.
1985: Geology of the Yellowknife-Hearne Lake area, District of Mackenzie: a segment across an Archean basin; Geological Survey of Canada, Memoir 414, 135 p.

Henderson, J.B. and Thompson, P.H.
1982: Geology of Healey Lake map area; Geological Survey of Canada Open File 860 (1:125,000 scale map and marginal notes).

Henderson, J.B., Thompson, P.H., and James, D.T.
1982: The Healey Lake map area and the Thelon Front problem, District of Mackenzie; *in* Current Research, Part A, Geological Survey of Canada, Paper 82-1A, p. 191-195.

Henderson, J.B., van Breemen, O., and Loveridge, W.D.
1987: Some U-Pb zircon ages from Archean basement, supracrustal and intrusive rocks, Yellowknife-Hearne Lake area, District of Mackenzie; *in* Radiogenic Age and Isotopic Studies: Report 1; Geological Survey of Canada, Paper 87-2.

Lambert, M.B.
1978: The Back River volcanic complex — a cauldron subsidence structure of Archean age; *in* Current Research, Part A, Geological Survey of Canada Paper 78-1A, p. 153-157.

Lambert, M.B. and Henderson, J.B.
1980: A uranium-lead age of zircons from volcanics and sediments of the Back River volcanic complex, eastern Slave Province, District of Mackenzie; *in* Loveridge, W.D., Rubidium-Strontium and Uranium-Lead Isotopic Age Studies, Report 3; *in* Current Research, Part C, Geological Survey of Canada, Paper 80-1C, p. 239-242.

Roddick, J.C., Loveridge, W.D., and Parrish, R.R.
1987: Precise U/Pb dating of zircon at the sub-nonogram Pb level; Isotope Geoscience, v. 66, p. 111-121.

Silver, L.T.
1969: A geochronological investigation of the anorthosite complex, Adirondack Mountains, New York; *in* Origin of Anorthosite and Related Rocks, ed. Y.W. Isachsen; New York State Museum and Science Service, Memoir 18, p. 233-251.

Thompson, P.H., Culshaw, N., Thompson, D.L., and Buchanan, J.R.
1985: Geology across the western boundary of the Thelon Tectonic Zone in the Tinney Hills — Overby Lake (west half) map area, District of Mackenzie; *in* Current Research, Part A, Geological Survey of Canada, Paper 85-1A, p. 555-572.

van Breemen, O. and Hanmer, S.
1986: Zircon morphology and U-Pb geochronology in active shear zones: studies on syntectonic intrusions along the northwest boundary of the Central Metasedimentary Belt, Grenville Province, Ontario; *in* Current Research, Part B, Geological Survey of Canada, Paper 86-1B, p. 775-784.

van Breemen, O., Davidson, A., Loveridge, W.D., and Sullivan, R.W.
1986: U-Pb zircon geochronology of Grenville tectonites, granulites and igneous precursors, Parry Sound, Ontario; *in* The Grenville Province, ed. J.M. Moore, A. Davidson and A.J. Baer; Geological Association of Canada, Special Paper 31, p. 191-207.

van Breemen, O., Henderson, J.B., Loveridge, W.D., and Thompson, P.H.
1987: U-Pb zircon and monazite geochronology and morphology of granulites and granite from the Thelon Tectonic zone, Healey Lake and Artillery Lake map areas, N.W.T.; *in* Current Research, Part A, Geological Survey of Canada, Paper 87-1A, p. 783-801.

York, D.
1969: Least squares fitting of a straight line with correlated errors; Earth and Planetary Science Letters, v. 5, p. 320-324.

Some U-Pb zircon ages from Archean basement, supracrustal and intrusive rocks, Yellowknife-Hearne Lake area, District of Mackenzie

J.B. Henderson[1], O. van Breemen[1], and W.D. Loveridge[1]

Henderson, J.B., van Breemen, O., and Loveridge, W.D., Some U-Pb zircon ages from Archean basement, supracrustal and intrusive rocks, Yellowknife-Hearne Lake area, District of Mackenzie; in Radiogenic Age and Isotopic Studies: Report 1, Geological Survey of Canada, Paper 87-2, p. 111-121, 1987.

Abstract

A cataclastic granitoid gneiss intruded by a weakly foliated granodiorite yields ages of 2819 +40/−31 Ma and 2683.5 ± 2 Ma respectively. These two units are part of the the Sleepy Dragon Complex, considered to be possibly basement to the Yellowknife Supergroup. Two Yellowknife felsic volcanic units are dated at 2663 +7/−5 Ma for the Turnback Rhyolite on the east side of the Archean Yellowknife supracrustal basin at Yellowknife, and 2684 +16/−24 for the Townsite felsic volcanics within the main volcanic sequence at Yellowknife. Ages on plutons intrusive into the Yellowknife Supergroup units include two dates on the Defeat Plutonic Suite at 2618 +7/−20 and 2620 ±8 Ma east and west of Yellowknife respectively, and ages of between 2.55 and 2.63 Ga for the Awry and 2581 +29/−24 Ma for the Stagg Plutonic Suite west of Yellowknife.

Résumé

On a évalué l'âge d'un gneiss granitoïde cataclastique traversé par une granodiorite faiblement feuilletée à 2819 ± 40/31 Ma et celui de la granodiorite à 2683,5 ± 2 Ma. Ces deux unités font partie du complexe de Sleepy Dragon que l'on estime constituer le socle du supergroupe de Yellowknife. Deux unités volcaniques felsiques du supergroupe de Yellowknife ont donné les âges suivants: 2663 + 7/− 5 Ma pour la rhyolite de Turnback, sur le côté est du bassin supracrustal de Yellowknife, d'âge archéen, à Yellowknife, et 2684 +16/−24 Ma pour les roches volcaniques felsiques de Townsite au sein de la principale séquence volcanique de Yellowknife. On a évalué l'âge des plutons traversant le supergroupe de Yellowknife de la façon suivante: série plutonique de Defeat, 2618 +7/−20 Ma à l'est et 2620 ± 8 Ma à l'ouest de Yellowknife; série plutonique d'Awry, entre 2,55 et 2,63 Ga; série plutonique de Stagg à l'ouest de Yellowknife, 2581 +29/−24 Ma.

INTRODUCTION

The Yellowknife-Hearne Lake area is located in the southern part of the Slave Structural Province and contains the most complete segment across an Archean depositional basin in the province. The city of Yellowknife and its important gold mines are situated on a volcanic sequence at the western margin of the mainly sediment-dominated basin, itself the location of numerous gold prospects. The basin (Fig. 1), considered by some to be the remnant of an ensialic rift (Henderson, 1985), consists mainly of metagreywacke-mudstone turbidites that range in grade from greenschist to middle amphibolite in the lower pressure facies series. Thick sequences of mainly basaltic to andesitic subaqueous volcanic rocks occur locally at the margin of the preserved basin. Smaller amounts of felsic volcanics occur with the mafic volcanics in varied amounts and as separate centres. A largely volcanic mantled basement area, the Sleepy Dragon Complex, consisting of metamorphosed and deformed granitoid plutons and gneisses occurs in the northest. Both the basement and Yellowknife Supergroup supracrustal rocks have been intruded by a varied assemblage of plutonic rocks ranging from quartz diorite to granite. These are particularly abundant west of Yellowknife (Fig. 1).

Eight U-Pb zircon geochronological case studies are presented in this report. These include two units in the Sleepy Dragon Complex, 70 km northeast of Yellowknife, the Turnback Rhyolite immediately southeast of the Sleepy Dragon Complex, the Townsite felsic volcanics at Yellowknife, and four granitoid units intrusive into the Yellowknife Supergroup including two samples of the Defeat Plutonic Suite east and west of Yellowknife and one each from the Awry and Stagg plutonic suites west of Yellowknife. These data build on the pioneering geochronological work of Green (1968), Green et al. (1968), and Green and Baadsgaard (1971) in the Yellowknife area. Further information on the geological context of this data can be found in Henderson (1985).

[1] Geological Survey of Canada, 601 Booth Street, Ottawa, Ontario K1A 0E8.

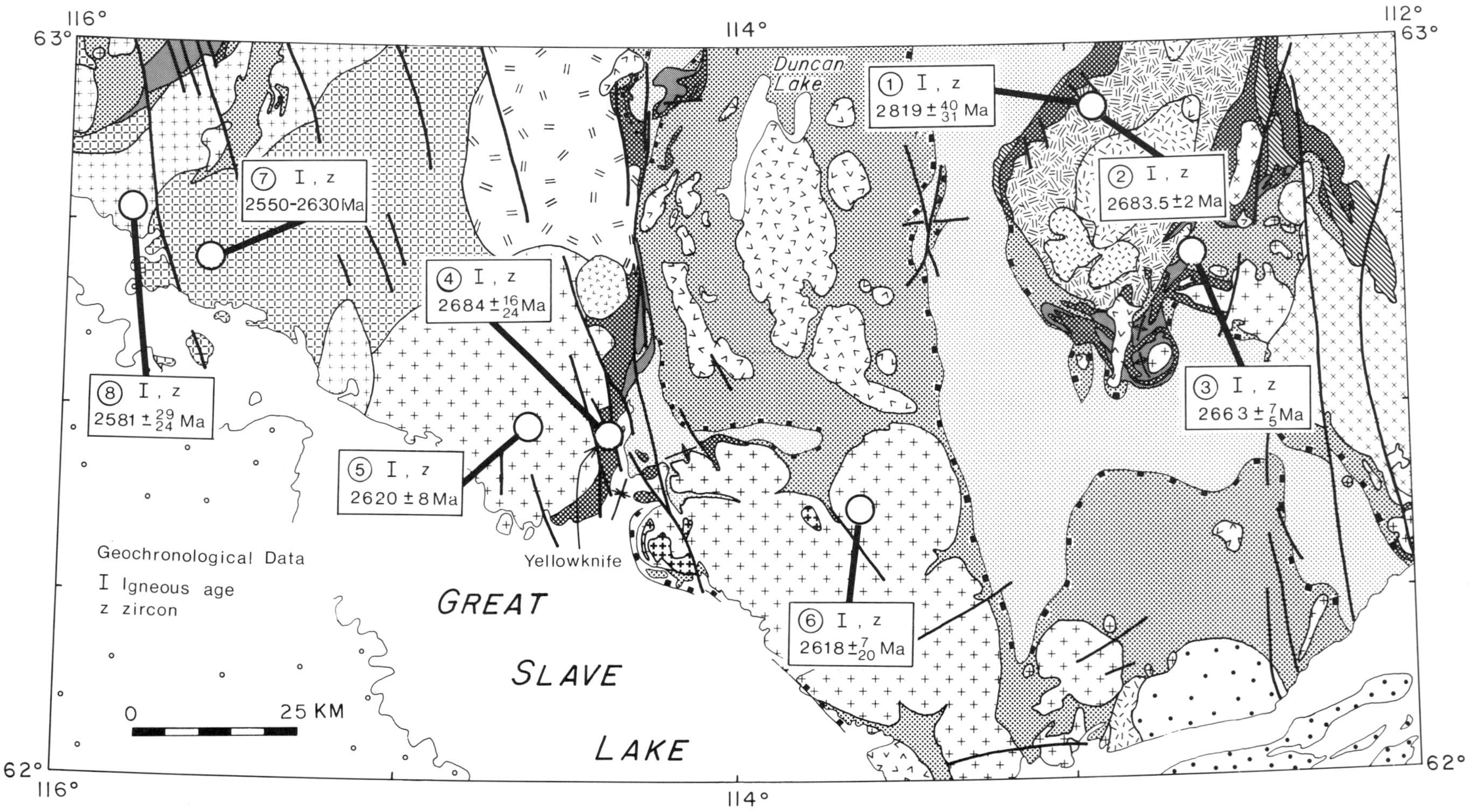
116°
114°
112°
63°
62°
Duncan Lake
① I, z
2819 $^{+40}_{-31}$ Ma
② I, z
2683.5 ± 2 Ma
③ I, z
2663 $^{+7}_{-5}$ Ma
④ I, z
2684 $^{+16}_{-24}$ Ma
⑤ I, z
2620 ± 8 Ma
⑥ I, z
2618 $^{+7}_{-20}$ Ma
⑦ I, z
2550-2630 Ma
⑧ I, z
2581 $^{+29}_{-24}$ Ma
Yellowknife
GREAT
SLAVE
LAKE
Geochronological Data
I Igneous age
z zircon
0
25 KM

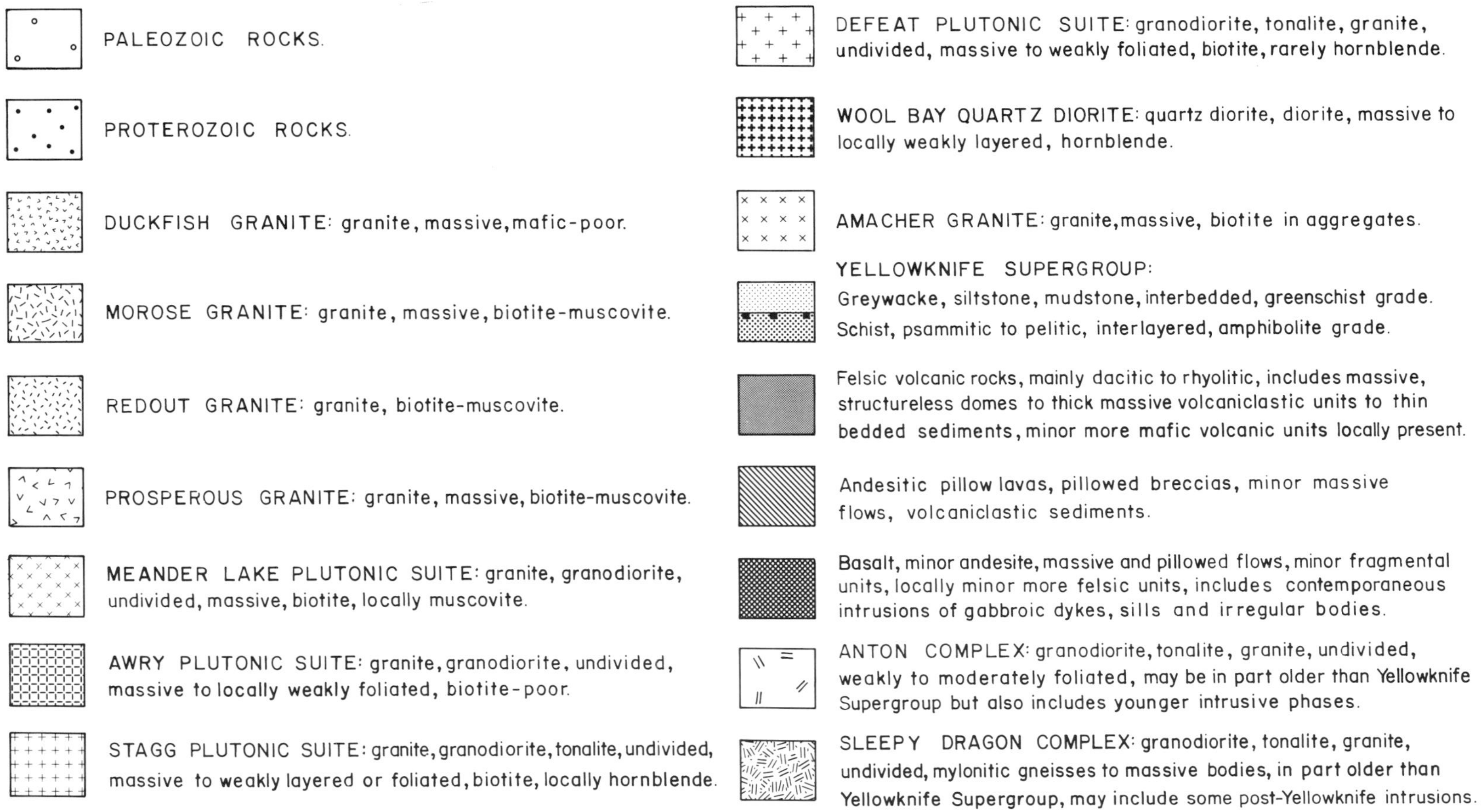

Figure 1. Geology of the Yellowknife — Hearne Lake area showing the location on the samples analyzed (circled numbers) and summarizing the geochronological data for each sample.

ANALYTICAL METHODS

General techniques of zircon concentration, chemical dissolution and isotopic analysis (Finnigan MAT 261 mass spectrometer) are similar to those used by van Breemen et al. (1987). Results are presented in Table 1 and Figures 3 to 6. Two sigma uncertainties in the isotopic ratios are 0.5% for ^{206}Pb ^{238}U and $^{207}Pb/^{235}U$ and 0.1% or 0.15% for $^{207}Pb/^{206}Pb$. Zircon samples (2 and 3) were analyzed more recently by improved techniques and this is reflected by the smaller amounts of analyzed material (Roddick et al., 1987; Parrish et al., 1987) as were repeat fractions from the remaining samples. Uncertainties for each data point are indicated on the concordia diagrams. Regression calculations for linear alignments of data points were done according to the methods of York (1969).

CASE STUDIES

Sleepy Dragon Complex

The Sleepy Dragon Complex consists of a heterogeneous assemblage of generally metamorphosed, and in some cases highly deformed, granitoids to locally gneissic units that are considered to be older than the Yellowknife Supergroup. This relationship was first suggested by Baragar (1966) on the basis of the locally developed extensive swarm of mafic intrusions in the west margin of the complex that occur both within the complex and in the adjacent volcanic sequence. He considered them to be possible feeders to the volcanics. The only published zircon geochronology prior to this study is that of Green and Baadsgaard (1971) who determined a $^{207}Pb/^{206}Pb$ zircon age of 2595 Ma from a locality 22 km south southwest of the sample locations for this study near Upper Ross Lake.

The two samples for this study were collected in the immediate vicinity of Sleepy Dragon Lake, an area that has been mapped at 1:50 000 scale (Davidson, 1972; unpublished map) and is remote from post- Yellowknife plutonic bodies. In that area Davidson (1972) outlined strongly foliated cataclastic to mylonitic granitoid rocks that are intruded by a metamorphosed and weakly foliated granodiorite. Both are intruded by metamorphosed mafic dykes. The granitoid units were both sampled for geochronology.

Sleepy Dragon Complex cataclastic granite (sample 1)

The cataclastic to strongly foliated to mylonitic granitoid occurs at the west end of Sleepy Dragon Lake between the Cameron River volcanics of the Yellowknife Supergroup and the lake (Davidson, 1972; *see* also Fig. 4 in Henderson, 1985). The unit includes megacrystic rocks with coarse pink microcline in a medium grained grey matrix, mylonitic granite and deformed pegmatite. The unit is heterogenous and contains local veins and dykes of weakly foliated aplitic granite. Layers of coarsely recrystallized amphibolite are also present locally. In thin section the rock analyzed is a medium- to coarse-grained somewhat annealed, cataclastic granite consisting of blocky, abraded, equant to slightly elongate microcline and plagioclase with granulated margins in a matrix of coarsely polygonized quartz and finer microcline and plagioclase. Mafic minerals form about 10 -15% of the rock, occur in ragged aggregates in the matrix and consist mainly of somewhat altered brown to olive-green biotite and lesser amounts of strongly coloured blue-green hornblende. Accessory minerals include sphene, epidote, apatite and zircon.

Zircons are pale brown, moderately clear and prismatic; L:B (length to breadth) = 1:1 to 1:5. Polished and etched grain mounts reveal inner crystals with subhedral igneous zoning and outer mantles characterized by usually irregular and discordant zoning (Fig. 2A). Synneusis twins in this outer mantle can be interpreted as dynamic impingement of the growing zircons with crystals in a crystal-liquid mush (van Breemen and Hanmer, 1986). Uranium concentrations are high (Table 1). Fractions 1c and 1f are strongly abraded zircons repicked from unanalyzed portions of fractions 1b and 1e. From Figure 3 it is apparent that the four most discordant data points show significantly more scatter than the three more concordant data points. A regression analysis of all seven fractions yields upper and lower intercept ages of 2829 +40/−31 Ma and ca. 820 Ma respectively with a mean square of weighted deviates (MSWD) of 38. The best fit line does not, however, pass through the most concordant point 1f. A regression line through points 1a, 1c, 1d, and 1f corresponds to a MSWD of 3.1 and yields an upper intercept age of 2819 +13/−12 Ma. The latter age with the larger uncertainty, 2819 +40/−31 Ma, is interpreted as the age of syntectonic granite emplacement and crystallization.

Sleepy Dragon Complex granodiorite (sample 2)

The granodiorite sampled is an irregularly banded, north-trending unit in the west-central part of Sleepy Dragon Lake (Davidson, 1972; *see* also Fig. 4 in Henderson, 1985). It sharply truncates the east trending cataclastic granitoid gneiss that occurs at the west end of the lake (described above). The rock sampled is a weakly foliated, grey, medium grained, equigranular granodiorite. It consists of anhedral to subhedral weakly altered plagioclase with irregular to weakly granulated margins, minor microcline and highly irregular masses of coarsely polycrystalline quartz. Ragged flakes of olive-green bioitite are partly replaced by chlorite. Accessory minerals are mainly epidote with minor zircon and sphene. The original igneous texture has been replaced by a metamorphic granoblastic texture with the rock having a weakly crushed aspect that is much less than that of the cataclastic granitoid gneiss it intrudes.

Zircons are brown and mostly clear. The euhedral to subhedral equidimensional to prismatic (L:B of 5:1) crystals have sharp terminations. All four fractions analyzed (Table 1) have been abraded strongly. Fraction 2d was repicked from 2c and consists of flat crystals. Data points can be fitted to a regression line with a MSWD of 0.3 yielding upper and lower intercept ages of 2683.5 ± 2.0 Ma and ca. 610 Ma (Fig. 3). The upper intercept age is interpreted as the time of emplacement and crystallization of the original pluton.

Turnback Rhyolite

The Turnback Rhyolite occurs in a complexly folded part of the Beaulieu volcanic belt south of the Sleepy Dragon Complex (Lambert, in press). This volcanic complex is unusual in the Yellowknife region in that it consists of about

Table 1. U-Pb zircon isotopic data

Sample and zircon fraction	Weight (mg)	U (ppm)	Pb* (ppm)	Measured $^{206}Pb/^{204}Pb$	Isotopic abundances (^{206}Pb=1000) ^{204}Pb	^{207}Pb	^{208}Pb	Isotopic ratios $^{206}Pb/^{238}U$	$^{207}Pb/^{235}U$	Age, Ma $^{207}Pb/^{206}Pb$
1. Cataclastic granitic gneiss, Sleepy Dragon Complex (62° 55.5′ 112° 56.0′)										
a, −149+105,N0,A	0.081	1279	651.6	1943	0.481	196.5	133.3	0.45234	11.8904	2747.8
b, −105+74,N-1	0.12	467.1	229.5	949	0.992	202.1	147.7	0.43709	11.4578	2743.2
c, −105+74,N-1,A	0.015	1796	940.4	2414	0.360	195.7	154.5	0.45600	12.0310	2753.8
d, −74+62,N-2	0.10	1165	572.2	1198	0.804	198.4	158.1	0.43183	11.2330	2730.5
e, −62+44,N-3	0.12	1130	567.1	829	1.180	201.0	192.2	0.43442	11.1829	2713.3
f, −62+44,N-3,A	0.015	1473	781.1	1135	0.820	202.6	142.3	0.47091	12.5107	2765.2
g, −44	0.087	1112	549.2	842	1.151	201.7	180.9	0.43047	11.1451	2722.8
2. Granodiorite, Sleepy Dragon Complex (62° 55.0′ 112° 56.6′)										
a, −160+149,N-4,A	0.11	485.3	270.1	948	0.965	193.5	170.8	0.48913	12.2540	2668.4
b, −160+149,N-4,A	0.12	477.0	268.7	4812	0.120	183.3	155.8	0.49074	12.3042	2669.8
c, −149+105,N-3,A	0.11	446.9	254.7	1800	0.456	187.8	162.4	0.49787	12.5075	2673.0
d, −149+105,N-3,A	0.035	390.1	229.2	2373	0.326	187.0	165.6	0.51012	12.8722	2680.3
3. Turnback Rhyolite (62° 42.6′ 112° 38.2′)										
a, −149+105,N3,A	0.13	366.5	161.8	3318	0.225	181.9	138.8	0.39128	9.6662	2645.2
b, −105+74,N1	0.11	379.4	150.7	2289	0.340	182.3	140.0	0.35294	8.6699	2635.8
c, −105+74,N1,A	0.12	301.9	143.4	3096	0.231	182.8	129.8	0.42375	10.5136	2652.4
d, −105+74,N1,A	0.029	361.7	211.0	752	1.171	195.2	175.3	0.51393	12.8128	2660.4
e, −74+62,N1	0.12	298.5	137.2	2671	0.283	183.0	142.4	0.40673	10.0683	2648.6
f, −74+62,N1,A	0.027	434.2	220.2	1338	0.581	188.4	151.5	0.44839	11.2090	2664.8
4. Townsite felsic volcanics (62° 27.4′ 114° 23′)										
a, −149+105,N1,A	0.11	951.1	468.0	3864	0.229	175.7	106.2	0.44970	10.7183	2585.5
b, −149+105,N1,A	0.049	907.3	470.7	4542	0.189	179.5	107.5	0.47157	11.5179	2626.3
c, −105+74,N1	0.11	1104	530.9	6180	0.136	174.0	102.8	0.43986	10.4528	2580.6
d, −105+74,N1,A	0.020	1077	573.6	4708	0.148	180.0	121.5	0.47803	11.7429	2635.8
e, −74+62,N1	0.12	1196	580.3	6408	0.135	173.9	113.2	0.44026	10.4560	2579.6
f, −74+62,N1,A	0.015	891.4	473.1	3286	0.202	181.4	112.6	0.48002	11.8402	2642.6
5. Defeat tonalite (west) (62° 27.7′ 114° 36.7′)										
a, +105,A	0.11	335.3	155.4	2401	0.315	177.7	114.8	0.42134	10.0954	2594.3
b, −105+74	0.11	335.6	160.4	2086	0.378	178.7	133.5	0.42880	10.2847	2596.1
c, −105+74,A	0.11	316.5	156.0	1784	0.456	180.7	144.4	0.43889	10.5906	2606.1
d, −105+74,A	0.007	332.6	179.2	1773	0.191	178.1	145.0	0.47641	11.5431	2613.0
e, −105+74,A	0.014	273.5	141.8	1253	0.414	181.1	140.3	0.46227	11.2137	2614.9
f, −74+62	0.12	357.8	164.1	2254	0.354	177.7	127.6	0.41330	9.8777	2590.1
6. Defeat tonalite (east) (62° 21.7′ 113° 37.6′′)										
a, +105,A	0.13	588.4	269.2	1304	0.716	177.6	142.2	0.41274	9.5970	2544.2
b, −105+74	0.12	569.6	253.2	842	1.134	182.0	107.5	0.41733	9.6594	2536.5
c, −105+74	0.10	553.7	255.8	1304	0.702	177.8	124.4	0.42256	9.8490	2548.2
d, −105+74,A	0.11	570.5	261.7	2095	0.419	174.7	100.2	0.42459	9.9203	2552.3
e, −105+74	0.021	430.5	222.8	4027	0.160	176.5	88.0	0.47850	11.5142	2601.5
f, −105+74,long	0.020	736.5	348.9	2231	0.389	171.8	108.6	0.43593	10.0307	2526.6
7. Awry granite (62° 42.7′ 115° 35.7′)										
a, +105,A	0.13	1477	501.5	1137	0.854	158.9	109.5	0.32089	6.5488	2323.1
b, −105+74	0.054	1668	653.7	1786	0.522	158.6	134.2	0.34962	7.3236	2367.7
c, −74+62	0.10	1529	519.1	1053	0.935	162.1	130.9	0.31553	6.5353	2348.4
d, −62	0.006	828.4	332.0	1281	0.594	167.4	77.2	0.38264	8.4372	2454.8
8. Stagg granite (62° 45.6′ 115° 49.2′)										
a, +105,A	0.10	430.9	181.2	10510	0.052	164.2	95.75	0.38875	8.7659	2492.6
b, −105+74,A	0.13	564.6	212.8	1624	0.586	168.4	113.9	0.34888	7.7433	2465.9
c, −105+74	0.023	212.1	106.4	2401	0.024	172.9	146.8	0.44509	10.4250	2556.4
d, −105+74	0.012	283.6	138.0	1682	0.334	172.0	151.1	0.43219	10.0015	2536.2
e, −105+74,cores,A	0.13	258.3	111.1	9365	0.049	168.2	119.8	0.38888	8.9878	2534.1

* radiogenic lead
N, nonmagnetic at given angle
M, magnetic at given angle
A, abraded

equal proportions of felsic and mafic volcanics. The felsic volcanics probably represent relics of local high relief and probably emergent centres superimposed on a background of subaqueous mafic volcanoes. The sample analyzed came from the southeast shore of Turnback Lake. In thin section the rock consists of a fine granoblastic quartzofeldspathic matrix with a strong foliation defined by well oriented muscovite and brown biotite laths. Lenses to roughly square or rectangular areas of coarsely polygonized quartz and plagioclase, up to 1.5 mm, represent recrystallized quartz and plagioclase phenocrysts. The original rhyolite has been metamorphosed at middle amphibolite grade based on the mineralogy of the nearby pelitic and psammitic metasediments.

Zircons are colourless, clear, have few facets and have L:B of 2:1 to 3:1. Fractions 3d and 3f consist of repicked and strongly abraded zircons (Table 1). A regression analysis of all six data points yields an upper intercept age of 2668 ± 7 Ma and a lower intercept age of ca. 180 Ma with a MSWD OF 15 (Fig. 3). The MSWD is reduced to 2.3 if data point 3f is not included in which case the upper and lower intercept ages are 2664 ± 3 Ma. As fraction 3d is essentially concordant, the age and uncertainty of crystallization of the zircon are estimated at 2663 + 7/ − 5 Ma.

Townsite felsic volcanics

The Townsite felsic volcanics occur within the central part of the dominantly mafic volcanic sequence of the Yellowknife Supergroup at Yellowknife, immediately north of the city (Henderson and Brown, 1966). It is a 400 m thick unit consisting of porphyritic fragmental volcanics ranging from coarse breccias to tuffs, massive flows and intrusive gabbro sills. The massive flow unit analyzed consists of about 30% phenocrysts and glomerocrysts of euhedral to subhedral, moderately to strongly altered plagioclase up to 7 mm, and rare rounded quartz phenocrysts in a fine, homogeneous quartzofeldspathic matrix. Fine biotite occurs evenly distributed throughout the matrix as well as in decussate aggregates, with minor quartz and feldspar, commonly associated with an opaque phase. Minor carbonate is also present. This rock contrasts strongly with the previously described Turnback Rhyolite in that it is unfoliated, at low metamorphic grade and primary textures are well preserved.

Zircons are euhedral with L:B of 2:1 to 3:1 and have a simple morphology. Crystals are brown, cloudy and contain from 5 to 10% inclusions. Fractions 4b, 4d and 4f are strongly abraded zircons repicked from unanalyzed remainders of fraction 4a, 4c, and 4e respectively (Table 1). Of the three most concordant fractions, 4f appears to have an older component which in this fine grained cloudy fraction may reflect either cores or xenocrystic zircon. There may also be an inherited component in fractions 4a, 4c and 4e, although the scatter of these data points may have resulted from variable times of Pb loss. A regression analysis of the five data points 4a-4d yields an upper intercept age of 2684 + 16/ − 14 Ma, a lower intercept age of ca. 1170 Ma with a MSWD of 31 (Fig. 4). The upper intercept age is interpreted as the time of zircon crystallization shortly prior to lava extrusion. In view of the suprisingly old lower intercept age, the zircon crystallization may be overestimated although it is not likely to be younger than 2660 Ma. Further work is clearly required on this felsic unit.

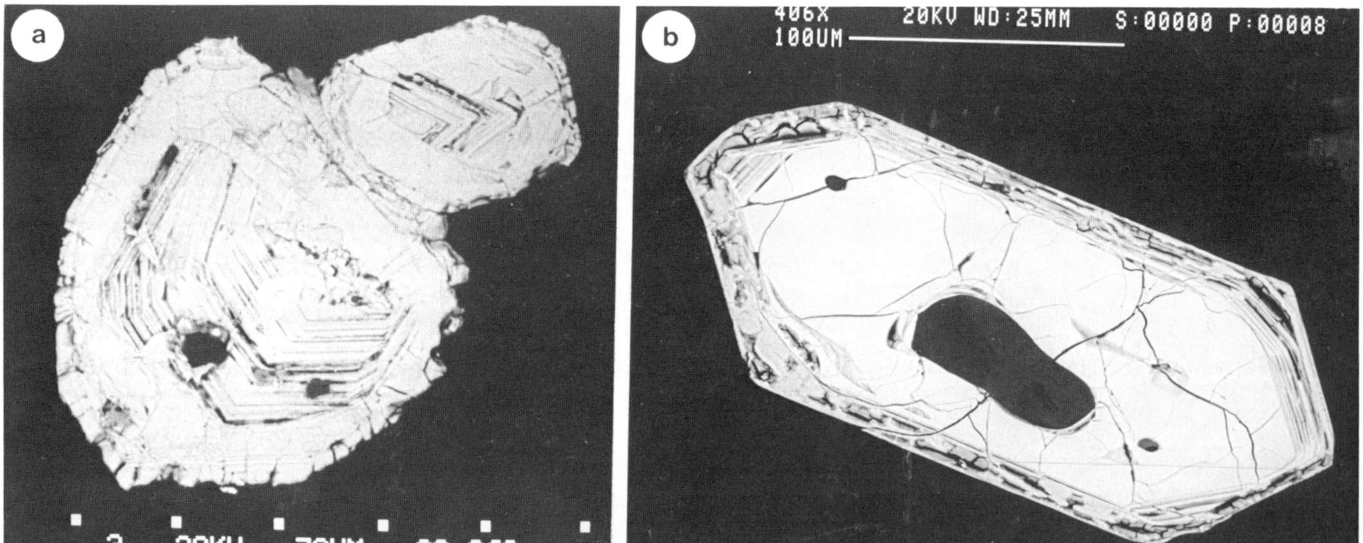

Figure 2. Scanning electron microscope images (back scattered electron mode) of polished and etched (HF) zircons from: **A.** Cataclastic granite gneiss, Sleepy Dragon Complex, showing synneusis twins which developed late in the growth history of the zircons. Inner resorbed and/or broken zircon centres have euhedral igneous zoning. **B.** Stagg granite. Inner zircon is low in U content (light shade) and shows simple euhedral igneous outline. Some irregular zonation is evident near the outer margin of the inner crystal as well as in the outer rim of the high U (darker shade) zircon, suggesting a continuous growth history. Outer perimeter is nearly euhedral. Larger dark inclusion or embayment consists of quartz and K-feldspar.

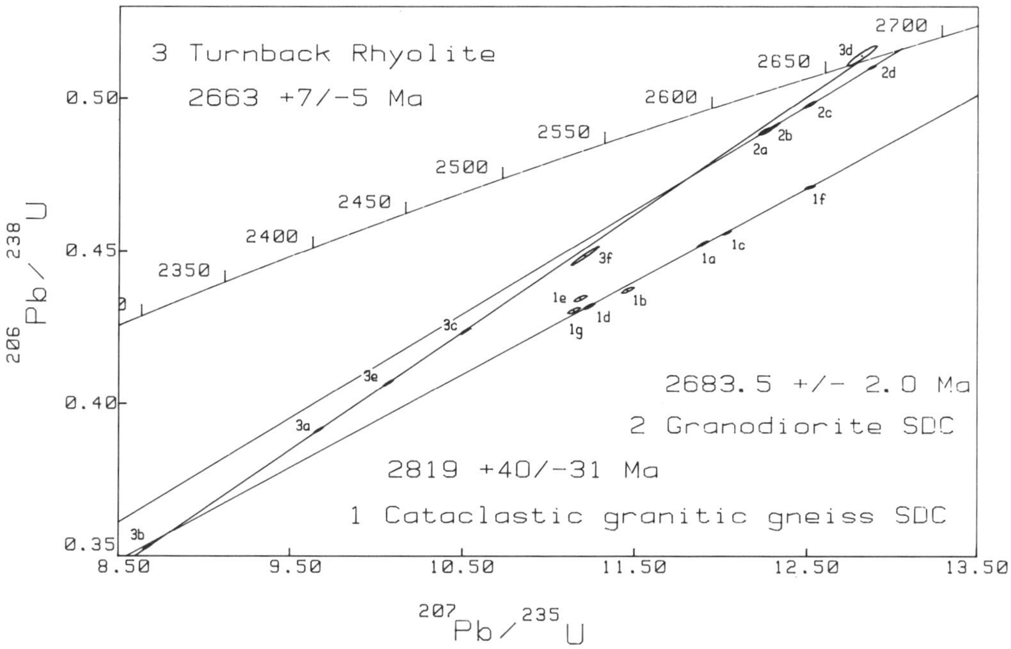

Figure 3. Isotope ratio plot for zircons from 1, Sleepy Dragon Complex (SDC) cataclastic granitic gneiss; 2, Sleepy Dragon Complex granodiorite and 3, Turnback Rhyolite. Numbers and letters with data points correspond to those of Table 1.

Defeat Plutonic Suite

The Defeat Plutonic Suite forms large plutonic complexes east and west of Yellowknife (Henderson, 1985), smaller intrusions in the south-central part of the Yellowknife basin and in the Beaulieu volcanic belt (Lambert, in press) southeast of the Sleepy Dragon Complex. These plutons have sharp, regionally concordant contacts with the Yellowknife Supergroup rocks they intrude; the one notable exception being the sharply discordant contact with the Kam Group volcanics at Yellowknife. To the west of Yellowknife the Defeat is intruded by the Awry Plutonic Suite. The plutons are typically massive except for a local, moderately developed marginal foliation. West of Yellowknife the Defeat has a weak, northeast-trending foliation that is roughly parallel to regional geological trends in the area west of Yellowknife. The Defeat plutons range in composition from granodiorite to tonalite. They are typically homogeneous, medium grained, equigranular rocks with a distinctive weathered texture displaying lath-shaped plagioclase and coarse, almost euhedral, biotite, particularly in the smaller bodies.

Two samples of the Defeat Plutonic Suite were analyzed: one west and one east of Yellowknife.

Defeat tonalite (west of Yellowknife) (sample 5)

This tonalite comes from the western body of the Defeat Plutonic Suite about 12 km west of Yellowknife on the highway. It consists primarily of subhedral to anhedral, weakly zoned plagioclase with moderately developed patchy to zoned alteration, elongate polycrystalline aggregates of quartz and ragged flakes to aggregates of olive-green biotite. Microcline is rare. Epidote is abundant, forming several per cent of the rock and is typically associated with biotite. Accessory minerals include sphene, apatite and zircon. The rock has a moderate foliation expressed by the orientation of biotite, plagioclase and elongation of the polycrystalline quartz masses. The mineral boundaries are sutured to granulated in some cases. The original igneous texture (which is evident in sample 6) is all but lost.

Zircons are pale pinkish brown and have L:B ratios of 2:1 to 4:1. Zoning is prominent and it is difficult to recognize cores. Fractions 5d and 5c are repicked and strongly abraded zircons from fractions 5c and 5b respectively. A regression analysis of all 6 data points yields an upper intercept age of 2626 ± 8 Ma, a lower intercept age of c. 380 Ma and a MSWD of 14 (Fig. 4). Assuming a small inherited component in fractions 5c and 5e (Fig. 4), a concordant regression line can be fitted to the remaining 4 data points yielding an upper intercept age of 2620 ± 3 Ma, a lower intercept age of c. 340 Ma and a MSWD of 0.66. In view of the scatter of data points the latter uncertainty is considered inadequate and so the age of crystallization and emplacement of the western Defeat tonalite is placed at 2620 ± 8 Ma.

Defeat tonalite (east of Yellowknife) (sample 6)

The tonalite was sampled at Defeat Lake, 45 km east of Yellowknife, a few kilometres from the margin is one of the large plutonic lobes of the eastern plutonic complex. This equigranular medium grained rock consists mainly of stubby euhedral to less commonly subhedral, zoned, moderately to weakly altered plagioclase, and irregular masses of strained but not polygonized quartz. Microcline is present as a rare interstitial phase. Euhedral lath-like olive-green biotite is the only mafic mineral and forms less than 10% of the rock. Chlorite is associated with it only rarely. Epidote is an abundant accessory mineral associated with biotite and the alteration products of plagioclase. Relatively coarse white mica is also associated with fine plagioclase alteration products. Other accessory minerals include apatite and zircon. The rock has a well preserved panidiomorphic igneous texture.

Zircons are clear and colourless and have L:B ratios ranging from 2:1 to 7:1. Zoning is strong and cores are common, as are transverse cracks. There are also a number of synneusis twins. Data points corresponding to the four fractions analyzed originally (6a-6d) are tightly clustered (Table 1). Two small and strongly abraded fractions are more concordant.

Of these, 6f consists of long thin crystals, whereas 6e consists of morphologies similar to 6a-6d. A regression line through data points 6a-6e yields an upper intercept age of 2618 ± 7 Ma a lower intercept age of ca. 780 Ma and a MSWD of 14 (Fig. 5). Data point 6f plots above this line. A regression analysis for all six data points yields an upper intercept age of 2620 + 41/ −29 Ma and a MSWD of 330. The anomalous position of data point 6f can be interpreted either in terms of similar amounts of inheritance in all fractions 6a-6e or in terms of earlier average lead loss from fraction 6f. Such earlier lead loss from long thin zircons has been demonstrated in a separate study on Archean plutonic rocks (unpublished data) and is also more likely in view of the higher U concentration of this fraction. The five point regression analysis is therefore considered more valid, especially since the consistency of the data does not support a hypothesis of significant inheritance. However, the possibility of a somewhat younger zircon crystallization age cannot be excluded. Hence the age of emplacement of the southeast granodiorite is placed at 2618 + 7/ −20 Ma.

Awry granite (sample 7)

The Awry Plutonic Suite is a northeast-trending granite to granodiorite unit that occupies the central part of the extensive granitoid terrane west of Yellowknife. It intrudes the

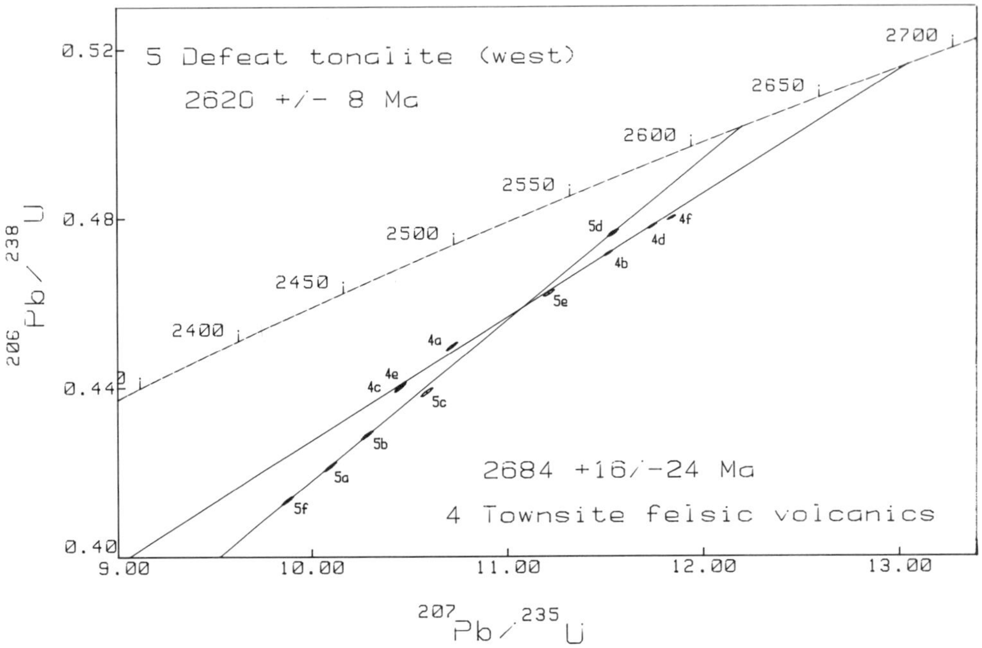

Figure 4. Isotope ratio plot for zircons from 4, Townsite felsic volcanics and 5, Defeat tonalite, west of Yellowknife. Numbers and letters with data points correspond to those of Table 1.

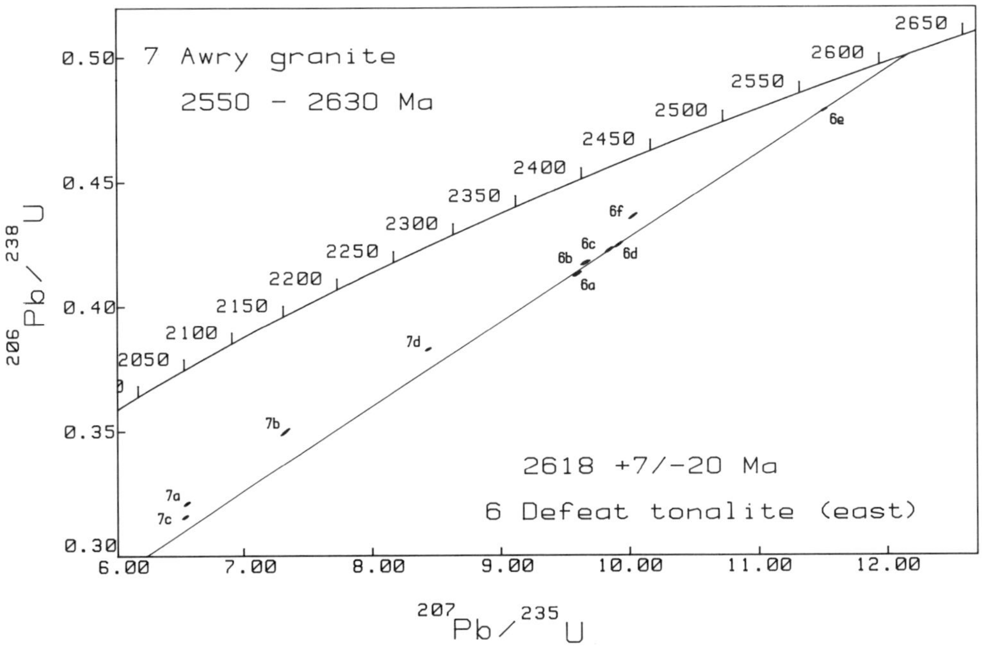

Figure 5. Isotope ratio plot for zircons from 6, Defeat tonalite, east of Yellowknife and 7, Awry granite. Numbers and letters with data points correspond to those of Table 1.

Defeat Plutonic Suite to the east, the Stagg Plutonic Suite to the west, and the Anton Complex to the northeast (Henderson, 1985). The contacts in general are not sharp and have extensive inclusions of the older units in the Awry, and dykes of Awry in the various older units. The granite sample analyzed is massive, light pink, medium-to medium coarse-grained, biotite poor, and consists mainly of equidimensional, anhedral, unzoned plagioclase with moderate patchy alteration that forms the coarsest phase in the rock. Slightly altered microcline of similar form but smaller size contains inclusions of plagioclase of varied size. Rare coarse subhedral Carlsbad twinned laths of microcline are also present. Abundant large polycrystalline quartz masses have irregular sutured boundaries. Minor brown to greenish brown biotite in ragged irregular flakes to small aggregates are slightly to moderately altered. Minor muscovite is also present. Accessory minerals include epidote, relatively coarse apatite and zircon. The rock has a granitic igneous texture modified due to strain.

Zircons have L:B ratios of 2:1 to 5:1. The crystals are mainly cloudy with many inclusions and cores. Twins are common and some are clearly of synneusis origin. Four zircon data points are strongly discordant and scattered, and a regression analysis yields an upper intercept age of 2570 + 117/−83 Ma, a lower intercept age of ca. 850 Ma and a MSWD of 305. Comparing the distribution of data points relative to the regression line for the southeast granodiorite, it is tentatively suggested that the Awry granite is younger than 2630 Ma or older than 2550 Ma.

Stagg Plutonic Suite (sample 8)

The Stagg Plutonic Suite has a northeasterly trend along the western margin of the extensive granitoid terrane west of Yellowknife. It is transected by a wide belt of high grade Yellowknife metasediments that it intrudes. It is itself intruded by the Awry Plutonic Suite to the east. The Stagg Plutonic Suite is a massive to locally weakly foliated, dark, typically mafic-rich, commonly megacrystic granitoid unit. It consists of two compositional phases; a tonalitic end member which is more common in the northern part of the unit and a more granite-like phase to the south which, except for the addition of abundant coarse microcline megacrysts, is similar texturally and compositionally to the tonalitic phase. The tonalitic member in places grades into the granitic member and in some cases is intruded by the granitic member although the reverse was never seen.

The sample analyzed is from the granitic phase of the Stagg collected 82 km west-northwest of Yellowknife on the highway. The granite is a massive, coarse grained quartz-rich rock in which the large and continuous masses of quartz are coarsely and irregularly polycrystalline. Coarse grained plagioclase is subhedral to anhedral, not zoned, and is moderately altered. Microcline occurs either as coarse subhedral elongate megacrysts or as fine anhedral grains. Both feldspars contain inclusions of plagioclase and minor chloritized biotite. Chlorite occurring as coarse individual flakes to aggregates has replaced the original biotite in the rock although the fine biotite texture is preserved. Accessory minerals include coarse apatite and zircon with in one case zircon occurring as inclusions within a plagioclase. Relatively coarse white mica is associated with the fine plagioclase alteration products.

A heterogeneous population of zircons, some twinned, was obtained from this sample. L:B ratios range from equidimensional to 2:1. Rounded cores are abundant and in many grains these cores are only partly overgrown by prismatic zircon of igneous origin. Polished and etched zircon sections show, however, that growth zoning in many of the cores is continuous with zoning of the high U mantle (Fig. 2B). In many grains irregular and discordant zoning occurs near, and on both sides of, the interface, and rounding of the inner crystal is apparently the result of magmatic corrosion.

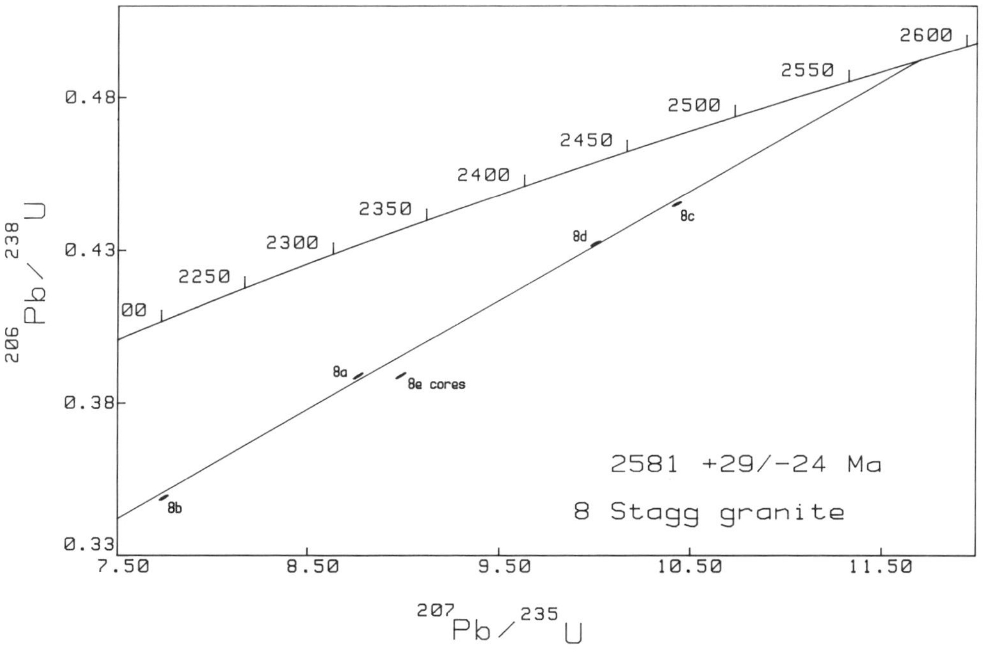

Figure 6. Isotope ratio plot for zircons from 8, Stagg granite. Numbers and letters with data correspond to those of Table 1.

Four zircon fractions 8a-8d were picked avoiding cores, and one fraction (8e) was picked for cores and strongly abraded (Table 1). A regression analysis of the four fractions 8a-8d yields an upper intercept age of 2581 + 29/−24 Ma, a lower intercept age of ca. 650 Ma and a MSWD of 171 (Fig. 6). The upper intercept age and uncertainty are tentatively interpreted in terms of zircon crystallization and emplacement of the Stagg granite. Sample 8e shows evidence for an older component, indicating that not all the rounded cores seen during hand picking can be interpreted in terms of continuous growth.

DISCUSSION

The 2819 + 40/−31 Ma age of the cataclastic granite gneiss of the Sleepy Dragon Complex (sample 1) supports the interpretation originally proposed by Baragar (1966) that at least part of the complex is older than the Yellowknife Supergroup. Although apparently 150 Ma older than the Yellowknife, it is relatively young compared to other basement ages determined in the Slave Province (i.e. 3.15 Ga basement at Point Lake (Krogh and Gibbins, 1978); ca. 2.9-3.2 Ga clasts in a diatreme in Yellowknife volcanics at Yellowknife (Nikic et al., 1980); 2.99 Ga basement near Indin Lake, western Slave Province (Frith et al., 1986)).

The 2683.5 ± 2 Ma age of the granodiorite in the Sleepy Dragon Complex (sample 2) that intrudes the cataclastic granite gneiss is interesting in that it is only about 20 Ma older than the age of volcanism determined at Turnback Lake, and is essentially identical to that of the Townsite felsic volcanics at Yellowknife. It is also very close to the age of meta quartz diorite at Healey Lake, which is part of a complex of granitoids, migmatitic granitoids and gneisses, parts of which on geological grounds (but not supported by present geochronological data) were also interpreted as basement to the Yellowknife Supergroup in that area (van Breemen et al., 1987).

The two ages of Yellowknife Supergroup volcanic units were determined at 2663 + 7/−5 Ma for the Turnback Rhyolite (sample 3) and 2684 + 16/−24 Ma for the Townsite felsic volcanics (sample 4) at Yellowknife. The Turnback Rhyolite age, although on the other side of the basin, compares very closely with the 2667 + 4.4/−4.0 Ma age of a synvolcanic intrusion in the Banting Group at Yellowknife, a felsic volcanic complex that occurs high in the stratigraphic section, above the unconformity that separates it from the main volcanic sequence (S.A. Bowring *in* Padgham (1985)). Padgham (1985), quoting S.A. Bowring, also reported an age of 2678 + 8.7/− 8.0 Ma for a series of irregular felsic dykes that cut the main volcanic sequence including the Townsite felsic volcanics, but are also cut off at the unconformity. They are thought to represent the conduit system to a felsic volcanic sequence like the Banting or Townsite that has not been preserved. These dyke ages are not at variance with the 2684 + 16/− 24 Ma age of the Townsite volcanics. The new Townsite zircon date is also close to the age of the Yellowknife felsic dome in the Back River volcanic complex at 2692 ± 2 Ma (van Breemen et al., 1987).

The two Defeat Plutonic Suite ages for the intrusions west and east of Yellowknife are consistant at 2620 ± 8 and 2618 + 7/− 20 Ma for samples 5 and 6. This compares favourably with the lead isochron data of Cumming and Tsong (1975) at 2635 ± 15 Ma and 2625 ± 40 Ma respectively. An earlier zircon concordia of Green and Baadsgaard (1971) for the granitoid terrane west of Yellowknife is significantly younger at 2557 ± 16 Ma although only 1 of the 4 points comes from the Defeat Plutonic Suite. Rb-Sr isochrons for the unit west and east of Yellowknife indicate an age of 2555 ± 58 Ma and 2585 ± 37 Ma respectively (Green et al., 1968; Green and Baadsgaard, 1971).

The Awry and Stagg plutonic suites (samples 7 and 8) from the western part of the granitoid complex west of Yellowknife have ages of between 2550 and 2630 Ma and 2581 + 29/− 24 Ma respectively. They are consistent with Green and Baadsgaard's (1971) earlier aggregate age determined in the granitoid terrane west of Yellowknife at 2557 ± 16 Ma. The apparent older age of strongly abraded cores from one zircon fraction from the Stagg Plutonic Suite supports to some extent the suggestion of Henderson (1985) that the Stagg may represent a partially metasomatized, and to some extent remobilized, equivalent of an older unit.

ACKNOWLEDGEMENTS

Fieldwork for the study was greatly facilitated by the use of a Department of Indian and Northern Affairs truck made available to us by W.A. Padgham of the Geology Division, Northern Affairs Program. Figure 1 was drafted by E. Hurdle. The paper has been improved by critical reviews by M.B. Lambert and R.R. Parrish.

REFERENCES

Baragar, W.R.A.
1966: Geochemistry of the Yellowknife volcanic rocks; Canadian Journal of Earth Sciences, v. 3 p. 9-30.

Cumming, G.L. and Tsong, F.
1975: Variations in the isotopic composition of volatilized lead and the age of the western granodiorite, Yellowknife, Northwest Territories; Canadian Journal of Earth Sciences, v. 12, p. 558-573.

Davidson, A.
1972: Granite studies in the Slave Province (parts of 85I); in Report of Activities, Part A, Geological Survey of Canada, Paper 72-1A, p. 109-115.

Frith, R.A., Loveridge, W.D., and van Breemen, O.
1986: U-Pb ages on zircon from basement granitoids of the western Slave Structural Province, northwestern Canadian Shield; *in* Current Research Part A, Geological Survey of Canada, Paper 86-1A, p. 113-119.

Green, D.C.
1968: Precambrian geology and geochronology of the Yellowknife area, N.W.T.; unpublished Ph.D. thesis, Department of Geology, The University of Alberta, 166 p.

Green, D.C. and Baadsgaard, H.
1971: Temporal evolution and petrogenesis of an Archean crustal segment at Yellowknife, N.W.T., Canada; Journal of Petrology, v. 12, p. 177-217.

Green, D.C., Baadsgard, H., and Cumming, G.L.
1968: Geochronology of the Yellowknife area, Northwest Territories, Canada; Canadian Journal of Earth Sciences, v. 5, p. 725-735.

Henderson, J.B.
1985: Geology of the Yellowknife-Hearne Lake area, District of Mackenzie: a segment across an Archean basin; Geological Survey of Canada, Memoir 414, 135 p.

Henderson, J.F. and Brown, I.C.
1966: Geology and structure of the Yellowknife Greenstone Belt, District of Mackenzie; Geological Survey of Canada, Bulletin 141, 87 p.

Krogh, T.E. and Gibbins, W.
1978: U-Pb isotopic ages of basement and supracrustal rocks in the Point Lake area of the Slave Structural Province, Canada; *in* Abstracts with Programs, Geological Association of Canada — Mineralogical Association of Canada, v. 3, p. 438 (abstract).

Lambert, M.B.
— The Cameron and Beaulieu River volcanic belts, District of Mackenzie, Northwest Territories; Geological Survey of Canada, Bulletin (in press).

Nikic, Z., Baadsgaard, H., Folinsbee, R.E., Krupicka, J., Leech, A.P., and Sasaki, A.
1980: Boulders from the basement, the trace of an ancient crust?; *in* Selected Studies of Archean Gneisses and Lower Proterozoic Rocks, Southern Canadian Shield, ed. G.B. Morey and G.N. Hanson; Geological Society of America, Special Paper 182, p. 169-175.

Padgham, W.A.
1985: Observations and speculations on supracrustal successions in the Slave Structural Province; *in* Evolution of Archean supracrustal sequences, ed. L.D. Ayres, p. C. Thurston, K.D. Card, and W. Weber; Geological Association of Canada, Special Paper 28, p. 133-151.

Parrish, R.R., Roddick, J.C., Loveridge, W.D., and Sullivan, R.W.
1987: Uranium-lead analytical techniques at the geochronology laboratory, Geological Survey of Canada; *in* Radiogenic Age and Isotopic Studies: Report 1; Geological Survey of Canada, Paper 87-2.

Roddick, J.C., Loveridge, W.D., and Parrish, R.R.
1987: Precise U/Pb dating of zircon at the sub-nanogram Pb level; Isotope Geoscience, v. 66, p. 111-121.

van Breemen, O. and Hanmer, S.
1986: Zircon morphology and U-Pb geochronology in active shear zones: studies on syntectonic intrusions along the nothwest boundary of the Central Metasedimentary Belt, Grenville Province, Ontario; *in* Current Research, Part B, Geological survey of Canada, Paper 86-1B, p. 775-784.

van Breemen, O., Davidson, A., Loveridge, W.D., and Sullivan, R.W.
1986: U-Pb zircon geochronology of Grenville tectonites, granulites and igneous precursors, Parry Sound, Ontario; *in* The Grenville Province, ed. J., Moore, A. Davidson and A.J. Baer; Geological Association of Canada, Special Paper 31, p. 191-207

van Breemem, O., Henderson, J.B., Sullivan, R.W., and Thompson, P.H.
1987: U-Pb zircon and monazite Ages from the eastern Slave Province, Healey Lake area, N.W.T.; *in* Radiogenic Age and Isotopic Studies Report 1, Geological Survey of Canada, Paper 87-2.

York, D.
1969: Least squares fitting of a straight line with correlated errors; Earth and Planetary Science Letters, v. 5, p. 320-324.

U-Pb zircon ages of felsic volcanic rocks in the Kaminak Lake area, District of Keewatin

J.K. Mortensen[1] and R.I. Thorpe[1]

Mortensen, J.K. and Thorpe, R.I., U-Pb zircon ages of felsic volcanic rocks in the Kaminak Lake area, District of Keewatin; in Radiogenic Age and Isotopic Studies: Report 1, Geological Survey of Canada, Paper 87-2, p. 123-128, 1987.

Abstract

Archean mafic and felsic metavolcanic rocks are well exposed in the Kaminak Lake portion of the Rankin-Ennadai belt. Earlier U-Pb zircon dating studies had provided minimum crystallization ages of about 2680 Ma for these rocks. In this study U-Pb zircon ages of 2697.5 ± 1.4 Ma and 2692.0 ± 1.0 Ma were obtained for felsic volcanic rocks in the middle and near the top, respectively, of the stratigraphic section. These ages are in good agreement with a previously published age of 2682 ± 6 Ma for rhyolitic volcanic rocks in northern Saskatchewan near the southwestern end of the Rankin-Ennadai belt.

Résumé

Les roches métavolcaniques mafiques et felsiques de l'Archéen affleurent nettement dans la portion de la zone de Rankin-Ennadai faisant partie du lac Kaminac. D'après les premières datations par la méthode U-Pb appliquée au zircon, l'âge minimum de la cristallisation de ces roches est d'environ 2680 Ma. Dans la présente étude, on a aussi évalué, par la méthode U-Pb appliquée au zircon, à 2697,5 ± 1,4 Ma et à 2692 ± 1 Ma respectivement l'âge des roches volcaniques felsiques des parties médianes et presque sommitales de la coupe stratigraphique. Ces âges présentent une bonne concordance avec l'âge de 2682 ± 6 Ma publié antérieurement dans le cas des roches volcaniques rhyolitiques du nord de la Saskatchewan, près de l'extrémité sud-ouest de la zone de Rankin-Ennadai.

INTRODUCTION

Well exposed Archean volcanic rocks in the Kaminak Lake area, Northwest Territories, form part of the extensive Rankin-Ennadai greenstone belt. This belt extends from Hudson Bay near Rankin Inlet for more than 700 km southwest to Ennadai Lake, N.W.T., and northern Saskatchewan, and is flanked by rocks, of both Archean and Proterozoic ages, that were much more highly deformed and metamorphosed during the Hudsonian Orogeny.

Following subdivision of the volcanic rocks in the vicinity of Kaminak Lake into a number of cyclic units by Ridler (1973) and Ridler and Shilts (1974), felsic members from the different cycles were sampled for geochronological study by R.H. Ridler, R.K. Wanless, and R.D. Stevens in 1972. The objectives were to establish the general age of the succession and the time span over which volcanism occurred to produce such a major accumulation of volcanic rocks. Although U-Pb zircon geochronology techniques were not sufficiently refined at the time for the second objective to be attained, analyses from several felsic volcanic units established a very approximate 2677 Ma age (Wanless, 1979).

The present study was undertaken to obtain more precise ages for two of the volcanic cycles by analyzing a number of zircon fractions from concentrates prepared during the earlier study. A zircon age for the volcanic rocks in the vicinity of Spi Lake can be subsequently used, in combination with lead isotope data for the massive sulphide deposit located there, in the refinement of lead evolution models for Archean ores.

GEOLOGICAL SETTING

Regional helicopter reconnaissance mapping (Wright, 1967) of southern District of Keewatin outlined the general geological framework of the area. More detailed mapping was subsequently done in the Kaminak Lake region by Davidson (1970a, b) and Bell (1971). Ridler (1971a, b, 1972, 1973, 1974. Ridler and Shilts, 1974) studied the regional stratigraphy of the supracrustal rocks between Padlei and Hudson Bay in the summers of 1970 to 1972. Stratigraphic relationships in the Kaminak Lake area are presently being reexamined by F.C. Taylor (personal communication, 1987).

Archean rocks of the Kaminak Supergroup in the Kaminak Lake region (Fig. 1) consist of a number of mafic-to-felsic volcanic cycles with intervening sedimentary units. This volcanic-dominant greenstone sequence constitutes part of the Rankin-Ennadai Belt, the largest Archean "outlier" that

[1] Geological Survey of Canada, 601 Booth Street, Ottawa, Ontario K1A 0E8

has been defined within the Churchill Structural Province. Only a narrow belt of Proterozoic rocks of the Hurwitz Group, that extends from near Carr Lake to Quartzite Lake, has been preserved in the area.

Rocks of the Rankin-Ennadai greenstone terrane are particularly well exposed in a wide belt from Spi Lake to Quartzite Lake (Fig. 1). Ridler and Shilts (1974) noted that even normally recessive rock units such as interflow black shales, sulphide and carbonate bodies, sericitic and chloritic "paper schists", and soapstone are exposed.

Sedimentary rocks are minor in the three stratigraphically lowest volcanic cycles, and include oxide-facies iron-formation and carbonate- and sulphide-rich units that Ridler (1971a, 1973) considered to be stratigraphically equivalent. Parts of the later cycles are predominantly sedimentary.

The volcanic rocks of the Kaminak Supergroup were subdivided by Ridler (1973, and in Ridler and Shilts, 1974) into one incomplete and four complete mafic to felsic cycles. Clastic sedimentary rocks are important associates of the upper, felsic volcanic portions of some cycles. Basement to the volcanic sequence has not yet been positively identified.

The intermediate to mafic units consist mainly of pillowed and massive flows, some of which are more than 120 m thick, that form the widespread "mafic plate" part of the cycles. Black shales, exhalite units, tuffs and fine grained clastic sediments occupy interflow positions. Andesites, in addition to,

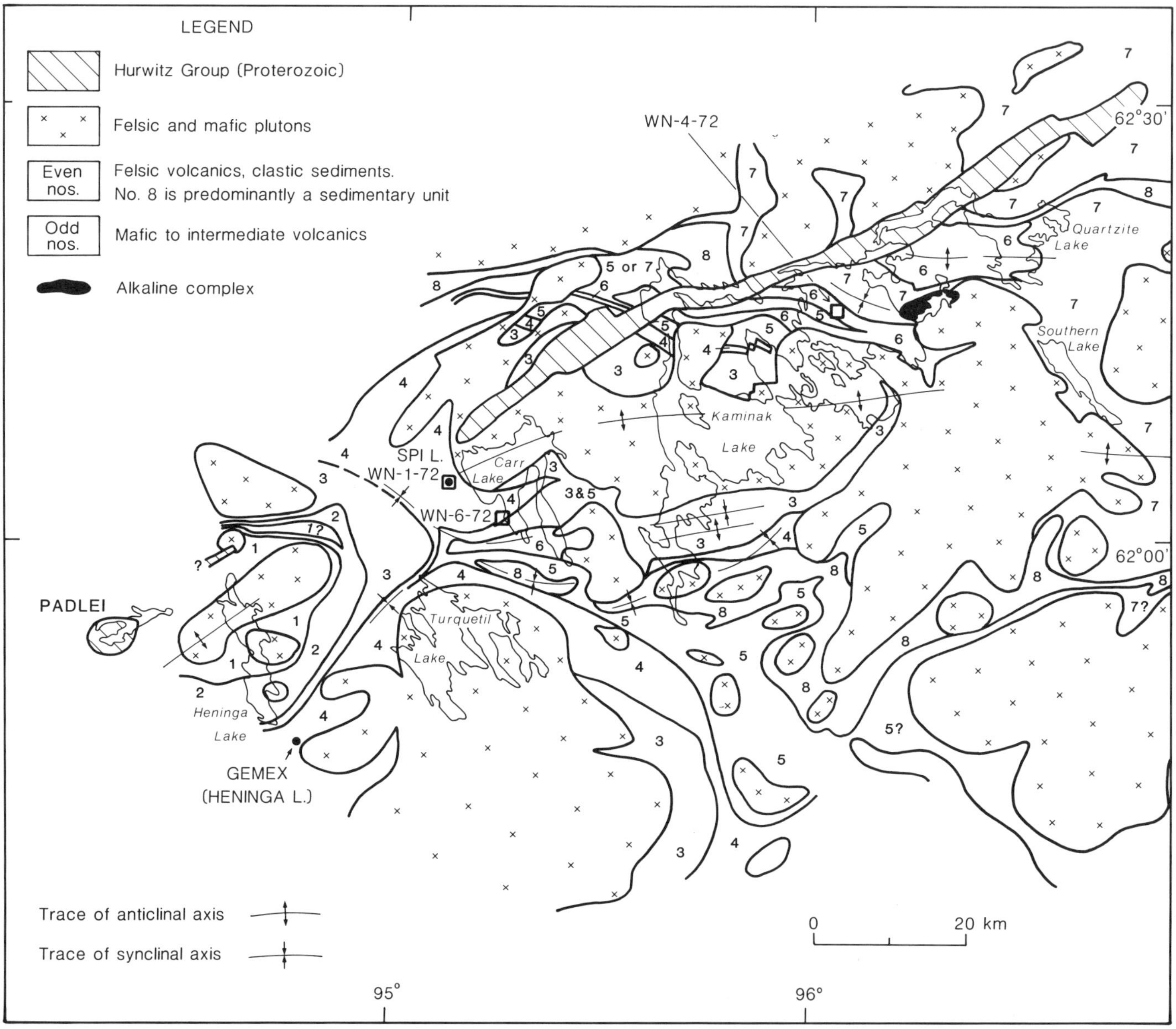

Figure 1. Simplified geology of the Kaminak Lake area (modified from Ridler and Shilts, 1974). The locations of the Kaminak Lake alkaline complex (Davidson, 1970c), the Spi Lake and Gemex (Heninga Lake) massive sulphide deposits (solid circles), and the zircon geochronology sites (open squares) are shown. Units 1 to 8 are, in stratigraphic sequence, the mafic (odd numbers) and felsic (even numbers) portions of volcanic cycles.

or in place of, the normal tholeiites, are present in units 3, 5, 7, and 9 (Fig. 1), especially toward the top of these units.

The felsic portions of the volcanic cycles are more lithologically variable. They include massive and fragmental flows, breccias, and tuffs, and locally thick sill-like masses of porphyry. Ridler and Shilts (1974) identified three definite and three possible centres of felsic volcanism. The centres at Maguse and Spi lakes contain coarse, angular breccias thought to represent proximal extrusive facies. The felsic units change thickness rapidly along strike, and in many cases grade laterally into sedimentary sequences.

The volcanic-sedimentary strata of the Kaminak region have been deformed into major, east-trending, doubly plunging, isoclinal folds. Some major domal folds are cored by large, semiconformable plutons of generally granitic composition (Davidson, 1970d). According to Ridler and Shilts (1974), a later period of folding is related to the emplacement of smaller, more discordant, mafic to felsic plutons and resulted in the development of interference fold structures.

Metamorphism is generally at greenschist facies (Davidson, 1970a, b, 1972) in the region studied by Ridler, except for higher grade contact aureoles flanking many of the larger stocks and batholiths. However, regional metamorphic grade related to Hudsonian metamorphism rises rapidly both to the north and south. Post-Hudsonian extension resulted in fault troughs that controlled Aphebian sedimentation (Ridler, 1973).

The main massive sulphide deposits in the Rankin-Ennadai belt include the Spi Lake, Gemex (Heninga Lake), and Rochon Lake occurrences, the latter near Ennadai Lake (Geological Survey of Canada, 1980; Franklin and Thorpe, 1982). The geology in the vicinity of the Spi Lake deposit has been mapped by Ridler (1974). Mineralization occurs both *in situ* on a small peninsula in Spi Lake (Fig. 1) and as Zn-rich boulders on an island in the lake (Shilts, 1974). The Gemex deposit, located southeast of Heninga Lake, nearly 37 km to the south-southwest, is also hosted by felsic volcanic rocks of the second cycle (unit 4 of Fig. 1). This deposit consists of three massive sulphide lenses, the largest of which is about 180 m long and 10.5 m wide, in a zone 830 m long. The mineralized zone has been incompletely explored, but probably contains less than 1 000 000 tonnes of base-metal-rich material in the three lenses.

PREVIOUS GEOCHRONOLOGY

Earlier U-Pb zircon studies of volcanic rocks in the Kaminak Lake area (Wanless, 1979, and unpublished data) yielded strongly discordant analyses which reflected the effects of substantial post-crystallization Pb-loss from the zircons. Although in some cases the data formed linear arrays defining discordia or Pb-loss lines, upper intercept ages were poorly constrained. Minimum ages for the volcanic units, however, were given by the oldest $^{207}Pb/^{206}Pb$ ages obtained for zircon from individual units, which ranged from 2674 to 2683 Ma.

Rhyolite from the Ennadai Group in northern Saskatchewan, near the southwestern end of the Rankin-Ennadai belt, was recently dated by the U-Pb zircon method at 2682 $\pm$ 6 Ma by Chiarenzelli and Macdonald (1986).

An alkaline intrusive complex on the northeastern shore of Kaminak Lake (Davidson, 1970c) has recently yielded a 2540 $\pm$ 76 Ma Sm-Nd age (Cavell et al., 1987; Cavell, personal communication, 1987), which is in agreement with a $^{207}Pb/^{206}Pb$ age of 2540 Ma (Wanless, unpublished data) on a discordant zircon fraction from the same complex. A whole-rock Sm-Nd age of 2671 $\pm$ 146 Ma has been obtained (Cavell et al., 1987; and Cavell, personal communication, 1987) for the volcanic rocks which belong to Ridler's third cycle (unit 7; Ridler and Shilts, 1974).

U-Pb ZIRCON DATING RESULTS

In this study, zircons from two felsic volcanic units were analyzed. Two samples are from unit 4 of Ridler and Shilts (1974) (Fig. 1); one taken at the Spi Lake sulphide deposit (sample WN-1-72), and the second taken from the same unit about 8 km along strike to the southeast (sample WN-6-72). The third sample (WN-4-72) is from a higher stratigraphic level within unit 6 of Ridler and Shilts (1974; Fig. 1). Samples WN-1-72 and WN-4-72 are from massive rhyolitic flows, whereas sample WN-6-72 is from a coarse rhyolitic breccia.

Original zircon concentrates obtained from these samples by Wanless were re-examined and found to still contain sufficient high quality material for further U-Pb analyses. In each case, the zircons form stubby, euhedral, well faceted grains with square cross-sections and simple prismatic terminations. The grains are typically very pale pink. Clear tubes and bubbles are the only types of inclusions present, and there is no visible evidence for inherited core material.

Several small subsamples were handpicked from the least magnetic zircon fraction, and were further subdivided using a high-magnetic-gradient "pin". Most fractions were subsequently strongly abraded prior to dissolution. Zircon dissolution and analysis and data reduction techniques employed in this study are described in detail by Parrish et al. (1987). The analytical data are given in Table 1.

Three analyses from the sample at Spi Lake (WN-1-72, Fig. 1) are slightly discordant and do not define a linear array (Fig. 2). The $^{207}Pb/^{206}Pb$ ages for individual fractions range from 2700 to 2707 Ma. The scatter of data cannot easily be explained by multiple episodes of Pb-loss, and is more likely due to the presence of a small amount of inherited core material in each of the fractions analyzed. Three analyses from the second sample from this stratigraphic unit (Fig. 2), however, yield more easily interpretable results. One fraction is concordant at 2697.5 $\pm$ 1.4 Ma, and provides the best estimate for crystallization age for the rock. Two additional analyses form a linear array with a calculated lower intercept age of 1100 Ma. However, the probability of fit to the line is low (9.9 %). This excess scatter outside of analytical error may reflect either a complex history of post-crystallization Pb-loss from the zircon, or minor inheritance of an older zircon component in some of the grains. This latter explanation is preferred in view of the evidence for inheritance observed in sample WN-1-72. The age for this older component can only be constrained to greater than the oldest $^{207}Pb/^{206}Pb$ age obtained from any of the zircon fractions (2707 Ma).

Sample WN-4-72 was taken from a stratigraphically higher unit than samples WN-1-72 and WN-6-72. Four zir-

Table 1. U-Pb zircon analytical data

Sample and zircon fraction	Weight (mg)	U (ppm)	Pb* (ppm)	Measured $^{206}Pb/^{204}Pb$	Isotopic abundances (^{206}Pb = 100) ^{204}Pb	^{207}Pb	^{208}Pb	Isotopic ratios $^{206}Pb/^{238}U$	$^{207}Pb/^{235}U$	Age, Ma $^{207}Pb/^{206}Pb$
				SAMPLE WN-1-72						
+74, N1, <M, A	0.0340	38.5	22.3	2617	0.0000	18.60	14.23	0.50707	13.0043	2707.1
+74, N1, >M, A	0.0450	71.8	43.3	5496	0.0000	18.52	17.31	0.51621	13.1815	2700.3
−74+44, N1, A	0.0380	78.0	46.6	5436	0.0000	18.55	15.79	0.51630	13.2061	2702.8
				SAMPLE WN-6-72						
+105, N1, <M, A	0.0420	59.5	34.9	3432	0.0035	18.54	12.92	0.51906	13.2341	2697.5
+105, N1, >M, A	0.0700	67.8	39.5	6622	0.0015	18.51	12.52	0.51664	13.1713	2697.3
−74+62, N1, A	0.1170	69.0	40.0	7826	0.0047	18.49	13.09	0.51348	13.0524	2692.5
				SAMPLE WN-4-72						
+105, N1, <M, A	0.0880	66.3	39.3	10390	0.0000	18.42	15.45	0.51493	13.0776	2691.0
+105, N1, >M, A	0.0380	68.7	40.9	3189	0.0067	18.50	16.00	0.51583	13.1007	2691.0
−74+62, N1, A	0.0440	79.1	47.8	3599	0.0094	18.53	17.29	0.51799	13.1547	2690.9
−74+62, N1	0.2270	79.6	42.1	1902	0.0485	18.54	22.49	0.4417	10.9268	2647.4

1) N1 = nonmagnetic at one degree side tilt on Frantz separator
2) A = abraded
3) <M, >M = nonmagnetic, magnetic with high-magnetic gradient "pin"

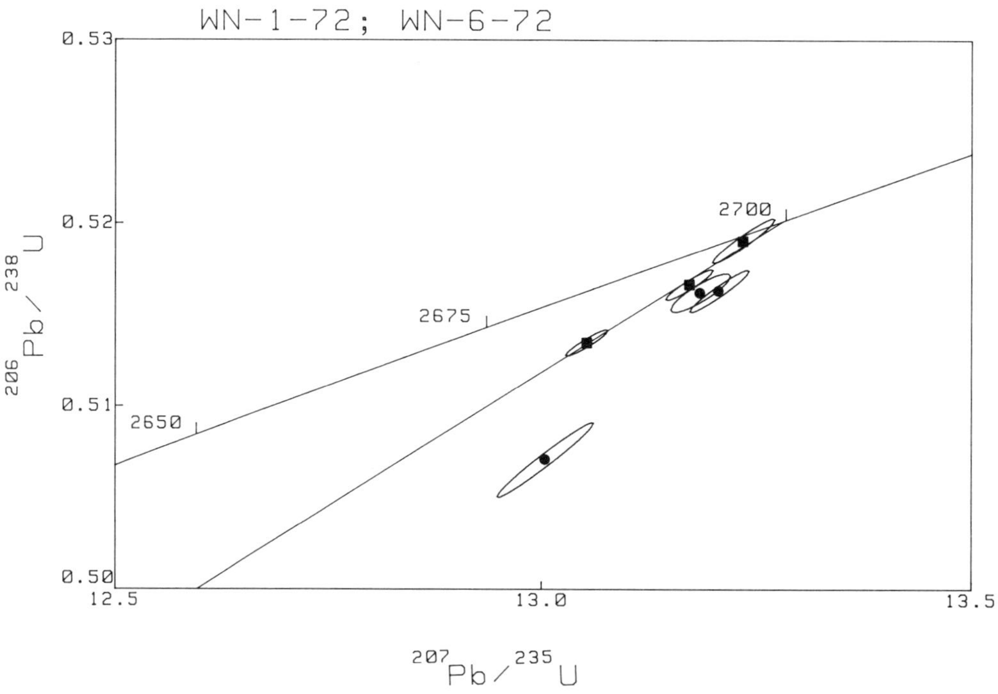

Figure 2. U-Pb concordia plot for zircons from sample WN-1-72 (circles) and WN-6-72 (squares). Regression line for WN-6-72 analyses plotted. Errors are shown at 2σ level.

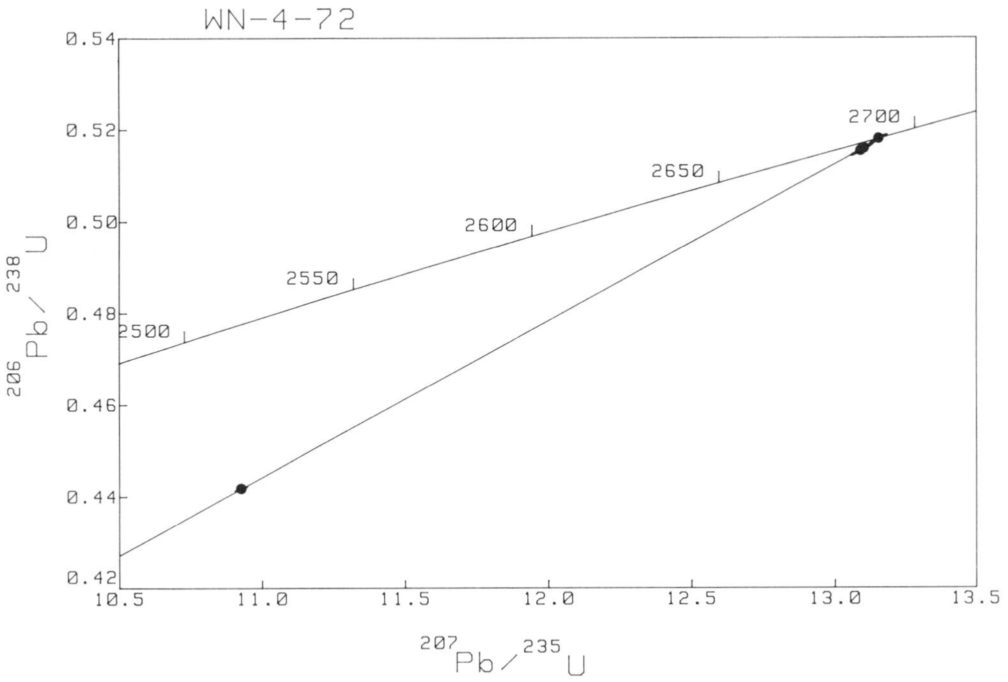

Figure 3. U-Pb concordia plot for zircons from sample WN-4-72. Errors are shown at 2σ level.

con fractions range from 0.1 to 18.1 % discordant. The calculated upper intercept age, which is taken as the crystallization age of the rock, is 2692.0 +1.0/-0.9 Ma. The calculated lower intercept of 589 Ma reflects relatively recent Pb loss.

DISCUSSION

U-Pb zircon data presented here provide the first precise ages for volcanism in the Rankin-Ennadai belt in the District of Keewatin, and give an indication of the duration of volcanism in the belt. The dated samples were from the middle and near the top of the stratigraphic sequence at Kaminak Lake (as recognized by Ridler and Shilts, 1974), and differ in age by 3 to 8 Ma (limits of 2σ errors on calculated ages). If the rhyolite unit from the Ennadai Group in northern Saskatchewan dated by Chiarenzelli and Macdonald (1986) is related to those studied at Kaminak Lake, volcanism in the Rankin-Ennadai belt may have persisted for as much as 23 Ma.

ACKNOWLEDGMENTS

Figure 1 was drafted by Nguyen Khanh Kim. The manuscript was improved as a result of critical reading by C.W. Jefferson and comments by F.C. Taylor.

REFERENCES

Bell, R.T.
1971: Geology of Henik Lakes (east half) and Ferguson Lake (east half) map-areas, District of Keewatin; Geological Survey of Canada, Paper 70-61, 31 p.

Cavell, P.A., Wijbrans, J.R. and Baadsgaard, H.
1987: Preliminary results of an isotopic study of the Kaminak Lake Archean terrane (N.W.T.); Geological Association of Canada, Program with Abstracts, 12, p. 29 (abstract).

Chiarenzelli, J.R. and Macdonald, R.
1986: A U-Pb zircon date for the Ennadai Group; *in* Summary of Investigations, 1986, Saskatchewan Geological Survey, Saskatchewan Energy and Mines, Miscellaneous Report 86-4, p. 112-113.

Davidson, A.
1970a: Precambrian geology, Kaminak Lake map-area, District of Keewatin; Geological Survey of Canada, Paper 69-51, 27 p.
1970b: Eskimo Point and Dawson Inlet map-areas (north halves), District of Keewatin; Geological Survey of Canada, Paper 70-27, 21 p.
1970c: Kaminak Lake alkalic complex, District of Keewatin; in Report of Activities, Part A, Geological Survey of Canada, Paper 70-1A, p. 135-137.
1970d: Plutonic complexes in the northeast part of the Ennadai-Rankin Inlet region; in Report of Activities, Part A, Geological Survey of Canada, Paper 70-1A, p. 133-134.
1972: The Churchill Province; *in* Variations in Tectonic Styles in Canada, Ed. R.A. Price and R.J.W. Douglas; Geological Association of Canada Special Paper 11, p. 382-433.

Franklin, J.M. and Thorpe, R.I.
1982: Comparative metallogeny of the Superior, Slave and Churchill provinces; *in* Precambrian Sulphide Deposits, Geological Association of Canada, Special Paper 25, p. 3-90.

Geological Survey of Canada
1980: Non-hydrocarbon mineral resource potential of parts of northern Canada; Geological Survey of Canada, Open File 716, 376 p.

Parrish, R.R., Roddick, J.C., Loveridge, W.D. and Sullivan, R.W.
1987: Uranium-lead analytical techniques at the geochronology laboratory; *in* Radiogenic Age and Isotopic Studies: Report 1; Geological Survey of Canada, Paper 87-2.

Ridler, R.H.
1971a: Relations of mineralization to stratigraphy in the Archean Rankin Inlet-Ennadai belt; Canadian Mining Journal, v. 92, no. 4, p. 50-53.
1971b: Volcanic stratigraphy and metallogeny of the Kaminak Group; *in* Report of Activities, Part A, Geological Survey of Canada, Paper 71-1A, p. 142-148.
1972: Volcanic stratigraphy and metallogeny of the Kaminak Group; *in* Report of Activities, Part A, Geological Survey of Canada, Paper 72-1A, p. 128-134.
1973: Volcanic stratigraphy and metallogeny, Rankin Inlet-Ennadai belt, District of Keewatin; *in* Report of Activities Part A, Geological Survey of Canada, Paper 73-1A, p. 165-174.
1974: Volcanic stratigraphy and metallogeny of the Kaminak Group, Spi Lake area, District of Keewatin; *in* Report of Activities, Part A, Geological Survey of Canada, Paper 74-1A, p. 181-185.

Ridler, R.H. and Shilts, W.W.
1974: Exploration for Archean polymetallic sulphide deposits in permafrost terranes: an integrated geological/geochemical technique; Geological Survey of Canada, Paper 73-34, 33 p.

Shilts, W.W.
1974: Drift prospecting in the Ennadai-Rankin Inlet greenstone belt, District of Keewatin; *in* Report of Activities, Part A, Geological Survey of Canada, Paper 74-1A, p. 259-261.

Wanless, R.K.
1979: Geochronology of Archean rocks of the Churchill Province; Geological Association of Canada, Program with Abstracts, v. 4, p. 85. (abstract).

Wright, G.M.
1967: Geology of the southeastern barren grounds, parts of the districts of Mackenzie and Keewatin; Geological Survey of Canada, Memoir 350, 91 p.

U-Pb zircon and sphene geochronology of Archean plutonic and orthogneissic rocks of the James Bay region and Bienville Domain, Quebec.

J.K. Mortensen[1] and A. Ciesielski[1]

Mortensen, J.K. and Ciesielski, A., U-Pb zircon and sphene geochronology of Archean plutonic and orthogneissic rocks of the James Bay region and Bienville Domain, Quebec; in Radiogenic Age and Isotopic Studies: Report 1, Geological Survey of Canada, Paper 87-2, p. 129-134, 1987.

Abstract

U-Pb zircon dates have been obtained on 4 samples of plutonic and orthogneissic Archean rocks from the east of James Bay, Quebec. Tonalitic gneiss from the northern part of the volcano-sedimentary belt of La Grande River has an age of 2811 ± 2 Ma. A late tectonic granodiorite pluton which intrudes the contact between the James Bay region to the south and the Bienville Domain to the north, yields an age of 2712 ± 3 Ma. Two samples of gneiss belonging to the Bienville Domain yield preliminary U-Pb zircon ages of about 2800 Ma; these have relatively large errors and may be correlative or synchronous with the pre-tectonic tonalitic gneisses from the La Grande belt farther south. U-Pb sphene ages from all four samples are younger than co-existing zircons and appear to reflect a range of cooling ages that may suggest progressive uplift to the north along major east-trending faults of uncertain age.

Résumé

Des datations à l'U-Pb (zircon et sphène) ont été effectuées sur quatre échantillons de roches plutoniques et orthogneissiques d'âge archéen provenant du côté est de la baie James au Québec. Les gneiss tonalitiques provenant du secteur nord de la zone volcano-sédimentaire de la rivière La Grande ont donné un âge de 2811 ± 2 Ma. Le pluton tardi-tectonique de granodiorite qui a pénétré le contact entre la région de la baie James au sud et le domaine de Bienville au nord donne un âge de 2712 ± 3 Ma. Deux échantillons de gneiss appartenant au domaine de Bienville au nord, donnent des âges aux environs de 2800 Ma. Bien que la marge d'erreur des âges soit grande, ces roches sont considérées comme synchrones des gneiss pré-tectoniques de la zone de la rivière La Grande. Les âges U-Pb (sphène) établis pour les quatre échantillons présentent des écarts appréciables et reflètent un refroidissement ainsi qu'une remontée progressive du socle, peut-être le long de failles est-ouest dont l'âge demeure indéterminé.

GEOLOGICAL SETTING

Archean basement on the east side of James Bay is a part of the Superior Structural Province and comprises orthogneiss, plutonic and metavolcanic rocks and the (meta)sedimentary basin of the Opinaca River (Fig. 1). Two different broad geological entities are present in the study area. To the north, the Bienville Domain is mostly orthogneissic. It covers a surface area of several thousand square kilometres. To the south, the James Bay geological region is composed of volcanic and sedimentary belts surrounding two large (meta) sedimentary basins, and plutonic and orthogneissic belts. The contact between the Bienville domain and the James Bay geological region is intruded by large late-tectonic granodioritic intrusions (Fig. 1).

The U-Pb dating study addressed three different types of orthogneiss.

A. Tonalitic orthogneisses immediately north of La Grande volcanic belt show a compositional spread in the trondhjemite and tonalite field. They are believed to be pre-tectonic and may be older than the volcanic rocks nearby.
B. Porphyritic granodiorite intrudes both the Bienville gneisses to the north and the volcanic rocks and gneisses of the James Bay region to the south.
C. Bienville gneisses range from tonalite to granite and consist mainly of an early tonalitic phase (blue phase) and a late granodioritic phase (pink phase).

The Bienville gneisses are mainly at amphibolite facies except for an area at granulite facies north of James Bay (Fig. 1). To the south, the granodiorite carries many amphibolite inclusions and may itself have reached the same metamorphic conditions. Farther south, the orthogneiss in contact with the metavolcanic rocks contains many metabasic

[1] Geological Survey of Canada, 601 Booth Street, Ottawa, Ontario K1A 0E8.

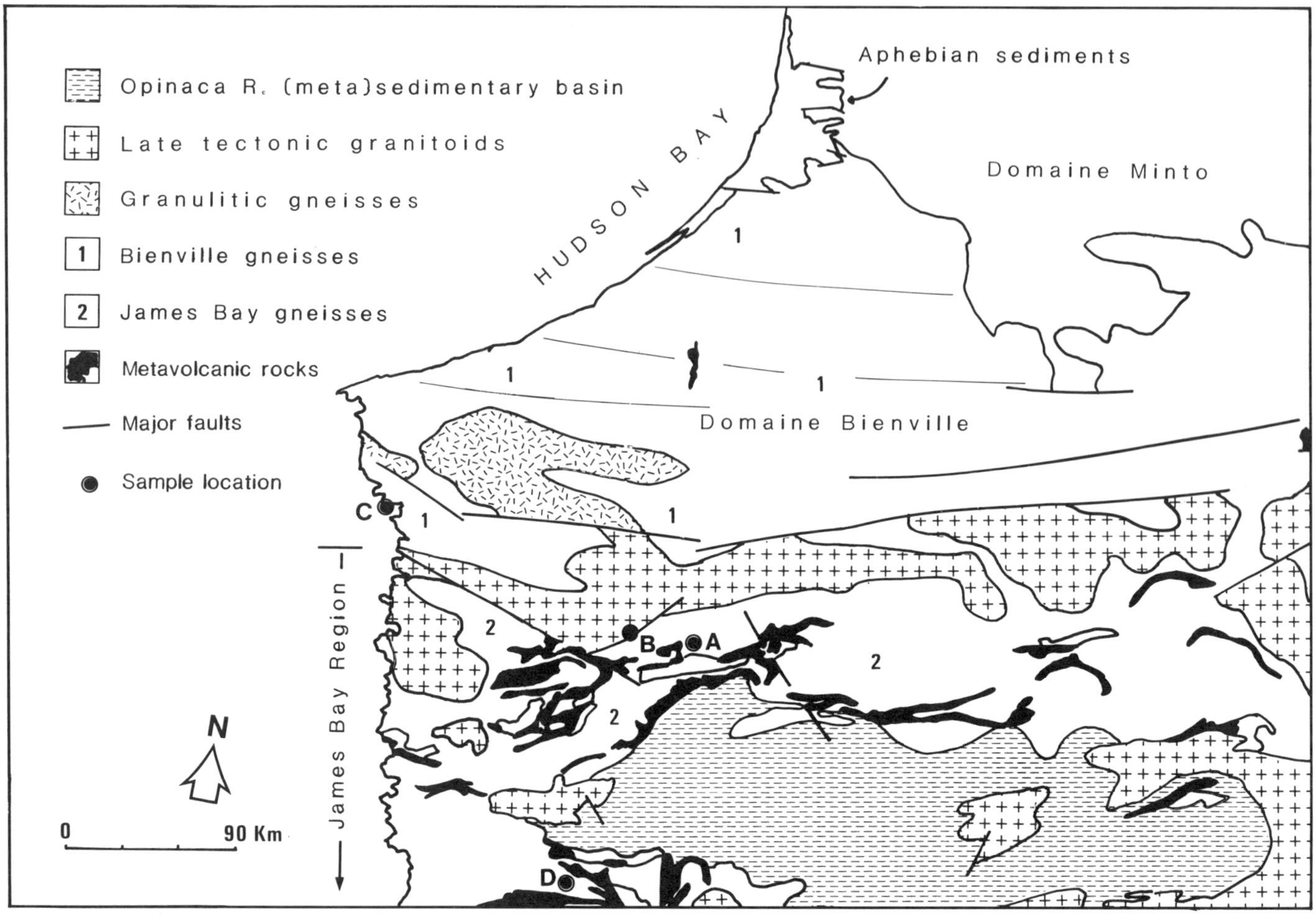

Figure 1. Schematic geological map of the study area, showing the boundary between (1) the volcano-sedimentary, gneissic and plutonic regions to the south, and (2) the dominantly gneissic domains to the north. This lithological or domain boundary is intruded by a late tectonic porphyritic granodiorite. D is the location of the Duxbury massif (Gauthier et al., 1982; Verpaelst et al., 1980). The map was modified after Card and Ciesielski (1986).

and metasedimentary inclusions. These are also believed to have been metamorphosed in the amphibolite facies. The associated volcanic rocks are at greenschist metamorphic facies, with the contact zone with plutons or gneisses of amphibolite facies. The rocks from the Opinaca River (meta) sedimentary basin farther south are in upper amphibolite metamorphic facies.

ANALYTICAL RESULTS

U-Pb analytical methods follow those outlined for zircon and sphene by Parrish et al. (1987). All zircon fractions have been abraded by the method of Krogh (1982). Data is presented in Table 1.

A. Tonalitic gneiss

(Leucocratic, recrystallized, fine-to medium-grained tonalite)

Zircons recovered from the sample are stubby to elongate, perfectly euhedral and range from almost colorless to pale pinkish brown. Polished and etched grain mounts of the zircon reveal strong internal igneous zonation with no evidence of inherited cores. Sphene recovered from the sample consists of broken fragments that are clear and pale yellow brown, with rare large flat crystal facets.

Five zircon fractions form a roughly linear array with some scatter at the lower end (Fig. 2). Two of the analyses, however, are concordant. One is relatively imprecise with a $^{207}Pb/^{206}Pb$ age of 2813.0 ± 7.4 Ma. The second concordant fraction, yields an identical $^{207}Pb/^{206}Pb$ age within the error limits but with a smaller associated uncertainty (2811.4 ± 2.4 Ma.) This is interpreted as the crystallization age of the tonalitic protolith for the gneiss. The discordance displayed by other zircon fractions is attributed to a possibly complex history of post-crystallization Pb-loss.

One sphene analysis from the gneiss (Fig. 2) plots on the zircon best-fit line. Since the sphene likely experienced the same Pb-loss history as the zircon, this implies that the sphene cooled through its closure temperature for the U-Pb system (approximately 600°C) at essentially the time of crystallization of the zircon, and has not been heated above this temperature since.

B. Porphyritic Granodiorite

(Massive equigranular, medium-to coarse-grained hornblende-biotite K-geldspar prophyritic granodiorite)

Zircons recovered from this sample are generally similar in appearance to those in the tonalitic gneiss. Polished and etched grain mounts reveal complex igneous zoning with no evidence for inherited cores. Sphene forms coarse euhedral tablets with rare fine opaque inclusions. The sphene is clear, unfractured, and pale yellow to yellow-brown.

Five fractions of zircon form a well-defined linear array with calculated upper and lower intercept ages of 2712.0 + 3.2/-2.3 Ma and 847.0 Ma (Fig. 3). The upper intercept is interpreted as the crystallization (emplacement) age of the pluton, and the lower intercept as an average age of Pb-loss from the zircons.

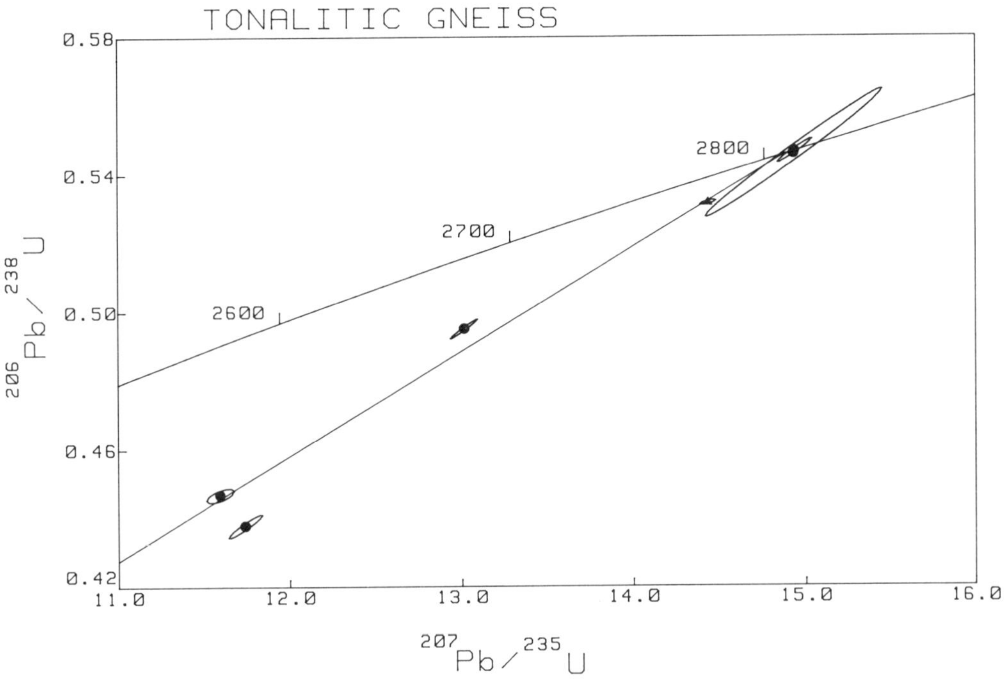

Figure 2. U-Pb concordia plot for zircon (circles) and sphene (triangle) from the tonalitic orthogneiss of the James Bay region, sample A. Errors are shown at the two sigma level.

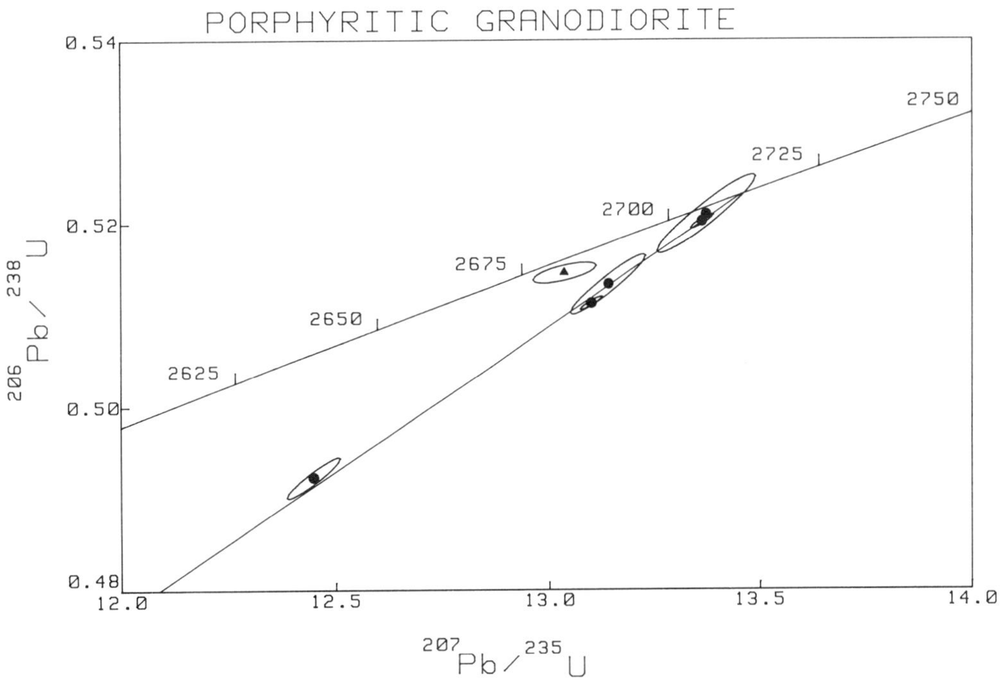

Figure 3. U-Pb concordia plot for zircon (circles) and sphene (triangles) from porphyritic granodiorite, sample B. Errors are shown at the two sigma level.

A single unabraded fraction of sphene from this sample was analyzed. The analysis is slightly discordant, and falls well above the discordia line defined by the zircons. Assuming that the sphene experienced the same relatively recent Pb-loss as did the zircon, the $^{207}Pb/^{206}Pb$ age of the sphene (2687 ± 8 Ma) is a good approximation of the cooling age for the sphene. This implies either that the intrusion cooled through the closure temperature of sphene (approximately 600°C) extremely slowly (over a period of about 25 Ma) following emplacement, or else the body cooled and was subsequently re-heated to above 600°C during some later thermal event.

C. Bienville gneiss.

Two compositionally distinct phases of this unit were studied. One is a well foliated equigranular, medium grained gneiss of bioitite-hornblende tonalitic composition (blue phase), while the other is a less foliated biotite granodiorite (pink phase). In the field, the two phases are in sharp but irregular contact; the pink phase is believed to be later.

Zircons recovered from the tonalitic phase are stubby to elongate, and display a considerable variety of crystal shapes, ranging from grains with square cross-sections and simple pyramidal terminations to multifaceted grains with rounded terminations. Many grains have smoothly rounded surfaces without visible facets. The zircons ranges from pale pink to deep reddish brown. The central regions of some grains appear slightly cloudy, and were initially thought to represent inherited cores; however, examination of polished and etched grain mounts reveals that these central regions display regular igneous zoning that is continuous with that in the bulk of the grains. The cloudiness is therefore probably due to more abundant fine inclusions and/or more extensive alteration and metamictization. Sphene from this sample occurs as flat euhedral tablets that are clear and yellow brown.

Zircons in the granodioritic pink phase are similar in appearance to those in the tonalitic phase except that multifaceted crystals are less abundant, and the grains have more abundant clear inclusions (clear tubes and bubbles). Sphene from the granodioritic phase occurs as clear, yellow-brown euhedral tablets.

Table 1. U-Pb zircon and sphene analytical data

Sample and zircon fraction	Weight (mg)	U (ppm)	Pb* (ppm)	Measured $^{206}Pb/^{204}Pb$	Isotopic abundances (^{206}Pb=100) ^{204}Pb	^{207}Pb	^{208}Pb	Isotopic ratios $^{206}Pb/^{238}U$	$^{207}Pb/^{235}U$	Age, Ma $^{207}Pb/^{206}Pb$
A: TONALITIC GNEISS										
+149, N1, A	0.0415	200.6	109.1	162.1	0.0450	20.05	10.32	0.49498	13.0126	2747.9
−149+105, N1, A	0.0312	24.8	15.1	556.8	0.1112	21.15	13.25	0.54669	14.9394	2811.4
−62, N1	0.0429	294.4	150.6	327.8	0.2938	22.38	23.39	0.44678	11.5944	2726.6
−149+105, N2, A	0.0320	38.2	23.3	298.7	0.2394	22.70	17.84	0.54601	14.9352	2813.0
+149, M2, A	0.0480	119.6	58.9	612.7	0.1367	21.09	15.67	0.43785	11.7404	2780.4
Sphene	0.315	74.1	47.7	645.3	0.1519	21.51	25.53	0.53162	14.4351	2800.9
B: PORPHYRITIC GRANODIORITE										
+105, N1, A	0.0312	147.6	94.3	2974	0.0000	18.89	23.33	0.52015	13.3622	2709.9
+105, N1, A (dup)	0.0381	195.4	120.2	748.9	0.1159	19.979	24.04	0.51329	13.1398	2704.1
−62, N1	0.0402	294.7	176.2	1119	0.0777	19.29	25.12	0.49227	12.4484	2683.9
+105, N2, A	0.0271	236.8	145.8	4115	0.0151	18.77	21.77	0.51124	13.0993	2705.6
+149, M2, A	0.0389	140.3	87.0	572.7	0.1512	20.45	24.23	0.52096	13.3734	2708.7
Sphene	0.3020	22.7	16.3	334.8	0.2897	21.92	51.34	0.51455	13.0365	2687.0
C: BIENVILLE GNEISS (tonalite)										
+177, N1, A, clr	0.2560	284.5	143.8	1991	0.0473	18.45	8.23	0.47170	11.6248	2641.2
+177, N1, A, brn	0.0386	433.4	235.1	1262	0.0709	19.09	12.21	0.49175	12.3566	2673.4
−62, N1, A	0.0420	3918.5	1775.0	5067	0.0183	17.03	5.49	0.43275	10.0281	2538.5
+149, M2, A	0.1690	1229.8	554.2	3403	0.0283	16.85	5.87	0.43147	9.8122	2506.9
Sphene	0.1640	43.8	26.1	509.1	0.1873	19.89	26.49	0.49953	12.116	2614.1
C: BIENVILLE GNEISS (granodiorite)										
+105, N1, A	0.0314	395.5	211.1	9182	0.0060	18.41	7.35	0.49445	12.4982	2683.2
−105+74, N1, A	0.112	460.5	240.9	6032	0.0153	18.22	7.68	0.48561	12.0772	2656.3
+105, N2, A	0.0419	443.1	233.3	9778	0.0069	18.30	8.69	0.48329	12.1360	2672.3
+149, M2, A	0.0421	1084.2	522.4	1764	0.0553	17.83	7.31	0.45676	10.7974	2571.8
Sphene	0.3410	53.2	42.0	582.6	0.1675	19.98	67.16	0.50495	12.4738	2645.1

1) N, M = non magnetic, magnetic at number of degrees side tilt on Frantz separator; A = abraded.
2) All zircon and sphene analyses corrected using an assumed composition for initial common Pb. (*see* Parrish et al., 1987)

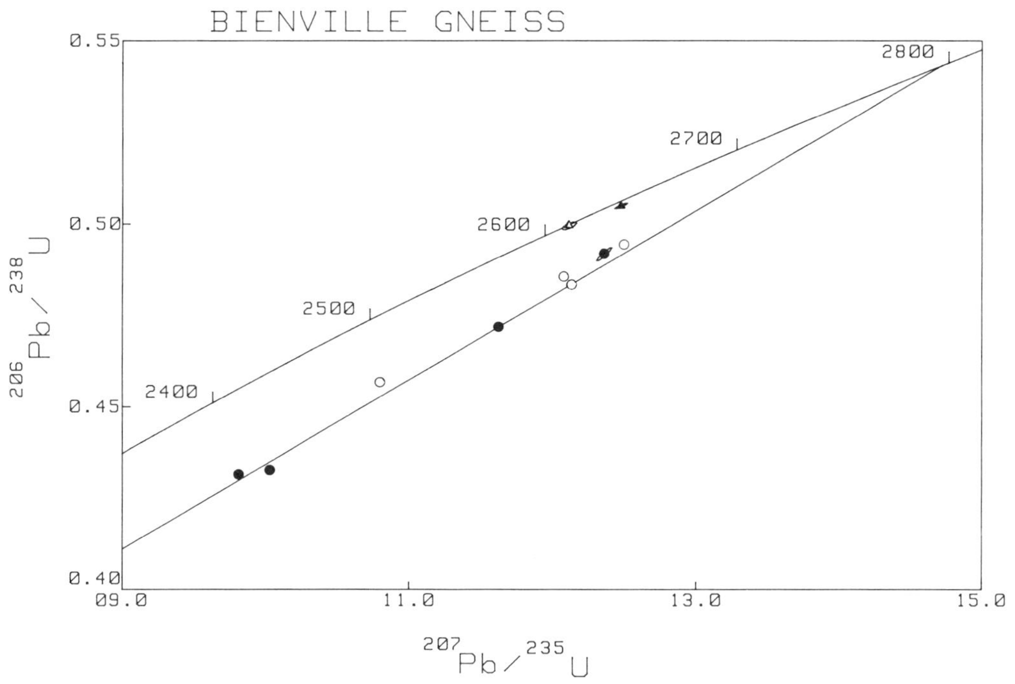

Figure 4. U-Pb concordia plot for zircon (circles) and sphene (triangles) from two phases of the Bienville gneiss. Analyses from the tonalitic phase are shown in solid, and those from the granodioritic phase as open symbols. Errors are shown at the two sigma level. Where not shown, errors ellipses are smaller than the symbols used.

Four fractions of zircon and one fraction of sphene were analyzed from each of the two phases. Zircons from both samples are high in uranium and quite discordant. Analyses for each sample form arrays with considerable scatter (Fig. 4). Upper intercept ages for best-fit regression lines through zircon data sets for the tonalitic and granodioritiic phases are 2797 + 122/-80 Ma and 2819 + 170/-80 Ma, respectively. Although the discordant nature of the zircon data precludes a more accurate estimate of emplacement age for the two samples, the calculated upper intercept ages suggest a possible correlation with the 2811 Ma tonalitic gneiss body farther south. Additional zircon analyses will be required to constrain the crystallization age of the Bienville gneiss.

Sphene from the tonalitic phase of the James Bay gneiss is essentially concordant at 2614 ± 5 Ma, while sphene from the granodioritic phase is slightly discordant and yields a considerably older $^{207}Pb/^{206}Pb$ age (Fig. 4). This date suggests that the sphene is probably igneous in origin, and that the tonalitic phase cooled through the sphene closure temperature at about 2615 Ma.

DISCUSSION

There are very limited reliable age data available from this part of Quebec. Gauthier (1981) and Gauthier et al. (1982) report U-Pb zircon ages of 2709 ± 2 Ma and 2708 ± 13 Ma for tonalitic and granodioritic phases (respectively) of the Duxbury Massif in the James Bay region (Fig. 1), and 2728 ± 3 Ma for gneissic tonalitic inclusions within the massif. It is probable that the porphyritic granodiorite pluton dated in this study (2712 + 3.2/ −2.3 Ma) is correlative with such plutonic phases found in the gneiss terrane of the James Bay region.

The U-Pb sphene ages obtained in this study have interesting implications for the history of metamorphism and subsequent uplift and cooling of the area. The most southerly sample dated (A, tonalitic gneiss), last cooled through approximately 600°C immediately following intrusion at about 2811 Ma. The porphyritic granodiorite (sample B), on the other hand, appears to have cooled through approximately 600°C about 25 Ma after intrusion. Since this body is mainly massive and late tectonic (there is no evidence for strong metamorphism after intrusion) the sphene age implies slow cooling of the body, presumably due to emplacement at considerable depth. Sample locations A and B are only 15 km apart but are separated by a major fault (the former Kanaaupscow River); the sense and magnitude of displacement of this fault is unknown, however the U-Pb cooling age from the sphene may indicate a significant northwest-side-up component. Sphenes from the Bienville gneiss record even younger cooling ages for sphene (as young as 2614 Ma), suggesting an even more protracted cooling (or reheating) history. Major east-trending faults present between sample locations B and C (Fig. 1), may also be north-side-up.

The regional metamorphism that affected the study area has not been directly dated. In the James Bay and Ungava region, the metamorphism is assumed to be Kenoran (2700 Ma) by analogy with the well documented metamorphism affecting the southern part of the Superior Province. Considering the late tectonic position of the porphyritic granodiorite (2712 Ma) and the post or late deformation growth of metamorphic minerals in the volcano-sedimentary belt of La Grande River, we assume the Kenoran Orogen in the study area to have culminated around 2700 Ma with slow uplift and post metamorphic cooling through approximately 600°C as late as 2615 Ma.

REFERENCES

Card, K.D., and Ciesielski, A.
1986: Subdivisions of the Superior Province of the Canadian Shield, Geoscience Canada, v. 13, no. 1, p. 5-13

Gauthier, G.
1981: Application de la méthode de datation Uranium/Plomb aux zircons du Massif Duxbury; mémoire de Maîtrise inédit, Université de Montréal, 149 p.

Gauthier, G., Brooks, C. and Krogh, T.E.
1982: The Duxbury Massif: a test case for dating a late Archean intrusion; Geological Association of Canada, Program with Abstracts, v. 7, p. 51

Krogh, T.E.
1982: Improved accuracy of U-Pb ages by the creation of more concordant systems using an air abrasion techniques; Geochimica et Cosmochimica Acta, v. 46, pp. 637-649

Parrish, R.R., Roddick, J.C., Loveridge, W.D. and Sullivan, R.W.
1987: U-Pb analytical techniques at the geochronology laboratory, Geological Survey of Canada; Geological Survey of Canada, Paper 87-2.

Verpaelst, P., Brooks, C. and Franconi, A.
1980: The 2.5 Ga Duxbury Massif, Quebec: a remobilized piece of pre-3.0 Ga sialic basement (?); Canadian Journal of Earth Science, v. 17, no. 1, p. 1-18.

Reconnaissance U-Pb zircon and monazite geochronology of the Lac Clairambault area, Ashuanipi complex, Quebec

J.K. Mortensen[1] and J.A. Percival[1]

Mortensen, J.K. and Percival, J.A., Reconnaissance U-Pb zircon and monazite geochronology of the Lac Clairambault area, Ashuanipi complex, Quebec; in Radiogenic Age and Isotopic Studies: Report 1, Geological Survey of Canada, Paper 87-2, p. 135-142, 1987.

Abstract

The Ashuanipi complex in northern Quebec consists mainly of granulite facies metasediments and metaplutonic rocks that have locally been retrograded to amphibolite facies. Zircon and monazite from five samples of the metamorphic rocks and late pegmatite from the Lac Clairambault area have been dated by U-Pb methods as part of a reconnaissance study of the geology of the complex. Metatonalite yields an emplacement age of 2692 ± 9 Ma. Metamorphic zircon in diatexite and retrograded diatexite is dated at about 2667 ± 1 Ma and a maximum of 2642 ± 3 Ma, respectively. A minimum age of 2786 Ma was obtained for detrital zircon in metasedimentary rocks. Crosscutting pegmatite is dated at 2654 ± 5 Ma. Monazite in the metamorphic rocks ranges in age from 2661 to 2668 Ma, reflecting cooling ages after peak granulite metamorphism.

These data suggest that felsic plutonism occurred during granulite facies metamorphism between about 2668 and 2649 Ma. Localized retrogression occurred after 2642 Ma. Age relationships and lithological associations in the Ashuanipi complex suggest correlation with the Quetico Subprovince in the western Superior Province.

Résumé

Le complexe d'Ashuanipi, dans le nord du Québec, comprend principalement des roches métaplutoniques et métasédimentaires à faciès des granulites qui ont été par endroits rétromorphosées dans le faciès des amphibolites. Du zircon et de la monazite provenant de cinq échantillons de roches métamorphiques et d'une pegmatite tardive prélevés dans le secteur du lac Clairambault, ont été datés par les méthodes U-Pb dans le cadre d'une étude de reconnaissance de la géologie du complexe. On a évalué l'âge de la mise en place de la métatonalite à 2692 ± 9 Ma. Le zircon métamorphique de la diatexite et de la diatexite rétromorphosée, a été respectivement daté à 2667 ± 1 Ma environ et à un maximum de 2642 ± 3 Ma. On a obtenu un âge minimum de 2786 Ma pour le zircon détritique de roches métasédimentaires. On a évalué l'âge d'une pegmatite traversant ces roches à 2654 ± 5 Ma. L'âge de la monazite des roches métamorphiques varie entre 2661 et 2668 Ma; il correspond, en outre, aux dates du refroidissement suivant le pic du métamorphisme associé au faciès des granulites.

Ces données semblent indiquer que le plutonisme felsique a eu lieu pendant l'épisode de métamorphisme associé au faciès des granulites, il y a 2268 à 2649 Ma environ. La rétromorphose localisée remonte à moins de 2642 Ma. Les relations d'âge et associations lithologiques caractérisant le complexe d'Ashuanipi suggèrent qu'une corrélation est possible avec la sous-province de Quetico dans la partie ouest de la province du lac Supérieur.

INTRODUCTION

Over the past decade the geochronology of Archean greenstone belts in the Superior Province has become increasingly refined. No less important in interpretation of the evolution of the Superior Province on a regional scale are the poorly known gneissic and plutonic regions. In order to understand how these high grade rocks relate to greenstone terranes, timing constraints are necessary. U-Pb zircon and monazite geochronology is a tool ideally suited to the study of felsic igneous terranes and has the potential for allowing regional correlation.

The Ashuanipi complex of the eastern Superior Province is a high grade gneiss region of some 300 by 300 km. Its origin with respect to the low grade easterly-trending belts

[1] Geological Survey of Canada, 601 Booth Street, Ottawa, Ontario K1A 0E8

of the western Superior Province is problematic. It may represent either a high grade basement to the belts, or alternatively, their highly metamorphosed extensions.

A reconnaissance survey in summer 1985 in the Lac Clairambault area returned a suite of five samples for geochronological work. The samples are representative of a spectrum of rock types ranging from tonalite to granite and from pre- to post-tectonic. In this paper, we report preliminary U-Pb zircon and monazite age data for these five samples and interpret these data in the light of known field and petrological relationships.

SUPERIOR PROVINCE

The Superior Province has been divided into four Archean "terrane types" by Card and Ciesielski (1986; Fig. 1):

1) volcano-plutonic
2) plutonic
3) metasedimentary
4) high grade gneiss

A large part of the Superior Province of Quebec is made up of high grade gneiss terranes, the Minto and Ashuanipi complexes (Fig. 1). Although both complexes are characterized by gneisses in the amphibolite and granulite facies, structural trends are different. The Minto block has a prominent northerly structural and aeromagnetic grain whereas the Ashuanipi complex has west and northwesterly trends. The two are separated by the Bienville plutonic domain (Fig. 1; Mortensen and Ciesielski, 1987).

ASHUANIPI COMPLEX

Most of the complex has been mapped at a scale of 1:500 000 (Eade, 1966; Stevenson, 1963) although recent maps of selected areas are available at 1:125 000 (Baragar, 1967; Percival, 1987), 1:100 000 (Sharma and Dubé, 1980) and 1:50 000 (Lapointe, 1986). Interest in the potential for gold mineralization has sparked exploration activity in the western part of the complex (Thomas and Butler, 1987).

A metamorphic map of New Quebec, including the Ashuanipi complex, was compiled by Herd (1978). Most of the complex is at granulite facies, with several patches of amphibolite facies overprint (Fig. 1).

In the Schefferville area, most of the granulites are of sedimentary origin, both metasedimentary migmatites and intrusive garnet-orthopyroxene diatexites (Percival, 1987). Early tonalites occur as abundant sills within the migmatites and later tonalites as irregular-shaped plutons. Late massive granite and syenite occur mainly as discrete oval plutons. On the basis of similarity of lithology, structural trend and

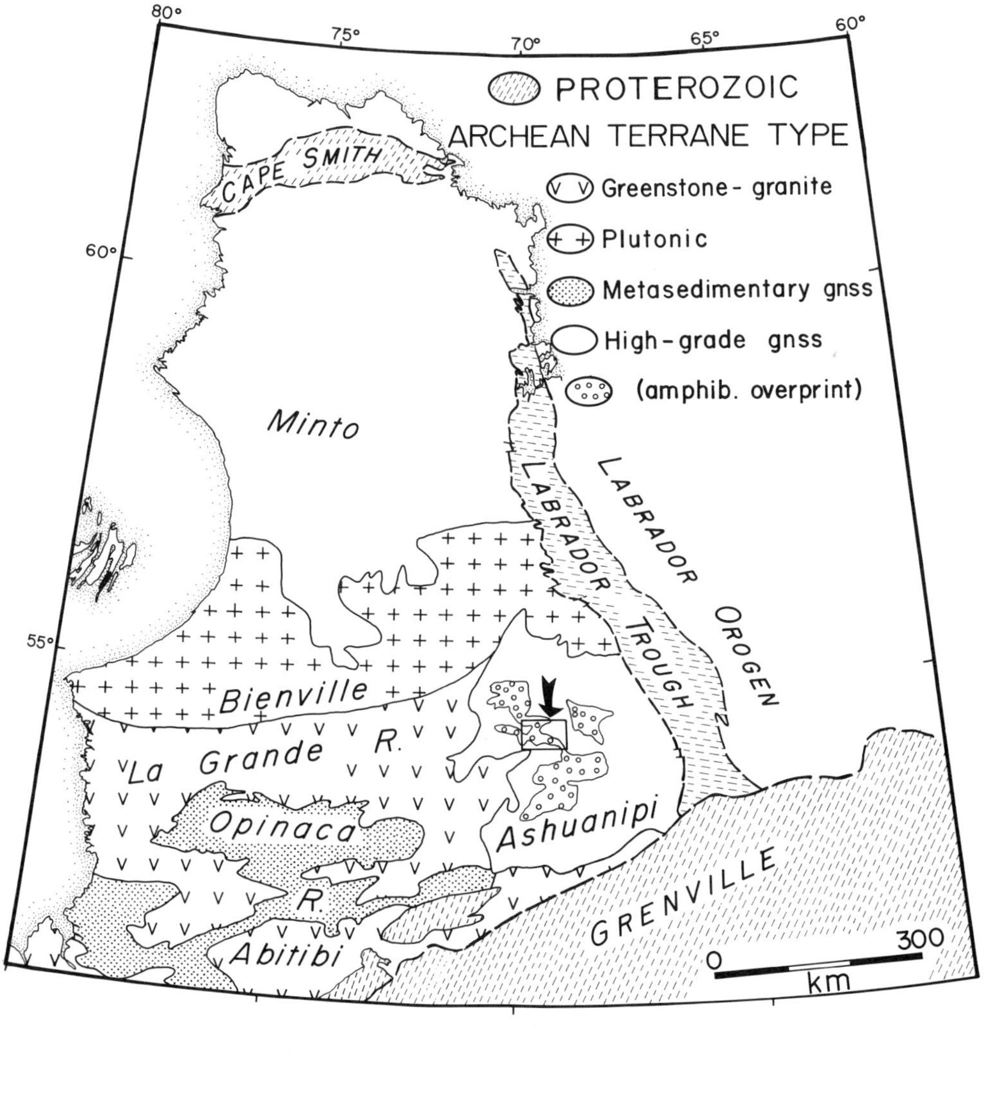

Figure 1. Generalized geology of northern Quebec, showing subdivisions of the eastern Superior Province after Card and Ciesielski (1986). Retrograde portions of the Ashuanipi complex (amphibolite facies overprint) after Herd (1978).

style, the Ashuanipi complex of the Schefferville area was correlated (Percival, 1987) with the metasedimentary gneiss belts of the western Superior Province.

LAC CLAIRAMBAULT REGION

Two mappable units were recognized in the Lac Clairambault area by Sharma and Dubé (1980): 1) metasedimentary rocks, and 2) a quartz diorite to granodiorite suite. In this study, the ''metasedimentary'' rocks are divided into two types of orthopyroxene-bearing diatexite, one with garnet and the other without. The quartz diorite-granodiorite unit is retained, and an additional rock type, massive to gneissic granite, is recognized.

The oldest recognizable rock type is garnet-biotite ± orthopyroxene paragneiss, which occurs as fist- to outcrop-sized inclusions in diatexite. The migmatitic rock contains up to 20% white coarse grained granitic leucosome and is cut by abundant pink and white pegmatite. Thin units of amphibolite and ultramafic rock also occur within diatexite and may represent early intrusions. Tonalite occurs as foliated to gneissic, medium grained clinopyroxene-orthopyroxene-hornblende-biotite rocks, cut by coarse grained pink granitic pegmatite.

Large parts of the region are underlain by diatexite (as defined by Brown, 1973): coarse grained, equigranular orthopyroxene-biotite-plagioclase-quartz-K feldspar ± garnet rocks. Garnet and non-garnet-bearing varieties are distinguished on Figure 2; both types contain paragneiss inclusions. Massive and gneissic textures are common in diatexite. Gneissic varieties have up to 20% diffuse, white leucocratic layers on the 1-5 cm scale and are cut by leucocratic pegmatites. The Lac Clairambault diatexites strongly resemble type-area charnockites from India (Pichamuthu, 1966). Similar rocks in the Schefferville area to the east form plutons intrusive into the gneissic sequence (Percival, 1987).

A large irregular body of pink leucocratic biotite granite occupies the southwest part of the map area (Fig. 2). It contains numerous inclusions of grey biotite schist and has pegmatitic phases and dykes. Most is massive, although gneissic textures are recognized locally (Fig. 2).

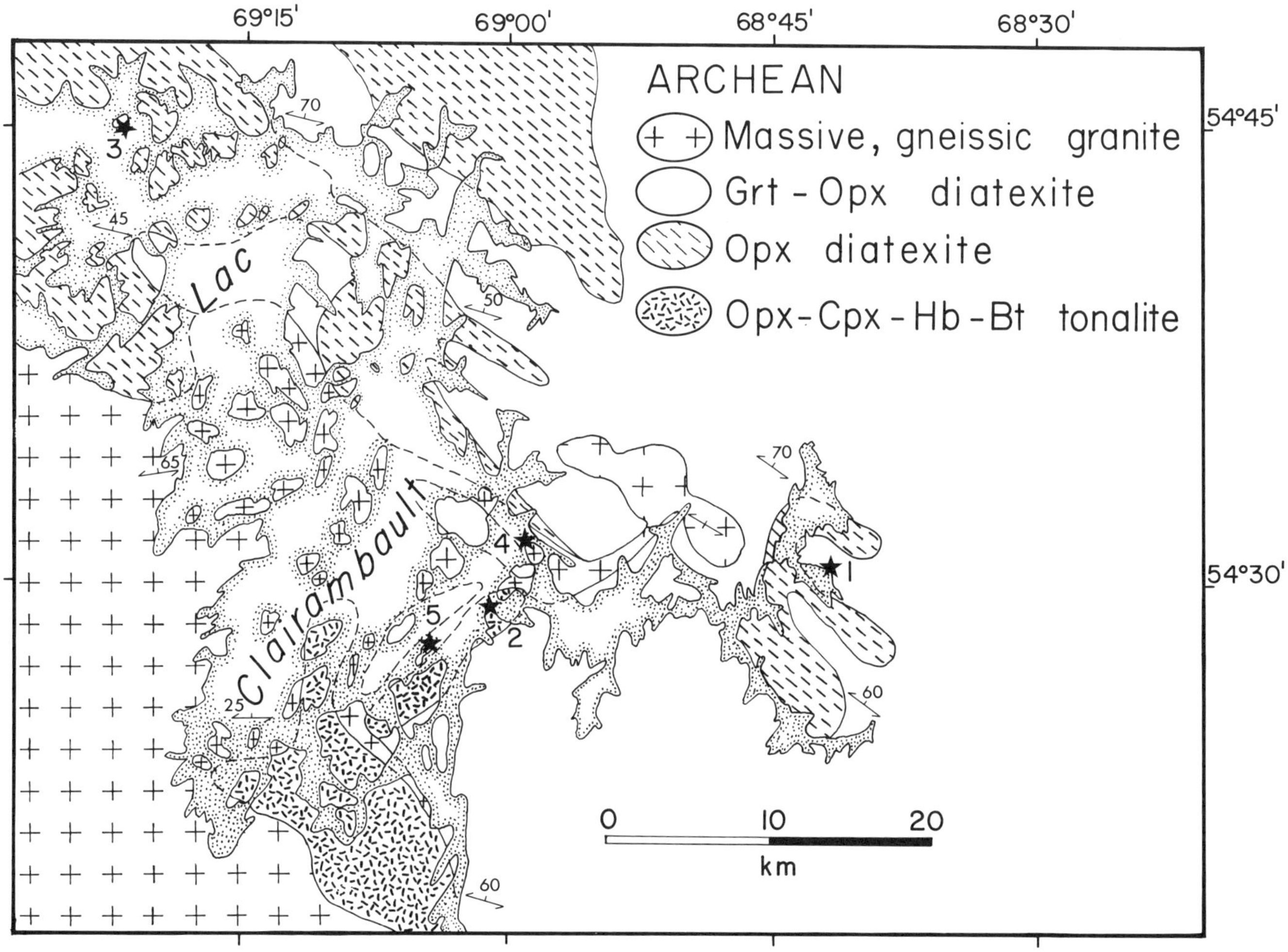

Figure 2. Generalized geological map of the Lac Clairambault area showing location of geochronology sampling sites (stars). Lake topography reflects the 5 m rise in water level caused by damming.

The retrogressed-granulite zone of Herd (1978) extends through the area. Small (1 km^2) retrograde patches were noted, associated with oval aeromagnetic highs. Although the transition from granulite to amphibolite was not observed, several transitional steps indicate that the progression is indeed retrograde rather than prograde. In the granulite facies, diatexite consists of dark grey, orthopyroxene-biotite-bearing matrix and inclusions, with white leucosome. The inferred retrogressed equivalent is a light pink or grey granodiorite with coarse biotite clots and some magnetite, fine grained grey biotite schist inclusions, and pink pegmatitic leucosome. Orthopyroxene relics are present locally in the cores of biotite clots in granodiorite and in the cores of inclusions. The retrogressive process included static addition of water and involved oxidation, judging by the pink colour and magnetite content of the retrograde rocks.

Structural trends are generally northwesterly; foliation dips moderately to the northeast. Lineations are rare. Much of the area is underlain by massive rocks without preferred orientation. A few small (2 m) fresh tholeiitic mafic dykes trend approximately 060°.

Five samples were chosen to represent the lithological spectrum in the Lac Clairambault area:

1) leucosome from migmatitc paragneiss;
2) foliated clinopyroxene-orthopyroxene-hornblende-biotite tonalite;
3) massive orthopyroxene-biotite diatexite;
4) retrogressed diatexite;
5) leuco-pegmatite.

SAMPLE DESCRIPTIONS AND PETROGRAPHY

1) Paragneiss leucosome. This sample was painstakingly extracted from a heterogeneous outcrop containing about 10% irregular layers of leucosome on the 1-5 cm scale. The paleosome is a foliated medium grained garnet-orthopyroxene-biotite-plagioclase-quartz schist, and the leucosome, a massive white, coarse grained plagioclase-quartz-K feldspar ± garnet ± orthopyroxene ± biotite granite. Small septa of biotite-rich paleosome could not be separated completely from the sample. Strongly pleochroic orthopyroxene is slightly altered to biotite and chlorite and the feldspars have minor sericitic alteration. Quartz has weak undulose extinction.

2) Foliated clinopyroxene-orthopyroxene-hornblende-biotite tonalite. This rock is a homogeneous, medium grained, foliated tonalite made up of clinopyroxene (5%), orthopyroxene (2%), hornblende (10%), biotite (7%), plagioclase (45%), quartz (30%), minor opaques and myrmeckite, and trace amounts of apatite, zircon and monazite. Ragged to subhedral pale green clinopyroxene contains inclusions of biotite, hornblende and plagioclase. Blocky, fresh orthopyroxene is nonpleochroic. Brown biotite is aligned in a foliation, parallel to aligned green hornblende. Twinned plagioclase has only incipient saussuritic alteration; quartz has weak undulose extinction. Zircon occurs as inclusions in biotite and plagioclase.

3) Orthopyroxene-biotite diatexite. This rock is coarse grained, massive, homogeneous and dark grey to green. Strongly pleochroic subhedral orthopyroxene to 8 mm (8%) is mildly to severely altered along cleavage fractures and grain boundaries to green biotite and chlorite (3%). Pleochroic green subhedral biotite (7%) also occurs as discrete, randomly oriented grains to 5 mm. Plagioclase (25%) and K feldspar (35%) have patchy sericitic alteration. Quartz (20%) has weak undulose extinction. Interstitial myrmeckite makes up about 1% of the rock. Medium grained apatite and opaques are associated with orthopyroxene. Traces of zircon and monazite occur as inclusions in biotite and plagioclase.

4) Retrogressed diatexite. In isolation this rock would probably be identified as a biotite granodiorite. However, in light of the evidence of regional retrogression and relict orthopyroxene in the outcrop, the unit is interpreted as a down-grade equivalent of sample 3. It is a massive, pink-grey, homogeneous, medium- to coarse-grained biotite granodiorite. Biotite schist inclusions in the outcrop have some orthopyroxene in their cores. Green biotite (15%) in the granodiorite occurs as randomly oriented 1-5 mm grains with only weak preferred orientation. Subhedral plagioclase (35%) and K feldspar porphyroblasts to 1 cm (20%) have minor to locally severe saussuritic and sericitic alteration. Anhedral to seriate quartz (30%) has weak undulose extinction. Minor amounts of an opaque mineral and apatite are associated with biotite, as are trace quantities of zircon and monazite.

5) Granite pegmatite. This rock is a coarse (to 10 cm) pink, leucocratic (< 1% biotite) granite, consisting of subequal amounts of plagioclase, K feldspar and quartz. Parts of the outcrop are aplitic.

U-Pb GEOCHRONOLOGY

Techniques employed in mineral separation and U-Pb analysis of zircons and monazites are discussed in detail in Parrish et al. (1987). Polished and etched grain mounts of zircons from each sample were prepared as an aid in interpretation of the U-Pb data. Complete tables of analytical data are not included in this report; these will be reported in full elsewhere when the study is complete.

Abundant zircon was recovered from all of the samples except the pegmatite, and monazite was recovered from three samples. The zircon grains range from round to very elongate (length: width ratios up to 5 in some samples). Grains with euhedral outlines were not observed in any of the concentrates. Etched grain mounts of zircon from each of the samples reveal clear core: rim relationships, with resorbed inner cores displaying strong igneous zoning and weakly zoned or unzoned rims. Examples of these relationships are shown in Figure 3. In some samples it was possible to isolate rounded, clear grains with no visible cores at all. The external morphology of the zircon grains and their internal structure is typical of grains which have been partially resorbed during a metamorphic event and subsequently overgrown in the solid state by a lower U-content metamorphic zircon rim (eg. van Breemen et al., 1987).

1) Paragneiss leucosome. Three fractions of zircon were analyzed. All are relatively U-rich (390 + ppm) and quite discordant (Fig. 4). The data form a crudely linear array, but since it is clear that a mixed zircon population is present, and subsequent Pb-loss has affected the grains, calculated concordia intercept ages are meaningless. The oldest

$^{207}Pb/^{206}Pb$ age obtained is 2786 Ma, for a fraction of zircon with very large cores and thin metamorphic rims. After abrasion, little of this rim material remained. The age of 2786 Ma is therefore a minimum for at least some of the detrital zircon grains in the metasedimentary protolith. Monazite is slightly discordant, and yields a $^{207}Pb/^{206}Pb$ age of 2668 Ma. This is interpreted to date the early stages of post-metamorphic cooling, and therefore could be similar to the age of the metamorphic zircon rims.

2) Foliated tonalite. Four fractions of zircon are all discordant, but form a linear array with some scatter (Fig. 4). Calculated upper and lower concordia intercept ages are 2691.7 + 9.7/ −7.8 Ma and 764.7 Ma. Metamorphic rims on zircon from this sample are relatively thin, and since all of the fractions were abraded prior to dissolution, little of the metamorphic component was present in any of the fractions. The upper intercept should therefore be close to the primary crystallization age of the tonalitic protolith.

3) Diatexite. In this sample, it was possible to separate a distinct population of zircon that appeared to be completely free of core material. Two fractions of this material were slightly discordant (Fig. 5), but yield identical $^{207}Pb/^{206}Pb$ ages of 2667.3 ± 1.2 Ma, which is considered to be close to the age of metamorphic zircon growth. These zircon frac-

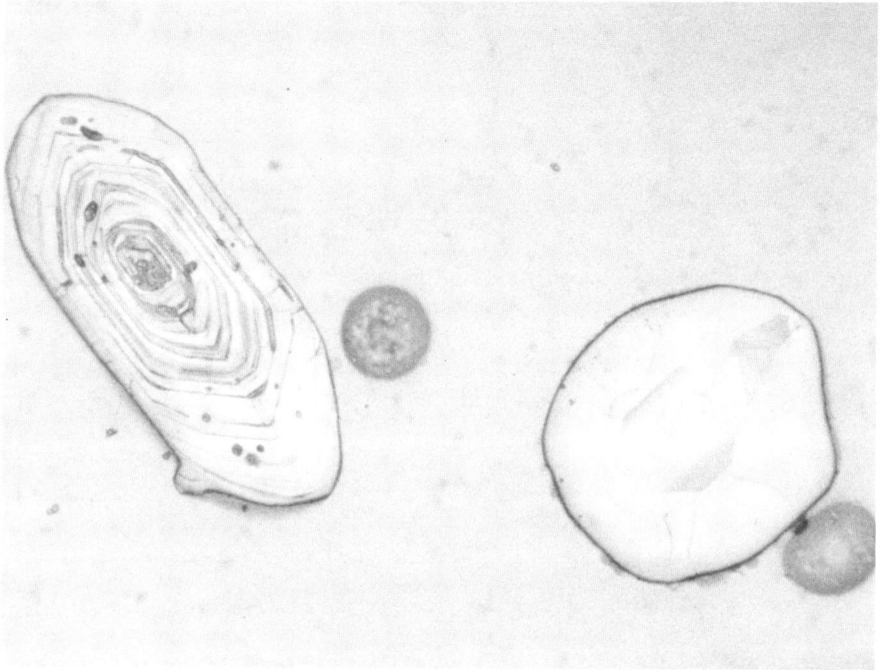

Figure 3. Etched polished grain mount of zircon from sample 3 (diatexite). Left-hand grain shows a strongly zoned igneous zircon core with a resorbed outline and a thin, discontinuous overgrowth of unzoned metamorphic zircon. Right-hand grain shows a small presumably igneous core with an irregular resorbed outline, and a thick overgrowth of unzoned metamorphic zircon. Width of the field of view is about 250 microns.

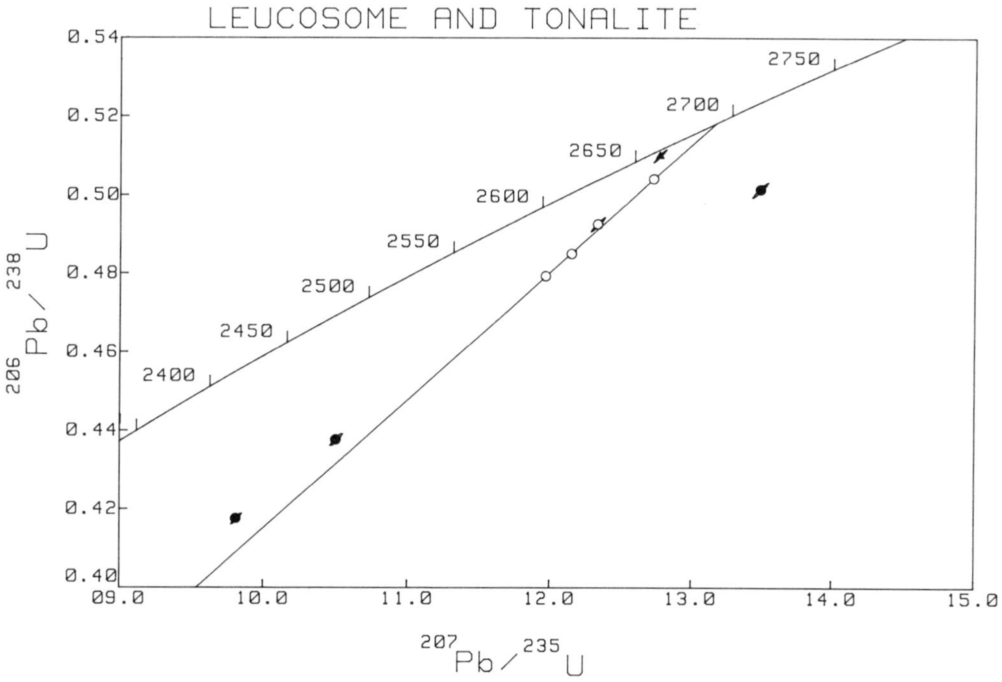

Figure 4. U-Pb concordia plot for zircons (circles) and monazites (triangles) for sample 1 (paragneiss leucosome; closed symbols), and sample 2 (foliated tonalite; open symbols). Errors are shown at the 2σ level (where not shown, error ellipses are smaller than the symbol used). Calculated regression line for foliated tonalite sample shown.

tions may have included a minor amount of core material, however, and 2667 Ma is therefore only a maximum possible age for metamorphic zircon growth. Monazite is slightly younger, with a $^{207}Pb/^{235}U$ age of 2660.8 ± 2.2 Ma, suggesting a slow post-metamorphic cooling through the closure temperature of monazite. The age of metamorphic zircon growth (probably coinciding with granulite facies metamorphism) is therefore bracketed between 2667.3 and 2660.1 Ma. The second population of zircon in this sample form elongate brownish grains. The oldest $^{207}Pb/^{206}Pb$ age obtained for zircon from this population is 2677 Ma, which is a minimum age for the igneous or sedimentary protolith of the rock.

4) Retrogressed diatexite. Three fractions form a nonlinear array (Fig. 5). One fraction of zircon which contained no visible core material yields a concordant age of 2641.9 ± 3.0 Ma. In the light of the complex zircon morphology and the inferred origin of this rock, some interpretation is required. This zircon date is younger than the regional monazite cooling ages, in the 2660-2670 Ma range, suggesting that this population of zircon may have grown during a post-regional metamorphism retrogressive event. This could have been fluid-controlled and localized, given the lack of evidence for a regional event at this time. If this is the case, then it is likely that two generations of metamorphic zircon

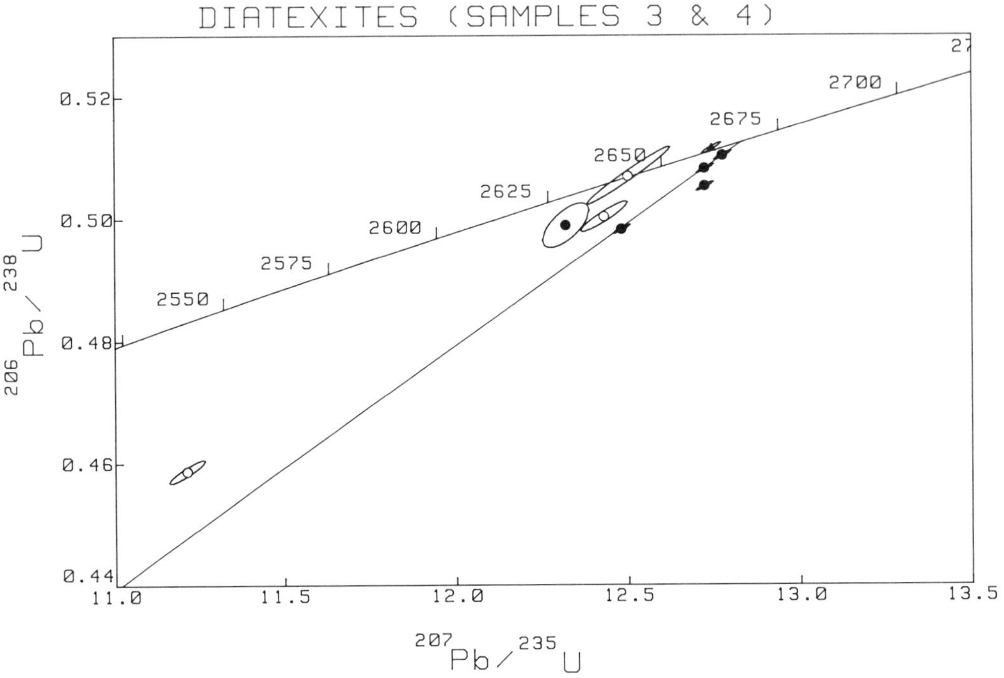

Figure 5. U-Pb concordia plot for zircons (circles) and monazites (triangles) from sample 3 (diatexite; closed symbols) and sample 4 (retrogressed diatexite; open symbols). Errors are shown at the 2σ level. Reference chord from 0 to 2667.3 Ma shown.

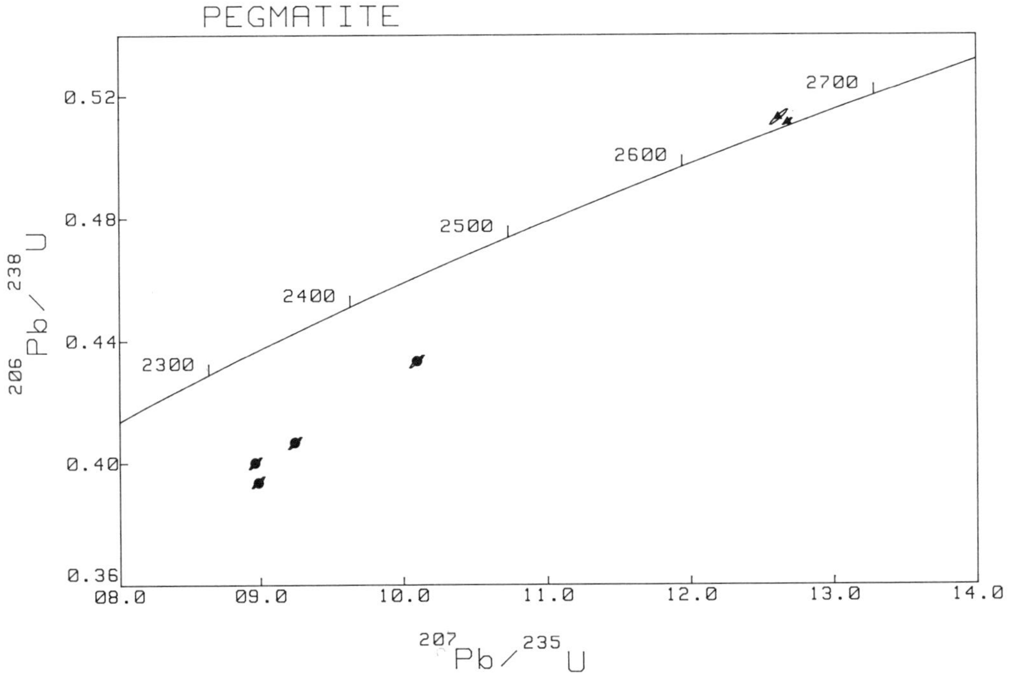

Figure 6. U-Pb concordia plot for zircons (circles) and monazites (triangles) from sample 5 (pegmatite). Errors are shown at the 2σ level.

are present in the rock, one of which grew during granulite facies metamorphism, and a second which grew during the retrogression. The age of the concordant zircon fraction should only be taken as a maximum age for the retrogression. There is insufficient data available from this rock as yet to place constraints on the protolith age.

5) Pegmatite. Four fractions of zircon from pegmatite have very high U-contents, and are extremely discordant in spite of abrasion (Fig. 6). Two fractions of monazite yield overlapping $^{207}Pb/^{235}U$ ages of 2651.6 ± 3.6 Ma and 2656.1 ± 2.4 Ma, suggesting a crystallization age for the pegmatite of about 2654 ± 5 Ma.

DISCUSSION

Rocks in the Lac Clairambault area have had a complex history. For example the diatexites are probably the result of partial or complete fusion of metasedimentary sources and crystallized under granulite facies conditions. Superimposed amphibolite facies metamorphism produces a non descript granodioritic rock with only vaguely preserved evidence of its complex evolution.

It is therefore not surprising that zircon growth patterns are also complex in the rocks with polyphase evolution, and it is possible that the zircon growth stages record the time of recrystallization events in the rocks. However, the ages of these events within single complex grains is generally beyond the means of conventional solid source U-Pb geochronology and will require ion microprobe study.

The strongest conclusion which can be made from the preliminary U-Pb results reported here is that major felsic plutonism occurred, under granulite facies conditions, in the time interval 2669-2650 Ma ago. This accounts for most of the intrusive rocks in the Lac Clairambault area. Foliated tonalite intrusions are slightly older, at 2692 ± 9 Ma. Protolith ages for the metasedimentary rocks are much more tenuous. Zircons from leucosome of paragneiss indicate a component at least 2786 Ma old, but it is likely that some lead loss occurred during the metamorphism and migmatization, and the true age of the detrital zircon is considerably greater than this.

Retrogression which affected sample 4 cannot be older than 2642 ± 3 Ma. The fact that this localized event is younger than the regional monazite ages suggests a relatively low-temperature process (600 < °C), probably driven by channelized fluid access.

On a regional basis, the age of major felsic plutonism and high grade metamorphism is younger that the post-tectonic plutonism in greenstone belts of the western Superior Province, at 2700-2680 Ma (Frarey and Krogh, 1986; Mortensen, 1987; Davis and Edwards, 1985). However, similar ages for felsic plutons and high grade metamorphism were reported by Percival and Sullivan (1986) for the Quetico belt of the western Superior Province. There, granites yield monazite ages of 2671 and 2667 Ma and pegmatite, 2670 Ma. In view of the similar lithology, structural style and orientation (Percival, 1987) and new evidence for synchroneity of intrusive and metamorphic event in the Quetico belt and Ashuanipi complex, the correlation appears compelling. Furthermore, it is necessary to consider some large-scale tectonic process which produced similar rock sequences at the same time in regions some 1600 km apart along strike.

ACKNOWLEDGMENTS

The paper benefited from a critical review by O. van Breemen

REFERENCES

Baragar, W.R.A.
1967: Wakuach Lake map-area, Quebec-Labrador (23O); Geological Survey of Canada, Memoir 344, 174 p.

Brown, M.
1973: The definition of metatexis, diatexis and migmatite; Proceedings of the Geological Association, v. 84, p. 371-382.

Card, K.D. and Ciesielski, A.
1986: Subdivisions of the Superior Province of the Canadian Shield, Geoscience Canada, v. 13, p. 5-13.

Davis, D.W. and Edwards, G.R.
1986: Crustal evolution of Archean rocks in the Kakagi Lake area, Wabigoon Subprovince, Ontario, as interpreted from high-precision U-Pb geochronology; Canadian Journal of Earth Sciences, v. 23, p. 182-192.

Eade, K.E.
1966: Fort George River and Kaniapiskau River (west half) map-areas, New Quebec; Geological Survey of Canada, Memoir 339, 84 p.

Frarey, M.J. and Krogh, T.E.
1986: U-Pb zircon ages of late internal plutons of the Abitibi and eastern Wawa subprovinces, Ontario and Quebec; *in* Current Research, Part A, Geological Survey of Canada, Paper 86-1A, p. 43-48.

Herd, R.K.
1978: Notes on metamorphism in New Quebec, *in* Metamorphism in the Canadian Shield, Ed. J.A. Fraser and W. W. Heywood; Geological Survey of Canada, Paper 78-10, p. 79-83.

Lapointe, B.
1986: Reconnaissance géologique de la région du lac Pailleraut, Territoire du Nouveau Québec; Ministère de l'Énergie et des Ressources du Québec, MB 85-73, 10 p.

Mortensen, J.K.
1987: Preliminary U-Pb ages for volcanic and plutonic rocks of the Noranda–Lac Abitibi area, Abitibi Subprovince, Quebec; *in* Current Research, Part A, Geological Survey of Canada, Paper 87-1A, p. 581-590.

Mortensen, J. K. and Ciesielski, A.
1987: U-Pb zircon and sphene geochronology of Archean plutonic and orthogneissic rocks of the James Bay region and Bienville Domain, Quebec; *in* Radiogenic Age and Isotopic Studies: Report 1; Geological Survey of Canada, Paper 87-2.

Parrish, R.R., Roddick, J.C., Loveridge, W.D. and Sullivan, R.W.
1987: Uranium-lead analytical techniques at the geochronology laboratory, Geological Survey of Canada; *in* Radiogenic Age and Isotopic Studies; Report 1; Geological Survey of Canada, Paper 87-2.

Percival, J.A.
1987: Geology of the Ashuanipi granulite complex in the Schefferville area, Quebec, *in* Current Research, Part A, Geological Survey of Canada, Paper 87-1A, p. 1-10.

Percival, J.A. and Sullivan, R.W.
1986: Age constraints on the evolution of the Quetico belt, Superior Province, Ontario, in Tectonic Evolution of Greenstone Belts; Lunar and Planetary Institute, Technical Report 86-10, p. 167-169.

Pichamuthu, C.S.
1966: The charnockite problem; Mysore Geological Association (Bangalore), 178 p.

Sharma, K.N.M. and Dubé, C.-Y.
1980: Region des Lacs Caniapiscau- Clairambault; Ministère de l'Énergie et des Ressources, DPV-755, 13 p.

Stevenson, I.M.
1963: Lac Bazil, Quebec; Geological Survey of Canada, Paper 62-37, 4 p.

van Breemen, O., Henderson, J.B., Loveridge, W.D. and Thompson, P.H.
1987: U-Pb zircon and monazite geochronology and zircon morphology of granulites and granitc from the Thelon Tectonic Zone, Healey Lake and Artillery Lake map areas, N.W.T.; *in* Current Research, Part A, Geological Survey of Canada, Paper 87- 1A, p. 783-801.

Thomas, A. and Butler, J.
1987: Gold reconnaissance in the Archean Ashuanipi complex of western Labrador, *in* Current Research, Newfoundland Department of Mines and Energy, Report 87-1, p. 237-255.

A compilation of K-Ar Ages
Report 17

P.A. Hunt[1] and J.C. Roddick[1]

Hunt, P.A. and Roddick, J.C., A compilation of K-Ar ages, Report 17; in Radiogenic Age and Isotopic Studies: Report 1, Geological Survey of Canada, Paper 87-2, p. 143-210, 1987.

Abstract

Two hundred and forty-five potassium-argon age determinations carried out by the Geological Survey of Canada are reported. Each age determination is accompanied by a description of the rock and mineral concentrate used; brief interpretative comments regarding the geological significance of each age are also provided where possible. The experimental procedures employed are described in brief outline. An index of all Geological Survey of Canada K-Ar age determinations published in this format has been prepared using NTS quadrangles as the primary reference.

Résumé

Les auteurs présentent 245 datations au potassium-argon effectuées par la Commission géologique du Canada. Chaque datation est accompagnée d'une description de la roche ou du concentré minéral utilisé ainsi que d'une brève interprétation touchant l'aspect géologique. Les méthodes expérimentales qui ont servi aux datations sont aussi résumées. De plus, à l'aide de quadrilatères du SRCN, un index de toutes les datations au potassium-argon a été publié par la Commission géologique du Canada.

INTRODUCTION

This compilation of K-Ar ages determined in the Geochronological Laboratories of the Geological Survey of Canada is the latest in a series of reports, the last of which was published in 1982 (Stevens et al; Paper 82-2). In this new contribution 245 determinations are reported. The format of this compilation is similar to the previous reports with data ordered by province or territory and subdivided by map sheet number. In addition to the GSC numbers, laboratory numbers (K-Ar xxxx) are also included for internal reference.

Experimental procedures

The data compiled here represents analysis over a period of time from 1981 to 1987 and some evolution has taken place in the methods of analysis, particularly of potassium. The laboratory K-Ar number may be correlated with the method of analysis. Up to 1983 (to K-Ar 3511) potassium was analyzed by X-ray fluorescence spectrometry (Lachance, in Wanless et al, 1965; Wanless et al, 1966). For a period in 1984-1985 (K-Ar 3512 to 3692) potassium was analyzed by isotope dilution on a Finnigan MAT 261 mass spectrometer using a ^{41}K-^{40}K double spike. More recent analysis ($>$K-Ar 3692) have been by atomic absorption spectrometry, which is described in more detail in Roddick and Souther (1987).

The techniques of argon analysis have also been modified from the previous procedures described by Stevens et al (1982) and are given in detail by Roddick and Souther (1987). The major changes involve the incorporation of the extraction line and mass spectrometer (A.E.I. MS-10) in a single vacuum line and a multi-sample loading system capable of holding six samples. An atmospheric Ar aliquot system is also incorporated to provide routine monitoring of mass spectrometer mass discrimination. While computer acquisition and processing of data has been in operation since 1978, complete computer control of the mass spectometer was initiated in 1986.

The complete series of reports including the present one is as follows:

		Determinations
GSC Paper 60-17,	Report 1	59-1 to 59-98
GSC Paper 61-17,	Report 2	60-1 to 60-152
GSC Paper 62-17,	Report 3	61-1 to 61-204
GSC Paper 63-17,	Report 4	62-1 to 62-190
GSC Paper 64-17,	Report 5	63-1 to 63-184
GSC Paper 65-17,	Report 6	64-1 to 64-165
GSC Paper 66-17,	Report 7	65-1 to 65-153
GSC Paper 67-2A,	Report 8	66-1 to 66-176
GSC Paper 69-2A,	Report 9	67-1 to 67-146
GSC Paper 71-2,	Report 10	70-1 to 70-156
GSC Paper 73-2,	Report 11	72-1 to 72-163
GSC Paper 74-2,	Report 12	73-1 to 73-198
GSC Paper 77-2,	Report 13	76-1 to 76-248
GSC Paper 79-2,	Report 14	78-1 to 78-230
GSC Paper 81-2,	Report 15	80-1 to 80-208
GSC Paper 82-2,	Report 16	81-1 to 81-226
GSC Paper 87-2,	Report 17	87-1 to 87-245

[1] Geological Survey of Canada, 601 Booth Street, Ottawa, Ontario, K1A 0E8.

REFERENCES

Wanless, R.K., Stevens, R.D., Lachance, G.R., and Rimsaite, J.Y.H.
1965: Age determinations and geological studies, Pt. 1 — Isotopic ages, Report 5; Geological Survey of Canada, paper 64-17, p. 1-126.
1966: Age determinations and geological studies, K-Ar isotopic ages, Report 6; Geological Survey of Canada, Paper 65-17.

Steiger, R.H. and Jager, E.
1977: Subcommission on Geochronology: Convention on the use of decay constants in Geo- and Cosmochronology; Earth and Planetary Science Letters, v. 36, p. 359-362.

Stevens, R.D., Delabio, R.N. and Lachance, G.R.
1982: Age determinations and geological studies, K-Ar Isotopic Ages, Report 16, Geological Survey of Canada, Paper 82-2.

Roddick, J.C. and Souther, J.G.
1987: Geochronology of Neogene Volcanic Rock in the norther Garibaldi Belt, B.C. (this volume).

NEWFOUNDLAND-LABRADOR
(GSC 87-1 TO GSC 87-13)

K-Ar Mineral Ages in the Michael Gabbro

GSC 87-1 Biotite, K-Ar age
984 ± 14 Ma
K = 7.41 %, radiogenic
Ar = 3770 × 10^{-7} cc/gm
K-Ar 3541 atmos. Ar = 0.9 %

From a metagabbro
(13 J) South shore of Groswater Bay, one km west of Nat's Discovery Pt. Labrador, 54°14′N, 58°04′W, Map-unit 8, Map 13-1969. Paper 69-48, Sample, EC82-12, collected and interpreted by R.E. Emslie.

The samples are metagabbros, related entirely or in part to the Michael gabbro suite (Emslie, 1983); the uncertainty arises because it is possible that gabbroic rocks of more than one age are present in the region. The biotite and all of the hornblendes but one appear to have attained closure temperatures during or at the close of the Grenville event. Hornblende EC82-39A (GSC 87-4), for unknown reasons, retains a significantly older K-Ar age of 1270 ± 17 Ma. Unfortunately, the least metamorphosed specimen, taken from the type locality at Lake Michael and having abundant igneous hornblende, entered the processing system and did not emerge.

The Michael gabbros have been dated by Rb-Sr whole rock samples at 1461 ± 96 Ma, Sr_i = 0.7033 ± 0.0021, an ''errorchron'' age (Fahrig and Loveridge, 1981). Recently, a more precise U-Pb zircon date of 1426 ± 6 Ma, considered to be a crystallization age, has been reported by Scharer et al. (1986).

All of the present K-Ar mineral samples lie within the Groswater Bay Terrane which Scharer et al. (1986) infer to have undergone partial melting with zircon and sphene growth at 970 Ma. All of the K-Ar mineral ages reported here are older than 970 Ma suggesting questions of interpretation remain to be answered. It is possible for example, that the 970 Ma event is associated with localized igneous activity rather than being regional in extent.

REFERENCES

Emslie, R.F.
1983: The coronitic Michael gabbros, Labrador: assessment of Grenvillian metamorphism in northeastern Grenville Province; *in* Current Research, Part A, Geological Survey of Canada, paper 83-1A, pp. 139-145.

Fahrig, W.F., and Loveridge, W.D.
1981: Rb-Sr study of the Michael Gabbro, Labrador; *in* Current Research, Part C, Geological Survey of Canada, Paper 81-1C, pp. 99-103.

Scharer, U., Krogh, T.E., and Gower, C.F.
1986: Age and evolution of the Grenville Province in eastern Labrador from U-Pb systematics in accessory minerals; Contributions to Mineralogy and Petrology, v94, pp. 438-451.

GSC 87-2 Hornblende, K-Ar age
988 ± 14 Ma
K = 1.03 %, radiogenic
Ar = 529.1 × 10^{-7} cc/gm
K-Ar 3542 atmos. Ar = 0.7 %

From a metagabbro separate
(13 J) For location, description and interpretation see GSC 87-1.

GSC 87-3 Hornblende, K-Ar age
1074 ± 15 Ma
K = .63 %, radiogenic
Ar = 359.6 × 10^{-7} cc/gm
K-Ar 3543 atmos. Ar = 3.5 %

From an amphibolite separate
(13 J) Ridge on south shore of small pond 11 km due north of Tom Luscombe Brook, Labrador, 54°27′N, 58°22′W. Map-unit 8, Map 13-1969, Paper 69-48. Sample EC82-33, collected and interpreted by R.F. Emslie.
For description and interpretation see GSC 87-1.

GSC 87-4 Hornblende, K-Ar age
1270 ± 17 Ma

K = 1.28%, radiogenic
Ar = 914.6 × 10^{-7} cc/gm
K-Ar 3544 atmos. Ar = 0.8%

From an amphibolite separate
(13 J) High hill 2.5 km south of Tom Luscombe Brook. Labrador, 54°20′N, 58°16′W, Map unit 8, Map 13-1969, Paper 69-48. Sample EC-82-39A, collected and interpreted by R.F. Emslie. For description and interpretation see GSC 87-1.

GSC 87-5 Hornblende, K-Ar age
1072 ± 15 Ma

K = .907%, radiogenic
Ar = 516.2 × 10^{-7} cc/gm
K-Ar 3557 atmos. Ar = 1.0%

From an amphibolite separate
(13 J) North shore of Double Mer, Newfoundland, 54°14′W, 58°27′W, Map-unit 8, Map GSC 13-1969, GSC Paper 69-48. Sample EC 82-10, collected and interpreted by R.F. Emslie. For description and interpretation see GSC 87-1.

GSC 87-6 Hornblende, K-Ar age
1389 ± 23 Ma

K = 1.03%, radiogenic
Ar = 834.1 × 10^{-7} cc/gm
K-Ar 3664 atmos. Ar = 1.7%

From a granite separate
(13M/6) Ridge south of tiny pond in NE trending valley, Labrador, 55°22′N, 63°19′W. Sample EC-79- 46, collected and interpreted by R.F. Emslie. See GSC 87-12 for description and interpretation.

GSC 87-7 Hornblende, K-Ar age
1394 ± 15 Ma

K = 1.16%, radiogenic
Ar = 943.6 × 10^{-7} cc/gm
K-Ar 3660 atmos. Ar = 1.7%

From a quartz monzonite separate
(13M/11) Hill top N. of lake. Labrador 55°44′N, 63°27′W. Sample EC-79-257, collected and interpreted by R.F. Emslie. See GSC 87-12 for description and interpretation.

GSC 87-8 Hornblende, K-Ar age
1386 ± 23 Ma

K = .998%, radiogenic
Ar = 809 × 10^{-7} cc/gm
K-Ar 3665 atmos. Ar = 1.5%

From a quartz monzonite separate
(13M/14) 1/4 W of hammer shaped pond, Labrador, 55°59′N, 63°14′N. Sample EC-79-452, collected and interpreted by R.F. Emslie.
See GSC 87-12 for description and interpretation.

GSC 87-9 Hornblende, K-Ar age
2449 ± 54 Ma

K = 1.14%, radiogenic
Ar = 2.307 × 10^{-4} cc/gm
K-Ar 3403 atmos. Ar = 1.7%

From a quartz monzonite separate.
(13 N/6) 11 km west-northwest of Hunt Lake, Labrador, NFL, 55°25′10″N, 61°22′12″W. Sample EE80-016, collected and interpreted by I.F. Ermanovics. See GSC 87-10 for description and interpretation.

GSC 87-10 Biotite, K-Ar age
2522 ± 31 Ma

K = 6.21%, radiogenic
Ar = 1.327 × 10^{-3} cc/gm
K-Ar 3389 atmos. Ar = 0.6%

From a gneiss separate.
(13 N/9) On southeast shore of bay 8 km south of Napatalik Island, Labrador, NFL, 55°30′N, 60°22′45″W. Sample EE80-151B, collected and interpreted by I.F. Ermanovics.

The mineral concentrates of those age determinations derive from metaplutonic rocks of Archean Hopedale block (southern Nain Province) that yield U-Pb zircon and Rb-Sr whole rock ages in the range 3200-2700 Ma (Grant et al., 1983).

The object of the present K-Ar determinations was to ascertain, in conjunction with other K-Ar ages available for the area, whether or not Aphebian tectonics affected rocks of Hopedale block, to the extent that K-Ar systematics can defect isotopic rejuvenation.

A number of K-Ar biotite and hornblende ages of rocks from Hopedale block range from 2400-2750 Ma (for a summary see Taylor 1979 and Stevens et al. 1982, p. 44). The average of nine age determinations that includes the present analyses is 2500 Ma. Two hornblende ages from this group are 2670 and 2750 Ma, and these are similar to Rb-Sr ages at about 2700 Ma that reflect deformation and metamorphism at that time. The remaining seven analyses fall in the range 2400-2520 ma and average 2450 Ma.

Accordingly, 2450 Ma can be considered to be the time of cooling of the crustal rocks in Hopedale block.

REFERENCES

Grant, N.K., Voner, F.R., Marzano, M.S., Hickman, M.H., and Ermanovics, I.F.

1983: A summary of Rb-Sr isotope studies in the Archean Hopedale block and the adjacent Proterozoic Makkovik Subprovince, Labrador: report 5; in Current Research, Part B, Geological Survey of Canada, Paper 83-1B, p. 127-134.

Taylor, F.C.

1979: Reconnaissance geology of a part of the Precambrian Shield, Northeastern Quebec, Northern Labrador and Northwest Territories; Geological Survey of Canada, Memoir 393, 99 p.

Stevens, R.D., Delabio, R.N. and Lachance, G.R.

1982: Age determinations and geological studies, K-Ar Isotopic Ages, Report 16, Geological Survey of Canada Paper 82-2, p. 44-45.

GSC 87-11 Hornblende, K-Ar age
1239 ± 44 Ma

K = 0.9%, radiogenic
Ar = 6.23×10^{-5} cc/gm
K-Ar 3387 atmos. Ar = 1.9%

From a granite separate.
(13 O/13) Farmyard (Nanuktok) Islands, 23 km east of Nunaksaluk Island, coast of Labrador, Newfoundland, 55°51′10″N, 59°57′30″W. Sample EE80-099, collected and interpreted by I.F. Ermanovics.

The sample is from a massive coarse-grained, pink, biotite-hornblende, perthite microcline granite that intrudes massive layered gabbroic rocks and metaplutonic rocks on a group of small islands 23 km offshore the mainland of the Archean Hopedale block. Rocks similar to this granite occur in the Nain Igneous Complex 60 km from the present locality, and are described by Hill (1982). Here, U-Pb zircon and Rb-Sr whole rock ages of massive granitoid rocks are in the range 1217-1270 Ma (Hill 1982).

Accordingly, the granite from Farmyard Islands is correlated with granitoid rocks of the Nain Igneous complex and the present K-Ar hornblende age is interpreted to approximate the time of intrusion of the granite on Farmyard Islands.

REFERENCE

Hill, J.D.

1982: Geology of the Flowers River-Notakwanon River area, Labrador; Newfoundland Department of Mines and Energy, Mineral Development Divison Report 82-6 140 p.

MISTASTIN BATHOLITH

R.F. Emslie

GSC 87-12 Hornblende, K-Ar age
1376 ± 23 Ma

K = 1.05%, radiogenic
Ar = 842.7×10^{-7} cc/gm
K-Ar 3662 atmos. Ar = 1.6%

From a quartz monzonite separate
(14D/3) Hill top between 2 lakes, Labrador 56°01′N, 63°16′W. Sample EC-79-251, collected and interpreted by R.F. Emslie.

Two groups of widely distributed samples are represented which comprise the two largest intrusive components of the 2,000 km^2 Mistastin batholith. These are pyroxene quartz monzonite (EC79-251, -257, -452, GSC 87-12, 7, 8) which is charnockitic in part, and biotite-hornblende granite (EC79-2, -46, -124, GSC 87-13, 6, 19). Pyroxene quartz monzonite is the older unit and is intruded by dykes of biotite-hornblende granite in many localities; together they form a typical rapakivi association and large ovoid perthite crystals (commonly 2 to 4 cm), mantled and unmantled by plagioclase, are abundant and widespread in both units.

The five hornblendes from widely separated locations in the batholith (e.g. EC79-124 and EC79-251 are about 70 km apart) are remarkably homogenous in cooling age. The mean time at which all samples cooled through the hornblende closure temperature was about 1387 Ma (Fig. 1). This homogeneity of hornblende ages is in good accord with the anorogenic tectonic setting of the batholith in which prolonged undisturbed cooling of such a large igneous complex might be anticipated. One other K-Ar hornblende age of 1301 ± 41 Ma from the batholith has been reported by Stockwell (in Wanless et al., 1974); this was a re- determination of an old sample that had previously yielded 1325 Ma.

Of particular interest is that the hornblende K-Ar closure ages averaging about 1387 Ma are some 25 to 30 Ma younger than the youngest U-Pb zircon ages from the batholith (J.C.M. Roddick, personal communication, 1986). Based on hydrothermal heating experiments on natural hornblendes Harrison (1981) suggests closure temperatures of 500-550°C for cooling rates of about 100°C/Ma, possibly extending to as low as 480°C for cooling rates near 5°C/Ma. Since the Mistastin batholith cooling rate seems to have been closer to the latter than the former, closure temperatures of about 500 ± 25°C may be a reasonable estimate. Harrison suggests that diffusivity of 40Ar in hornblendes is not sensitive to Mg/Mg + Fe in the mineral; his most iron- rich hornblende with Mg/Mg + Fe = 0.36 is, however, still relatively magnesian compared to these Mistastin hornblendes which have Mg/Mg + Fe = 0.22 and below (Fig. 1).

The biotite age of 1374 ± 15 Ma is roughly in accord with the hornblende cooling ages. One previously determined biotite has been reported by Stockwell (in Wanless et al., 1965) as 1340 ± 50 Ma. Like the hornblendes, biotites from the batholith are all very iron-rich (Mg/Mg + Fe = 0.30 to 0.08). Harrison et al. (1985) point out that, unlike hornblende, the thermal energy required for 40Ar diffusivity in biotite decreases strongly with decreasing Mg/Mg+Fe. These authors estimate closure temperatures for biotite with intermediate Mg/Mg + Fe to be in the range 310 to 280°C for cooling rates of 10°C/Ma to 1°C/Ma. Two biotite samples are insufficient for firm conclusions but they are consistent with very slow cooling of the batholith.

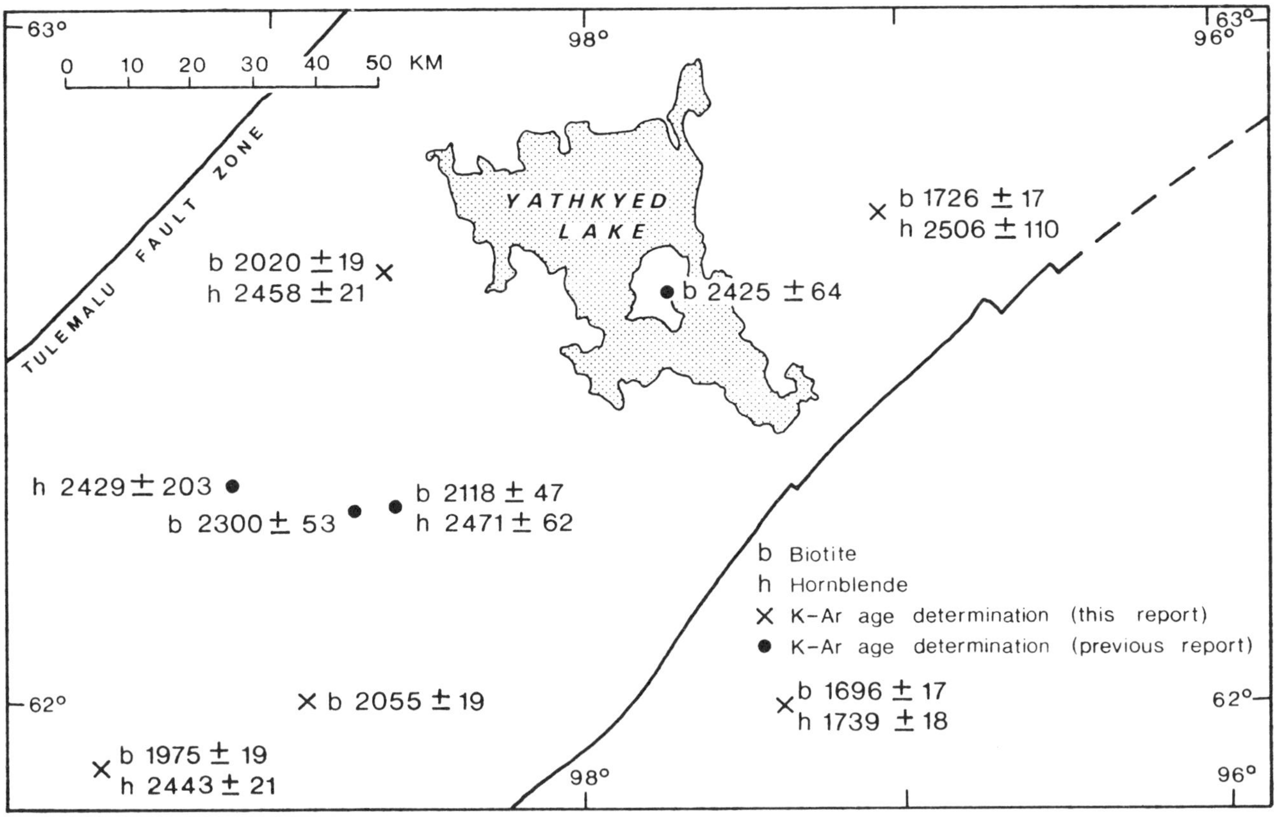

Figure 1. K-Ar mineral ages showing analytical uncertainties for the two major intrusive units of Mistastin batholith. Abbreviations: bi-biotite, hb-hornblende. Field sample number suffixes (2, 46, 124, etc.) are shown for identification.

The intrusions comprising Mistastin batholith crystallized from relatively dry magmas with little evidence of approach to fluid saturation even in the younger members so it is reasonable to expect that conductive heat loss strongly dominated over convective cooling of the sub-solidus batholith. Assuming conduction was the sole mechanism of cooling, the apparent extended cooling time is still much greater than would be anticipated even taking into account apparent crystallization depths of 10-12 km as suggested by the mineral assemblages. Cooling intervals of this magnitude, that is tens of Ma vs. several Ma are more commonly associated with crustal cooling following an orogenic event (e.g. Chamberlain and Karabinos, 1987) rather than simple cooling of a batholith. The implication is clear that cooling was retarded due to abnormally high heat flow in the crust. Abnormally high crustal heat flow is entirely consistent with the long-lived anorogenic magmatism associated with the Elsonian event, the younger limits of which are becoming less well defined because similar anorogenic magmatism, although on a much reduced scale, continued until at least 1305 Ma (Simmons et al., 1986).

An Rb-Sr whole rock isochron age of 1347 $\pm$ 15 Ma (^{87}Rb decay constant 1.39×10^{-11} yr^{-1}) for these same rock units has been published by Marchand and Crockett (1977). A relationship in which the Rb-Sr isochron whole rock age is substantially younger than K-Ar hornblende cooling ages is unusual (note that an ^{87}Rb decay constant of 1.42×10^{-11} yr^{-1} yields and even younger age). The implication would be that the Rb-Sr whole rock system remained open to lower temperatures than anticipated or was subjected to later disturbance; the much younger meteorite impact event at Mistastin Lake does not seem to be a plausible cause. Preliminary results of new Rb-Sr whole rock measurements on a suite of samples from the batholith suggest an age in better accord with the K-Ar mineral ages (J.C.M. Roddick, personal communication, 1986).

REFERENCES

Chamberlain, C.P., and Karabinos, P.
1987: Influence of deformation on pressure-temperature paths of metamorphism; Geology, vol. 15, pp. 42-44.

Harrison, T.M.
1981: Diffusion of ^{40}Ar in hornblende; contributions to Mineralogy and Petrology, vol. 78, pp. 324-331.

Harrison, T.M., Duncan, I., and McDougall, I.
1985: Diffusion of ^{40}Ar in biotite: temperature, pressure, and compositional effects; Geochimica et Cosmochimica Acta, vol. 49, pp. 2461-2468.

Marchand, M., and Crockett, J.H.
1977: Sr isotopes and trace element geochemistry of the impact melt and target rocks at the Mistastin Lake crater, Labrador; Geochimica et Cosmochimica Acta, vol. 41, pp. 1487-1495.

Simmons, K.R., Wiebe, R.A., Snyder, G.A.,and Simmons, E.C.
1986: U-Pb zircon age for the Newark Island layered intrusion, Nain anorthosite complex, Labrador; Geological Society of America, Abstracts with Programs, vol. 18, No. 6, P. 751.

Wanless, R.K., Stevens, R.D., Lachance, G.R., and Rimsaite, R.Y.H.
1965: Age determinations and geological studies, Part 1 -Isotopic ages, Report 5, Geological Survey of Canada Paper 64-17.

Wanless, R.K., Stevens, R.D., Lachance, G.R., and Dilabio, R.N.D.
1974: Age determinations and geological studies, K-Ar isotopic ages, Report 12, Geological Survey of Canada, Paper 74-2.

GSC 87-13 Biotite, K-Ar age
1374 ± 15 Ma

K = 6.08 %, radiogenic
Ar = 4864 × 10^{-7} cc/gm
K-Ar 3666 atmos. Ar = 1.4 %

From a granite separate
(23 P/8) SE shore, small round pond., Labrador 55°25′N, 64°00′W. Sample EC-79-2, collected and interpreted by R.F. Emslie. See GSC 87-12 for description and interpretation.

NOVA SCOTIA
(GSC 87-14)

GSC 87-14 Muscovite, K-Ar age
439 ± 10 Ma

K = 1.86 %, radiogenic
Ar = 3.593 × 10^{-5} cc/gm
K-Ar 3452 atmos. Ar = 30.8 %

From a quartzite separate
(11 F) Barymin Mine, Cape Breton Island, Nova Scotia, 45°51′45″N, 60°24′00″W. Sample SP-4075A, collected by D.F. Sangster.

NEW BRUNSWICK
(GSC 87-15 TO GSC 87-18)

GSC 87-15 Biotite, K-Ar age
390 ± 7 Ma

K = 6.73 %, radiogenic
Ar = 1.139 × 10^{-4} cc/gm
K-Ar 3467 atmos. Ar = 3.3 %

From a biotite granite separate
(21 G) Lake George antimony deposit, 3 km north of Lake George, New Brunswick, 45°51′31″N, 67°02′17″W. Sample SYA81-139 from Consolidated Durham Mines and Resources Limited diamond drill hole 81-24 at a depth of 513.5 to 518.2 m, collected and interpreted by W.D. Sinclair.

The sample is a non-foliated, medium-grained biotite granite with sericitized feldspar phenocrysts up to 1 cm long. It consists of 30 % quartz, 30 % K-feldspar, 25 % albite, 8 % biotite, 3 % sericite, 2 % chlorite, 1 % hornblende and 1 % carbonate with accessory sphene and rutile. Biotite ranges from unaltered to 50 % chloritized.

The sample is from an unexposed stock that has intruded Upper Silurian metagraywacke and slate. Scheelite- and molybdenite-bearing quartz veinlet stockworks in a calc- silicate hornfels zone that surround this stock are spatially and likely genetically related to it. An antimony-bearing quartz vein crosscuts the granite and the associated stockworks.

The date for this stock is the same as the age of crystallization estimated by McCutcheon et al. (1981) for the nearby Pokiok batholith. Compositional similarities between the stock and two-mica phases within the Pokiok batholith support a petrogenetic relationship. However, hydrothermal muscovites associated with the scheelite- and molybdenite- bearing quartz veinlet stockworks, which postdate the granite, yielded ages of 411 ± 8 and 413 ± 8 Ma (Seal et al., 1985). Because the biotite has been partially chloritized and may have suffered some postcrystallization argon loss, the age reported here must be viewed with caution.

REFERENCES

McCutcheon, S.R., Lutes, G., Gauthier, G., and Brooks, C.
1981: The Pokiok batholith: a contaminated Acadian intrusion with an anomalous Rb/Sr age; Canadian Journal of Earth Sciences, v. 18, p. 910-918.

Seal, R.R., II, Archibald, D.A., Clark, A.M., and Farrar, E.
1985: K-Ar evidence for pre-Devonian orogeny, Lake George area, Fredericton Trough, S.W. New Brunswick; Geological Association of Canada/Mineralogical Association of Canada Joint Annual Meeting, Program with Abstracts, v. 10, p. A55.

GSC 87-16 Whole rock, K-Ar age
339 ± 10 Ma

K = 1.75 %, radiogenic
Ar = 2.537 × 10^{-5} cc/gm
K-Ar 3508 atmos. Ar = 6.6 %

From hydrothermally-altered granite porphyry separate.
(21 G) Fire Tower zone, Mount Pleasant

tungsten-molybdenum deposit, 75 km north of St. George, N.B., 45°25′52″N, 66°48′47″W. Sample SYA82-95 collected from diamond drill hole B13 at 20.6 m. Collected and interpreted by W.D. Sinclair.
See GSC 87-17 for description and interpretation.

GSC 87-17 Whole rock, K-Ar age
334 ± 7 Ma

K = 4.77%, radiogenic
Ar = 6.803 × 10^{-5} cc/gm
K-Ar 3509 atmos. Ar = 7.5%

From hydrothermally-altered granite porphyry separate.
(21 G) Fire Tower zone, Mount Pleasant tungsten-molybdenum deposit, 75 km north of St. George, N.B., 45°25′52″N, 66°48′47″W. Sample SYA82-97 collected from diamond drill hole B13 at 29.6 m. Collected and interpreted by W.D. Sinclair.

Two samples of hydrothermally-altered granite porphyry were collected from the Fire Tower zone of the Mount Pleasant tungsten-molybdenum deposit in order to determine the age of this deposit. Both samples consist of very fine grained mixtures of biotite and chlorite with minor amounts of quartz, fluorite and trace amounts of sulphide minerals. Sample SYA82-95 (GSC 87-16) contains more biotite than chlorite; the reverse is true for SYA82-97.

Two distinct phases of mineralizaiton are present in the Fire Tower zone at Mount Pleasant. Porphyry tungsten-molybdenum mineralization is associated with extensive fracturing and brecciation which appear to be related to crystallization of a fine grained granite. Tin-bearing polymetallic zones, on the other hand, form small, irregular veins and replacement zones that are superimposed on the porphyry tungsten-molybdenum deposit and surrounding rocks. These polymetallic zones are spatially associated with irregular dykes of granite porphyry that crosscut the prophyry tungsten-molybdenum deposit. Hydrothermal alteration of the two samples dated appears to be related to the formation of the tin-bearing polymetallic zones. The ages obtained (334 ± 7 and 339 ± 10 Ma) therefore indicate that polymetallic mineralization occurred within the interval 329 to 341 Ma, based on the common overlap of the error limits for the two ages. Because the polymetallic zones are younger than the porphyry tungsten-molybdenum deposit, this interval represents a minimum age for the tungsten- molybdenum mineralization.

GSC 87-18 Biotite, K-Ar age
869 ± 13 Ma

K = 7.05%, radiogenic
Ar = 3059 × 10^{-7} cc/gm
K-Ar 3659 atmos. Ar = 1.9%

(21 O/4) On new highway westside Saint John River, New Brunswick. Sample PC1/81.

The sample was a glacial erratic boulder (Biotite- bearing 'granite') from New Brunswick, 47°02′; 67°48′30″W, collected and interpreted by V.K. Prest.

The boulder was one of several erratics of mixed lithologies cleared from newly broken ground (field) on the west side of the Trans Canada Highway southwest of Grand Falls.

It was collected for K/Ar analysis to identify its source area--whether from relatively nearby Appalachian intrusives or from the more distant Laurentians. The age of 869 ± 13 Ma suggests long distance transport from the north.

The dated boulder serves to confirm that Late Wisconsinan Laurentide ice advanced as far as the Grand Falls area but it remains to be proved that it made greater inroads. The complicated pattern of ice flow features in the Maritimes region suggests a dominance of locally developed ice during the late Wisconsinan.

QUEBEC
(GSC 87-19 TO GSC 87-31)

GSC 87-19 Hornblende, K-Ar age
1388 ± 15 Ma

K = 1.08%, radiogenic
Ar = 877.5 × 10^{-7} cc/gm
K-Ar 3661 atmos. Ar = 1.2%

From a granite separate
(13M/3) Hill top almost on Que-Lab Boundary Quebec. 55°14′N, 63°25′W. Sample EC-79-124, collected and interpreted by R.F. Emslie. See GSC 87-12 for description and interpretation.

GSC 87-20 Biotite, K-Ar age
2324 ± 21 Ma

K = 8.01%, radiogenic
Ar = 14750 × 10^{-7} cc/gm
K-Ar 3540 atmos. Ar = 0.1%

From a nepheline syenite separate
(23 O) 60 km NW of Schefferville, just west of the Labrador Trough. Quebec, 55°05′32″N, 67°38′40″W. Sample NF04981. Collected and interpreted by W.R.A. Baragar.

The sample dated is a nepheline syenite from one of a group of small (2-20 km long) syenite and nepheline syenite plutons that lie within the Archean craton just west of the Labrador Trough. A cluster of 3 of these was first discovered by geologists of the Iron Ore Company following the lead of earlier reports which had noted the presence of nepheline syenite boulders in the drift. These are now called the Goodwood River plutons and were described by Fumerton and Barry (1984). The sample is from the southern-most pluton of the cluster.

Subsequently Percival (1987) located more of these bodies in the map area to the south thus extending their north-south range to about 120 kms.

The syenite and nepheline syenite plutons are the latest felsic rocks in the Archean terrane (Percival, 1987) just west of the Labrador Trough but may have been emplaced prior to the cessation of late Archean deformation. Fumerton and Barry (1984) report the presence of well-developed foliation and lineation within some of the bodies which matches those of the surrounding host rock and Percival's map (1987) shows a strong tendency for country rock foliation to wrap around the plutons. The age obtained on the sample, therefore, could be expected to date the last phase of the Kenoran Orogeny if it were primary. However, at 2324 ± 21 Ma it is unlikely to be a primary age. Since the sample is from a locality within about 10-15 kms of lightly deformed rocks of the Labrador Trough it seems reasonable to attribute its somewhat younger than Kenoran age to updating by the late Aphebian deformation which affected the Trough. What is less in doubt is the interpretation of the nepheline syenite and syenite plutons as being of late Archean age.

REFERENCES

Fumerton, S.L. and Barry, A.P.
1984: Probable Archean nepheline syenite plutons in the Superior Province adjacent to the Labrador Trough; Canadian Journal of Earch Sciences, V. 21, p. 615-618.

Percival, J.A.
1987: Geology of the Ashuanipi granulite complex in the Schefferville area, Québec; Geological Survey of Canada Paper 87-1A, p. 1-10.

GSC 87-21 Biotite, K-Ar age
2413 ± 29 Ma

K = 7.97%, radiogenic
Ar = 1.570×10^{-3} cc/gm
K-Ar 3421 atmos. Ar = 0.6%

From a schist separate
(31 M) Northeast of Lac St-Amand, east of St. Edouard de Fabre, Québec, 47°15′N, 79°07′W. Sample DM-08V24-2, collected by J. van der Leeden.

GSC 87-22 Biotite, K-Ar age
2268 ± 28 Ma

K = 7.53%, radiogenic
Ar = 1.328×10^{-3} cc/gm
K-Ar 3423 atmos. Ar = 0.4%

From a gneiss separate
(31 M) South shore of Grassy Lake, South-southwest of Belleterre, Québec, 47°18′N, 78°48′W. Sample DM-GRAS, collected by J. van der Leeden.

GSC 87-23 Biotite, K-Ar age
1501 ± 22 Ma

K = 7.83%, radiogenic
Ar = 7.126×10^{-4}
K-Ar 3422 atmos. Ar = 0.9%

From a schist separate
(31 M/2) Northern part of Lac Ecarté, south-southeast of Belleterre, Québec, 47°13′N, 78°35′W. Sample DM-LEc, collected by J. van der Leeden.

GSC 87-24 Biotite, K-Ar age
1108 ± 17 Ma

K = 7.59%, radiogenic
Ar = 4.514×10^{-4} cc/gm
K-Ar 3426 atmos. Ar = 1.1%

From a schist separate
(31 M/2) West of Cox Lake, south of Belleterre, Quebec, 47°13′N, 78°46′W. Sample DM-08V1-20, collected by J. van der Leeden.

GSC 87-25 Biotite, K-Ar age
1755 ± 24 Ma

K = 7.64%, radiogenic
Ar = 8.814×10^{-4} cc/gm
K-Ar 3435 atmos. Ar = 0.8%

From a schist separate
(31 M/3) North of Lac St-Amand, east of St-Edouard de Fabre, Quebec, 47°15′N, 79°08′W. Sample DM-STA, collected by J. van der Leeden.

GSC 87-26 Biotite, K-Ar age
2494 ± 30 Ma

K = 7.44%, radiogenic
Ar = 1.557×10^{-3} cc/gm
K-Ar 3425 atmos. Ar = 0.4%

From a schist separate
(31 M/6) South of Lac Argentier; east-southeast of St. Placide de Béarn, Québec, 47°16′N, 79°07′W. Sample DM-6V13, collected by J. van der Leeden.

GSC 87-27 Biotite, K-Ar age
1245 ± 19 Ma

K = 8.0%, radiogenic
Ar = 5.576×10^{-4} cc/gm
K-Ar 3424 atmos. Ar = 0.9%

(31 M/7) From a schist separate
South of Lake Opogan; south of Belleterre, Quebec, 47°15′N, 78°41′W. Sample DM-OV7-11, collected by J. van der Leeden.

GSC 87-28 Biotite, K-Ar age
1312 ± 20 Ma

K-Ar 3427
K = 7.46%, radiogenic
Ar = 5.594 × 10^{-4} cc/gm
atmos. Ar = 0.7%

(31 M/7) From a schist separate
East of Lac Lavoie, southwest of Belleterre, Quebec 47°16′N, 78°51′W. Sample DM-OV9-2, collected by J. van der Leeden.

GSC 87-29 Muscovite, K-Ar age
1549 ± 22 Ma

K-Ar 3429
K = 6.58%, radiogenic
Ar = 6.274 × 10^{-4} cc/gm
atmos. Ar = 1.1

(31 M/7) From a granodiorite separate
At southern end of Lac des Bois, south-southeast of Latulipe, Québec, 47°18′N, 78°58′W. Sample DM-LdB, collected by J. van der Leeden.

GSC 87-30 Biotite, K-Ar age
1052 ± 16 Ma

K-Ar 3433
K = 7.95%, radiogenic
Ar = 4.413 × 10^{-4} cc/gm
atmos. Ar = 4.8%

(31 M/7) From a granodiorite separate
South of Lac Opogan, south of Belleterre, Quebec, 47°15′N, 78°41′W. Sample DM-0V7-11-G collected by J. van der Leeden.

GSC 87-31 Whole rock, K-Ar age
1809 ± 18 Ma

K-Ar 3537
K = 2.16%, radiogenic
Ar = 2616 × 10^{-7} cc/gm
atmos. Ar = 0.3%

(32 N) From a basalt separate
From Quebec, 51°29.5′N, 073°22.5′W. Sample FA-761008. Collected and interpreted by Walter Fahrig.

This analysis was carried out on chilled marginal material of a Mistassini dyke. Two earlier K-Ar ages from 20 km north on the same dyke (1925 ± 60 Ma, G.S.C. 70-97, Wanless et al. 1972 and 1959 ± 58 Ma, G.S.C. 66-140, Wanless et al., 1968) were obtained from chilled dyke material and from biotite less than two centimetres from the dyke margin. The three analyses suggest an age of at least 2 Ga, possibly as great as 2.1 Ga.

REFERENCES

Wanless, R.K., Stevens, R.D., Lachance, G.R., and Edmonds, C.M.
1968: Age Determinations and Geological Studies, K-Ar Isotopic Ages, Report 8: G.S.C. Paper 67-2, Part A

Wanless, R.K., Stevens, R.D., Lachance G.R., and Delabio, R.N.
1972: Age Determinations and Geological Studies, K-Ar Isotopic Ages, Report 10: G.S.C. Paper 71-2.

ONTARIO
(GSC 87-32 TO GSC 87-62)

GSC 87-32 Whole rock, K-Ar age
1444 ± 16 Ma

K-Ar 3521
K = .265%, radiogenic
Ar = 227.9 × 10^{-7} cc/gm
atmos. Ar = 2.7%

(41 J) From a diabase separate
North of Massey Town, Ontario; 46°27′50″N, 82°11′35″W. Sample KCA31, collected and interpreted by C. Kamineni.

This is an abundant map unit in the East Bull Lake area, Massey, Ontario. Mapped is Nipissing diabase. Fresh and contains olivine, pyroxenes and plagioclase as main constituents. The K-Ar age suggests that this unit intruded the East bull Lake area during Middle Proterozoic.

GSC 87-33 Whole rock, K-Ar age
1154 ± 13 Ma

K-Ar 3522
K = .448%, radiogenic
Ar = 281.5 × 10^{-7} cc/gm
atmos. Ar = 3.2%

(41 J) From a gabbro seperate
North of Massey Town, Ontario, 46°26′28″N, 82°14′30″W. Sample MCB1, collected and interpreted by C. Kamineni.
See GSC 87-40 for description.

GSC 87-34 Whole rock, K-Ar age
1723 ± 17 Ma

K-Ar 3523
K = .195%, radiogenic
Ar = 218.6 × 10^{-7} cc/gm
atmos. Ar = 3.1%

(41 J) From an olivine gabbro separate
North of Massey Town, Ontario, 46°25′28″N, 82°12′18″W. Sample MCB2, collected and interpreted by C. Kamineni.

Fresh olivine gabbro sample collected from the troctolite unit of the East Bull Lake anorthosite-gabbro layered complex. The age suggests a Middle Proterozoic age for the pluton.

GSC 87-35 Hornblende, K-Ar age
1983 ± 19 Ma

K = .258%, radiogenic
Ar = 361.9 × 10^{-7} cc/gm

K-Ar 3560 atmos. Ar = 2.5%

(41 J) From a gabbro separate
North of Massey Town, Ontario, 46°26′28″N, 82°14′22″W. Sample KCA17, collected and interpreted by C. Kamineni.

Hornblende separated from a mafic dyke occurring within the East Bull Lake anorthosite-gabbro layered complex. Hornblende occurs as an alteration product of clinopyroxene. No relicts of clinopyroxene are preserved. Ilmenite, sphene, plagioclase (An_{55}) are the other minerals present. The age suggests intrusion (alteration) during Middle Proterozoic for the dyke.

GSC 87-36 Hornblende, K-Ar age
1868 ± 18 Ma

K = .276%, radiogenic
Ar = 351.1 × 10^{-7} cc/gm

K-Ar 3562 atmos. Ar = 2.0%

(41 J) From a gabbro dyke separate
Ontario, 46°26′20″N, 82°13′28″W. Sample KCA94, collected and interpreted by C. Kamineni.

Hornblende occurs as visible pods of 1-2 mm size in a medium-grained mafic dyke. Clinopyroxene is absent. Plagioclase has a composition of $An_{50\text{-}55}$. Epidote, sphene and ilmenite are the other minerals present. The age suggests that the dyke intruded (altered) during Middle Proterozoic.

GSC 87-37 Hornblende, K-Ar age
1698 ± 57 Ma

K = .0764%, radiogenic
Ar = 83.77 × 10^{-7} cc/gm

K-Ar 3647 atmos. Ar = 20.5%

(41 J) From a mafic dyke separate
Whiskey Lake Area, north of Massey; Ontario 46°25′05″N, 82°10′50″W. Sample collected and interpreted by C. Kamineni.

Hornblende occurs as a randomly oriented acicular crystals in a fine-grained dyke. The general orientation of the dyke is NE-SW, in contrast to the NW-SE trend of the major swarm. Plagioclase composition is An_{55}. The age suggests that the dyke intruded (altered) during Middle Proterozoic.

GSC 87-38 Feldspar, K-Ar age
831 ± 18 Ma

K = 12.9%, radiogenic
Ar = 5283 × 10^{-7} cc/gm

K-Ar 3648 atmos. Ar = 0.8%

(41 J) From a K-feldspar separate
Whiskey Lake area, North of Massey, Ontario, 46°25′10″N, 82°10′25″W. Sample KCA34, collected and interpreted by C. Kamineni.

K-feldspar is identified as adularia variety. It occurs as fracture-filling within the gabbro. The age suggests that the K-feldspar formed during Grenville Orogeny.

GSC 87-39 Whole rock, K-Ar age
1623 ± 183 Ma

K = 0.24%, radiogenic
Ar = 2.455 × 10^{-5} cc/gm

K-Ar 3506 atmos. Ar = 3.0%

(41 J/8) From a diabase separate
East Bull Lake, north of Massey, Ontario, 46°26′10″N, 82°10′30″W. Sample L6-87, 6-3, collected and interpreted by C. Kamineni.

Mafic dyke, altered to amphibolite, occurs as intrusion in the East Bull Lake anorthosite-gabbro layered complex. The dyke represents Middle Proterozoic event.

GSC 87-40 Whole rock, K-Ar age
1170 ± 39 Ma

K = 0.97%, radiogenic
Ar = 6.208 × 10^{-5} cc/gm

K-Ar 3507 atmos. Ar = 2.7%

(41 J/8) From a norite separate
East Bull Lake, north of Massey, Ontario, 46°27′40″N, 82°10′15″W. Sample L8-244-5, collected and interpreted by C. Kamineni.

Mafic dyke intruding the East Bull Lake area. The dyke has a NE orientation. It is extremely fresh with olivine, pyroxenes and plagioclase as main constituents. The age indicates that the dyke intruded during Upper Proterozoic.

GSC 87-41 Whole rock, K-Ar age
1325 ± 15 Ma

K = .728%, radiogenic
Ar = 553.3 × 10^{-7} cc/gm

K-Ar 3681 atmos. Ar = 4.4%

(41 J/10) From a diabase. Diabase dyke
32 km N of Elliot Lake along highway 639. 46°37′32″N, 82°48′20″W. Sample BXA83/1720, collected by K.L. Buchan.

GSC 87-42 Biotite, K-Ar age
2265 ± 20 Ma

K = 7.59%, radiogenic
Ar = 13350 × 10^{-7} cc/gm
K-Ar 3551 atmos. Ar = 0.6%

From a dioritic gneiss separate
(410) West of Ivanhoe River, Ontario 47°52′10″N, 82°55′58″W. Sample P80-872 A, collected and interpreted by J.A. Percival.
See GSC 87-47 for description and interpretation.

GSC 87-43 Hornblende, K-Ar age
2404 ± 21 Ma

K = 1.06%, radiogenic
Ar = 2082 × 10^{-7} cc/gm
K-Ar 3552 atmos. Ar = 0.4%

From a dioritic gneiss separate
(410) See GSC 87-42 for location and GSC 87-47 for description and interpretation.

GSC 87-44 Hornblende, K-Ar age
2444 ± 64 Ma

K = 1.07%, radiogenic
Ar = 2.158 × 10^{-4} cc/gm
K-Ar 3479 atmos. Ar = 2.1

From a dioritic gneiss separate
(42 B) Boner Lake, Ontario, 48°22′18″N, 83°03′57″W. Sample P80-887D, collected and interpreted by J.A. Percival.
See GSC 87-47 for description and interpretation.

GSC 87-45 Biotite, K-Ar age
2256 ± 20 Ma

K = 7.57%, radiogenic
Ar = 13230 × 10^{-7} cc/gm
K-Ar 3547 atmos. Ar = 0.4%

From a tonalite gneiss separate.
(42 B) From the northern Ivanhoe River, Ontario, 48°19′57″N, 82°28′33″W. Sample P80-842, collected and interpreted by J.A. Percival.
See GSC 87-46 for description and interpretation.

GSC 87-46 Hornblende, K-Ar age
2446 ± 25 Ma

K = .616%, radiogenic
Ar = 1244 × 10^{-7} cc/gm
K-Ar 3548 atmos. Ar = 0.8%

From a tonalite gneiss separate.
(42 B) See GSC 87-45 for location.

The hornblende separate is from a medium grained, strongly foliated to streaky-layered hornblende-biotite plagioclase-quartz tonalite from the eastern Kapuskasing structural zone. Tonalite is interlayered with paragneiss, mafic gneiss and anorthosite and contains local mafic gneiss xenoliths. The 2446 ± 25 Ma date is comparable to previously determined K-Ar dates on hornblende from the eastern Kapuskasing zone: 2505 Ma on hornblende from the Shawmere anorthosite complex (Watkinson et al, 1972) and 2459 Ma on hornblende from mafic gneiss near the Ivanhoe Lake cataclastic zone (Percival, 1981). Hornblende dates on gneisses in the western Kapuskasing Zone are 2627 ± 88 Ma (Stevens et al., 1982), 2594 ± 151 Ma (Percival et al., 1981) and 2444 ± 64 Ma (GSC 87-44). All of the dates are interpreted to be the time at which the rocks cooled to below the Ar retention temperature in hornblende. Collected and interpreted by J.A. Percival.

REFERENCES

Percival, J.A.,
1981: Geological evolution of part of the central Superior Province based on relationships among the Abitibi and Wawa subprovinces and the Kapuskasing structural zone. Unpublished Ph.D. thesis, Queen's University, 300 p.

Percival J.A., Loveridge, W.D., and Sullivan, R.W.
1981: U-Pb zircon ages of tonalitic metaconglomerate cobbles and quartz monzonite from the Kapuskasing structural zone in the Chapleau area, Ontario; *in* Current Research, Part C, Geological Survey of Canada, Paper 81-1C, p. 107-113.

Stevens, R.D., Delabio, R.N., and Lachance, G.R.,
1982: Age determinations and geological studies: K-Ar isotopic ages, report 15; Geological Survey of Canada, Paper 81-2.

Watkinson, D.H., Thurston, P.C., and Shafigullah, M.
1972: The Shawmere anorthosite of Archean age in the Kapuskasing belt, Ontario; Journal of Geology, v. 80, p. 1736-739.

GSC 87-47 Biotite, K-Ar age
2366 ± 21 Ma

K = 7.53%, radiogenic
Ar = 14320 × 10^{-7} cc/gm
K-Ar 3549 atmos. Ar = 0.2%

From a diorite gneiss separate
(42 B) South of Robson Lake, west of Chaphau River, 48°04′55″N, 83°15′00″W. Sample P80-637B, collected and interpreted by J.A. Percival.

Hornblende and biotite from tonalitic rocks of the Kapuskasing structure were analyzed to study the cooling and uplift history of this granulite terrane of the central Superior Province. Metamorphic grade increases toward the structure from the west, from greenschist, through amphibolite, to granulite facies, over a 120 km distance. The tectonic model for uplift of the structure involves a southeast-verging crustal-scale thrust, bringing the high-grade rocks to the surface and exposing an oblique crustal cross-section (Percival and Card, 1983). This predicts (Fountain and Salisbury, 1981) a younging of isotopic ages toward the bounding sole thrust, the Ivanhoe Lake cataclastic zone, because the deeper levels cooled more slowly and passed through the blocking temperature for different isotopic systems later than the rocks at higher structural levels.

Results presented below are arranged from northwest to southeast, such that the numbers at the bottom come from the highest-grade rocks, closest to the Ivanhoe Lake structure.

There is a clear indication of younging toward the southeast in the hornblende dates, from 2569 ± 22 Ma (GSC 87-48), to 2404 ± 21 Ma (GSC 87-43). There is a suggestion of a concomitant decrease in biotite ages, from 2366 ± 21 (GSC 87-47), to 2265 ± 20 Ma (GSC 87-42).

In conjunction with U-Pb dates on metamorphic zircons from the same area (Percival and Krogh, 1983), the data indicate a slow cooling history, from metamorphic conditions in the 800°C range at 2650 Ma, to the hornblende blocking temperature at about 2400 Ma (250°C over 250 Ma = 1°C/Ma), to the biotite blocking temperature at about 2265 Ma ago (200°C over 135 Ma = 1.5°C/Ma). Uplift and final cooling of the granulites occurred after 2265 Ma ago.

REFERENCES

Fountain, D.M. and Salisbury, M.H.
1981: Exposed cross-sections through the continental crust: implications for crustal structure, petrology and evolution; Earth and Planetary Science Letters, 56, 263-277.

Percival, J.A.
1983: High-grade metamorphism in the Chapleau-Foleyet area, Ontario; American Mineralogist, 68, 667-686.

Percival, J.A. and Card, K.D.
1983: Archean crust as revealed in the Kapuskasing uplift, Superior Province, Canada; Geology, 11, 323-326.

Percival, J.A. and Krogh, T.E.
1983: U-Pb zircon geochronology of the Kapuskasing structural zone and vicinity in the Chapleau-Foleyet area, Ontario; Canadian Journal of Earth Sciences, 20, 830-843.

GSC 87-48 Hornblende, K-Ar age
2569 ± 22 Ma

K = .98%, radiogenic
Ar = 2167 × 10^{-7} cc/gm
K-Ar 3550 atmos. Ar = 0.3%

From a diorite gneiss separate
(42 B) See GSC 87-47 for location, description and interpretation.

GSC 87-49 Biotite, K-Ar age
2323 ± 21 Ma

K = 7.38%, radiogenic
Ar = 13570 × 10-7 cc/gm
K-Ar 3553 atmos. Ar = 0.3%

From a dioritic gneiss separate
(42 B) South of Mishionga Lake, Ontario, 48°16′54″N, 82°44′55″W. Sample P80-809, collected and interpreted by J.A Percival.
See GSC 87-47 for description and interpretation.

GSC 87-50 Hornblende, K-Ar age
2658 ± 67 Ma

K = 0.93%, radiogenic
Ar = 2.194 × 10^{-4} cc/gm
K-Ar 3438 atmos. Ar = 0.6%

From a porphyritic granite separate
(52 B) East of Dashwa Lake, near town of Atikokan, Ontario, 48°52′N, 92°42′W. Sample ATK-2, collected and interpreted by D.C. Kamineni.

Hornblende separated from an unaltered granodiorite sample collected from the view of the Eye-Dashwa Lakes pluton. The pluton is a late syn-kinematic intrusion in the Wabigoon Subprovince of the Superior Structural province. The age suggests that the granite intruded during the Kenoran event.

GSC 87-51 Hornblende, K-Ar age
2472 ± 71 Ma

K = 0.84%, radiogenic
Ar = 1.73 × 10^{-4} cc/gm
K-Ar 3439 atmos. Ar = 1.3%

From a porphyritic granite separate
(52 B) East of Dashwa Lake, near town of Atikokan, Ontario, 48°52′N, 92°40′W. Sample ATK-1, collected and interpreted by D.C. Kamineni.

Hornblende separated from a fault zone occurring within the rim part of the Eye-Dashwa lakes pluton (Wabigon subprovince). Note the age is younger by approx. 200 Ma, compared to sample GSC 87-50; this is attributed to resetting by late Kenoran brittle faulting.

GSC 87-52 Muscovite, K-Ar age
2655 ± 31 Ma

K = 8.87%, radiogenic
Ar = 2.088 × 10^{-3} cc/gm
K-Ar 3446 atmos. Ar = 0.4%

From an aplite separate
(52 B) Atikokan, Ontario Sample ATK-3, collected and interpreted by C. Kamineni.

Muscovite separated from an aplite dyke, within the Eye-Dashwa lakes pluton. The K-Ar date indicates that the dyke is related to the Kenoran event and intruded in the late stages of crystallization of the Eye-Dashwa pluton.

GSC 87-53 Muscovite, K-Ar age
2626 ± 30 Ma

K = 7.53%, radiogenic
Ar = 1.736 × 10^{-3} cc/gm
K-Ar 3447 atmos. Ar = 1.1%

From an aplite dike separate
(52 B) Atikokan, Ontario Sample ATK-4 (crust), collected and interpreted by C. Kamineni.

Muscovite separated from an epidote-bearing fault within the Eye-Dashwa lakes pluton. The K-Ar age indicates a late Kenoran age for the fault.

GSC 87-54 Muscovite, K-Ar age
2397 ± 28 Ma

K = 6.38%, radiogenic
Ar = 1.242 × 10^{-3} cc/gm
K-Ar 3448 atmos. Ar = 0.9%

From an aplite dike separate.
(52 B) Atikokan, Ontario Sample ATK-4 (matrix), collected and interpreted by C. Kamineni.

Muscovite separated from an aplite dyke located within the Eye-Dashwa lakes pluton, Atikokan, Ontario. K-Ar age is somewhat lower than the norm, possibly due to resetting during late Kenoran.

GSC 87-55 Whole rock, K-Ar age
1132 ± 27 Ma

K = 1.25%, radiogenic
Ar = 7.651 × 10^{-5} cc/gm
K-Ar 3458 atmos. Ar = 1.1%

From a diabase separate
(52 B) Near Atikokan, northwestern Ontario, 48°57′N, 91°45′W. Sample ATK-295B, collected and interpreted by C. Kamineni.

Fresh diabase dyke — occurs as intrusion in the Eye-Dashwa lakes pluton. The K-Ar age indicates that the dyke samples belong to the Keweenawan event.

GSC 87-56 Whole rock, K-Ar age
1143 ± 27 Ma

K = 1.26%, radiogenic
Ar = 7.813 × 10^{-5} cc/gm
K-Ar 3459 atmos. Ar = 0.9%

From a diabase separate
(52 B) Near Atikokan, northwestern Ontario, 48°54′N, 91°40′W. Sample ATK-6, collected and interpreted by C. Kamineni.
See GSC 87-55 for description.

GSC 87-57 Hornblende, K-Ar age
2487 ± 72 Ma

K = 0.83%, radiogenic
Ar = 1.728 × 10^{-4} cc/gm
K-Ar 3460 atmos. Ar = 1.0%

From a hornblende porphyry separate.
(52 B/13) Drill core near Atikokan, Ontario, 48°54′N, 91°42′W. Sample DEA79-ATK-1-192, collected and interpreted by C. Kamineni.

Hornblende separate from a hornblende porphyry. This occurs as a dyke within the Eye-Dashwa lakes pluton. The K-Ar age suggests that it is a late Kenoran dyke that intruded the Eye-Dashwa lakes pluton.

Sample EEGB-79-5: Redditt Granodiorite (Lount Lake batholith)

GSC 87-58 Biotite, K-Ar age
2341 ± 21 Ma

K = 7.25%, radiogenic
Ar = 13520 × 10^{-7} cc/gm
K-Ar 3556 atmos. Ar = 0.1%

From a granodiorite separate
(52E/10) On an outcrop on the west shore of Norway Lake, Ontario. 49°59′30″N, 94°23′25″W. Sample EEGB-79-5, collected and interpreted by Ermanovics.

The object of these age determinations is to continue to broaden observations made in the Wabigoon and English River subprovinces of Manitoba that K-Ar ages of biotites of plutonic rocks are 200 to 300 Ma younger than primary emplacement ages as determined from U-Pb zircon ages (Ermanovics and Wanless, 1983). The present samples yield K-Ar biotite ages of 2341 ± 21 Ma and 2443 ± 21 Ma (GSC 87-59) but U-Pb zircon ages of 2702 +4/−3 Ma and 2700 ± 2 Ma respectively.

It should be noted that hornblende concentrates from these samples were too impure (containing either a large proportion of biotite or too heavily chloritized) for K-Ar analysis.

The sample was selected from an outcrop on the southwest shore of Norway Lake approximately 1 kilometre north of the town of Redditt. The sample is a medium grained, microcline-megacrystic biotite granodiorite. Zircons separated from this sample have a U-Pb upper concordia intercept age of 2702 +4/−3 Ma (Beakhouse, 1983). Samples selected from this phase define a Rb-Sr whole rock isochron age of 2635 ± 56 Ma but anomalously radiogenic Sr in epidote suggests that the Rb-Sr system has been modified during deuteric alteration accompanying subsolidus cooling of the batholith (Beakhouse, 1983).

Both samples are from plutons that are interpreted to postdate the main regional deformation and metamorphism in the Winnipeg River belt of Ontario (English River Subprovince about 10 to 13 km north of Wabigoon Subprovince). They show no evidence of having been deformed or recrystallized subsequent to their crystallization and as such have an apparent geologic history that includes crystallization, subsolidus cooling (with accompanying deuteric alteration), uplift and erosion. The protracted history suggested by the U-Pb zircon, Rb-Sr whole rock and K-Ar biotite ages is interpreted to, in part, reflect the contrasting response of these systems to slow cooling of the crust. Based on available U-Pb zircon geochronology, the Winnipeg River belt contains some of the youngest granites (e.g. Lac du Bonnet granite — 2665 ± 20 Ma; Krogh et al, 1976a; Eastern Lac Seul pegmatitic granite — 2560 ± 40 Ma; Krogh et al., 1976b) yet found in the western Superior Province. Taken in conjunction with this observation, the relatively young K-Ar biotite ages suggests that this segment of the crust cooled more slowly than that exposed in adjacent subprovinces.

REFERENCES

Beakhouse, G.P.

1983: Geological, geochemical and Rb-Sr and U-Pb zircon geochronological investigations of granitoid rocks from the Winnipeg River belt, northwestern Ontario and southeastern Manitoba; Unpublished Ph.D. thesis, McMaster University, Hamilton, Ontario, 376 p.

Ermanovics, I. and Wanless, R.K.

1983: Isotopic age studies and tectonic interpretation of Superior Province in Manitoba; Geological Survey of Canada, Paper 82-12.

Krogh, T.E., Davis, G.L., Ermanovics I., and Harris, N.B.W.

1976a: U-Pb isotopic ages of zircons from the Berens block and English River gneiss belt; Proceedings of the 1976 Geotraverse Conference, Precambrian Research Group, University of Toronto, 46 p.

Krogh, T.E., Harris, N.B.W., and Davis, G.L.

1976b: Archean rocks from the eastern Lac Seul region of the English River gneiss belt, northwestern Ontario, part 2, Geochronology; Canadian Journal of Earth Sciences, 13, p. 1212-1215.

Sample EEGB-79-6: Quartz Diorite (Trout lake)

GSC 87-59 Biotite, K-Ar age
2443 ± 21 Ma

K = 7.48%, radiogenic
Ar = 15060 × 10^{-7} cc/gm
K-Ar 3558 atmos. Ar = 0.3%

From a quartz diorite separate
(52 E/15) Outcrop on the Minaki Rd. between the east end of Trout Lake and the south end of Meekin Lake, Ontario, 49°48′44″N, 94°46′15″W. Sample EEGB-79-6, collected by G. Beakhouse and interpreted by Ermanovics.

This sample was selected from an outcrop along highway 596 near the southeast end of Trout Lake. The sample is a medium grained, equigranular, biotite-hornblende quartz diorite. Zircons separated from this sample yield an upper concordia intercept age of 2700 ± 2 Ma and a suite of samples from this phase define a Rb-Sr whole rock isochron of 2652 ± 127 Ma (Beakhouse, 1983). See GSC 87-58 for interpretation.

GSC 87-60 Biotite, K-Ar age
2510 ± 30 Ma

K = 6.79%, radiogenic
Ar = 1.438 × 10^{-3} cc/gm
K-Ar 3417 atmos. Ar = 0.4%

From a granodiorite separate
(52 N) East shore of Perrigo Lake, 16 km east of Confederation Lake, Ontario. Map unit 7a, GSC Map 1200A. Sample FC-RL-5-78, collected and interpreted by M.J. Frarey.

The rock sampled is a medium-grained, pink, unfoliated granodiorite to diorite forming part of an ovoid pluton about 10 kilometres in long dimension intruding metavolcanic and metasedimentary rocks of Vehi Subprovince, Superior Province. The pluton is one of a group mapped as the youngest intrusives of the Trout Lake map-area (Map 1200A). The age is interpreted as the time of cooling of the pluton past the argon blocking temperature for biotite. In contrast, hornblende from the same rock gave a K-Ar age of 2661 ± 84 Ma (not published); depending on how the error limits are applied, there may have been a rather prolonged time interval between blocking temperatures of the two minerals.

GSC 87-61 Biotite, K-Ar age
2506 ± 31 Ma

K = 7.75%, radiogenic
Ar = 1.636 × 10^{-3} cc/gm
K-Ar 3410 atmos. Ar = 0.10%

From a granodiorite separate.
(53 B) Northwest shore of Weagamour Lake 40 km northwest of North Caribou Lake, northwestern Ontario, 52°56′N, 91°25′W. Map unit no. 8, GSC Map 19-1961. Sample FC-RL-2-78, collected and interpreted by M.J. Frarey.

The sample is from an igneous body intrusive into the North Caribou Lake greenstone belt, which forms part of Sachigo Subprovince of Superior Province. Where sampled, the rock is an unfoliated, fairly homogeneous, coarse-grained granodiorite consisting of potash feldspar, sodic plagioclase, quartz, biotite and minor accessories. The age is interpreted as the time of cooling of the intrusion to the closing temperature for the K-Ar system. The age of emplacement is probably considerably older; elsewhere in this subprovince, the ages of both volcanism and late plutonism are known to be pre-2700 Ma from U-Pb zircon dating.

GSC 87-62 Biotite, K-Ar age
2305 ± 27 Ma

K = 7.37%, radiogenic
Ar = 1.338 × 10^{-3} cc/gm

K-Ar 3416 atmos. Ar = 0.6%

From a granite separate
(53 C) 3.2 km south of Southtrout Lake, Ontario, 52°51′N, 93°41′W. Map unit 8, GSC map 1201A. Sample FC-RL-1-78. Collected and interpreted by M.J. Frarey.

The sample is from an extensive body of leucocratic granite (Setting Net Lake Batholith) intruding volcanic and sedimentary rocks of the Favourable Lake greenstone belt of Sachigo Subprovince, Superior Province. The age is interpreted as an updated cooling age for the batholith. The updating may be due to movement along adjacent faults or to regional reheating of the northwest part of Sachigo Subprovince during the Proterozoic.

MANITOBA
(GSC 87-63 TO GSC 87-64)

GSC 87-63 Biotite, K-Ar age
2416 ± 21 Ma

K = 7.91 %, radiogenic
Ar = 15630 × 10^{-7} cc/gm
K-Ar 3545 atmos. Ar = 0.2 %

From an amphibolite separate
(52 L/5) Collected from quartz southeast margin of Lac du Bonnet Batholith, Manitoba, 50°21′00″, 95°40′00″. Sample 81-MCB-6-313, collected and interpreted by G. McCrank and A. Brown.

The sample is from a strongly foliated and recrystallized mafic dyke, intrusive into an early granitic phase of the Lac du Bonnet Batholith (Rb-Sr age 2680 Ma), which is located in the English River Belt of Superior Province.

The dykes are black, medium grained, and composed of hornblende, biotite, quartz and plagioclase. The larger dykes, more than one cm wide, are oriented ESE (105°), and smaller ones, about 10 cm wide are oriented ENE (070°). The dykes are not traceable beyond their exposed locales in the foliated granite phase and thus their true dimensions are indeterminate.

Deformation, recrystallization and age reset of the dykes, along with foliation of the granite phase, are believed to be synchronous, and attributable to intrusion of the main phase of the batholith.

GSC 87-64 Hornblende, K-Ar age
2657 ± 22 Ma

K = .493 %, radiogenic
Ar = 1162 × 10^{-7} cc/gm
K-Ar 3546 atmos. Ar = 0.7 %

From an amphibolite separate
(52 L/5) See GSC 87-63 for description and interpretation.

DISTRICT OF KEEWATIN
(GSC 87-65 TO GSC 87-75)

GSC 87-65 Muscovite, K-Ar age
1666 ± 23 Ma

K = 6.56 %, radiogenic
Ar = 6.983 × 10^{-4} cc/gm
K-Ar 3432 atmos. Ar = 0.9 %

From a lamprophyre separate
(56 D/1) North of Baker Lake, District of Keewatin, 64°10′17″N, 94°28′59″W. Sample SMA6-0068, collected by M. Schau.

GSC 87-66 Biotite, K-Ar age
2032 ± 25 Ma

K = 6.84 %, radiogenic
Ar = 9.997 × 10^{-4} cc/gm
K-Ar 3434 atmos. Ar = 1.0 %

From a lamprophyre separate
(56 D/1) North of Baker Lake, District of Keewatin 64°08′19″N, 94°27′10″W. Sample SMA6-0744A, collected by M. Schau.

GSC 87-67 Biotite, K-Ar age
1975 ± 19 Ma

K = 6.82 %, radiogenic
Ar = 9510 × 10^{-7} cc/gm
K-Ar 3636 atmos. Ar = 0.5 %

From a gabbro separate
(65 G) 65 km north of Watterson Lake, N.W.T., 61°55′10″N, 99°30′45″, Map unit Ag. Ref: Eade & Chandler, 1975; Geology of Waterson Lake (W 1/2), District of Keewatin; GSC Paper 74-64. Sample EA-150-73, collected and interpreted by K. Eade.

See GSC 87-73 for description and interpretation.

GSC 87-68 Hornblende, K-Ar age
2443 ± 21 Ma

K = .617 %, radiogenic
Ar = 1244 × 10^{-7} cc/gm
K-Ar 3637 atmos. Ar = 0.5 %

From a gabbro separate
(65 G) See GSC 87-67 for location and GSC 87-73 for description and interpretation.

GSC 87-69 Biotite, K-Ar age
1726 ± 17 Ma

K = 6.6 %, radiogenic
Ar = 7422 × 10^{-7} cc/gm
K-Ar 3630 atmos. Ar = 0.1 %

From a paragneiss separate
(65 I) 30 km east of Yathkyed lake, N.W.T., District of Keewatin, 62°44′10″, 97°04′45″, Map unit Asn, Ref. Eade, 1986, Precambrian Geology of the Tukmalu Lake — Yathkyed Lake area, District of Keewatin; G.S.C. Paper 84-11. Sample EA-1828-73, collected and interpreted by K.E. Eade.
See GSC 87-73 for description and interpretation.

GSC 87-70 Hornblende, K-Ar age
2506 ± 110 Ma

K = 1.16%, radiogenic
Ar = 2456 × 10^{-7} cc/gm
K-Ar 3631 atmos. Ar = 0.5%

From a paragneiss separate
(65 I) See GSC 87-69 for location and GSC 87-73 for description and interpretation.

GSC 87-71 Biotite, K-Ar age
1696 ± 17 Ma

K = 7.08%, radiogenic
Ar = 7750 × 10^{-7} cc/gm
K-Ar 3634 atmos. Ar = 0.5%

From a granodiorite gneiss separate
(65 I) 12 km east of Imikula Lake N.W.T., 62°00′01″N, 97°19′40″W. Map unit Ang, Ref.: Eade, 1980; Precambrian Geology of the Tukmalu Lake-Yathkyed Lake area, District of Keewatin GSC paper 84-11. Sample EA-240-73, collected and interpreted by K.E. Eade.
See GSC 87-73 for description and interpretation.

GSC 87-72 Hornblende, K-Ar age
1739 ± 18 Ma

K = .396%, radiogenic
Ar = 450.2 × 10^{-7} cc/gm
K-Ar 3635 atmos. Ar = 1.6%

From a granodiorite gneiss separate
(65 I) See GSC 87-71 for location and GSC 87-73 for description and interpretation.

GSC 87-73 Biotite, K-Ar age
2055 ± 19 Ma

K = 7.71%, radiogenic
Ar = 11480 × 10^{-7} cc/gm
K-Ar 3629 atmos. Ar = 0.2%

From a granodiorite gneiss separate.
(65 J) 45 km ESE of Angikuni Lake; district of Keewatin, N.W.T., 62°01′10″N, 98°52′45″W. Map unit Ang, Ref.: Eade, 1986; Precambrian Geology of the Tulemalu Lake-Yathkyed Lake Area, District of Keewatin, G.S.C. Paper 84-11. Sample EA-260-75, collected and interpreted by K.E. Eade.

Seven of the nine determinations reported here are from well foliated gneisses, the other two from a massive gabbro pluton. All these rocks are part of the Archean craton of southern district of Keewatin. Throughout most of this region Archean metamorphic and plutonic rocks, similar to those dated here, contain hornblende and/or biotite whoe K-Ar isotopic systems have been re-set during the Hudsonian Orogeny. The determinations described here provide additional K-Ar isotopic data toward delineating the area of Archean rocks near Yathkyed Lake that have undergone little or no subsequent metamorphism associated with the Hudsonian Orogeny.

The 2055 ± 19 Ma age (GSC 87-73) on biotite from granodiorite gneiss southwest of Yathkyed Lake suggests there has been minor effects on the biotite system but less than to the southwest on sample GSC 87-67 biotite.

The biotite, 1726 ± 17 Ma (GSC 87-69) and hornblende, 2506 ± 110 Ma (GSC 87-70) pair from migmatized paragneiss east of Yathkyed Lake suggest that Hudsonian metamorphism was sufficient to re-set the isotopic system in biotite but that hornblende retains the cooling age set following the Kenoran Orogeny.

The biotite-hornblende pair (GSC 87-74, 87-75) from granodiorite orthogneiss present west of Yathkyed Lake gives ages of 2020 ± 19 Ma and 2458 ± 21 Ma respectively. As in the just described pair, the hornblende isotopic system is probably untouched but the biotite system has been affected. A previously published K-Ar on biotite from pyroxene- bearing granodiorite gneiss present approximately half way between the locations of GSC 87-69 — 87-70 and GSC 87-74 — 87-75 gives an age of 2425 ± 54 Ma. It is suggested that the K-Ar system in the biotite there has been undisturbed by metamorphism after closure following the Kenoran Orogeny.

A small Archean gabbro pluton approximately 105 km southwest of Yathkyed Lake contains biotite and hornblende (GSC 87-67, 87-68) that yields K-Ar ages of 1975 ± 19 Ma and 2443 ± 21 Ma respectively. The biotite here has been slightly more affected than is the adjacent, GSC 87-73, sample. The hornblende from gabbro retains the cooling age set following the Kenoran Orogeny.

Biotite and hornblende samples, GSC 87-71 and GSC 87-72, from granodiorite orthogneiss southeast of Yathkyed Lake gives ages of 1696 ± 17 Ma and 1739 ± 18 Ma respectively. In this orthogneiss the K-Ar systems of both minerals have been totally re-set as a result of the Hudsonian Orogeny. This granodiorite orthogneiss is east of a major northeast trending fault zone and it is suggested all Archean rocks east of the fault in this region, have been metamorphosed during the Hudsonian Orogeny.

Figure 2 shows the location of the presently reported dates as well as some previously reported data. In this area there are also a number of K-Ar ages on gabbro dykes and adjacent country rocks that have been discussed elsewhere (Fahrig et al., 1984; Wanless et al., 1979).

Approximately 50 km west of the Tulemalu Fault Zone (Fig. 2) Archean granodiorite gneiss has hornblende and biotite K-Ar ages of 1854 ± 39 Ma (GSC 81-146) and 1815 ± 25 Ma (GSC 81-147), respectively (Stevens et al., 1982; Tella and Eade, 1985). These closure dates are older than GSC 87-71 and GSC 87-72 which are closer to K-Ar ages elsewhere in the region and are thought to result from regional metamorphism associated with the Hudsonian Orogeny.

Approximately 150 km east southeast of Yathkyed Lake, in the Kaminak Lake area, Davidson (1970) recognized an area in which metamorphism associated with the Hudsonian Orogeny is slight or absent.

REFERENCES

Davidson, A.
1970: Precambrian geology, Kaminak Lake map-area, District of Keewatin; Geological Survey of Canada, Paper 69-51.

Fahrig, W.F., Christie, K.W., Eade, K.E. and Tella, S.
1984: Paleomagnetism of the Tulemalu dykes, Northwest Territories, Canada; Canadian Journal of Earth Sciences, Vol. 21, No. 5, p. 544-553.

Stevens, R.D., Delabio, R.N., and Lachance, G.R.
1982: Age determinations and geological studies K-Ar isotopic ages, Report 16; Geological Survey of Canada, Paper 82-2.

Tella, S. and Eade, K.E.
1985: Geology, Kamilukuak Lake, District of Keewatin, Northwest Territories; Geological Survey of Canada, Map 1629A.

Wanless, R.K., Stevens, R.D., Lachance, G.R., and Delabio, R.N.
1979: Age determinations and geological studies K-Ar isotopic ages, Report 14; Geological Survey of Canada, Paper 79-2.

GSC 87-74 Biotite, K-Ar age
2020 ± 19 Ma

K = 7.37%, radiogenic
Ar = 10670 × 10^{-7} cc/gm
K-Ar 3632 atmos. Ar = 0.2%

From a granodiorite orthogneiss separate
(65 J) Approximately 20 km west of Yathkyed Lake, District of Keewatin, N.W.T., 62°39′00″W, 98°38′30″W, map unit Ango. Ref: Eade, 1980; Precambrian Geology of the Tukemalu Lake-Yathkyed Lake area, District of Keewatin; G.S.C. Paper 84-11. Sample EA-1310-75, collected by G. Stott and interpreted by K. Eade.
See GSC 87-73 for description and interpretation.

GSC 87-75 Hornblende, K-Ar age
2458 ± 21 Ma

K = 1.1%, radiogenic
Ar = 2240 × 10^{-7} cc/gm
K-Ar 3633 atmos. Ar = 0.4%

From a granodiorite orthogneiss separate.
(65 J) See GSC 87-74 for location and GSC 87-73 for description and interpretation.

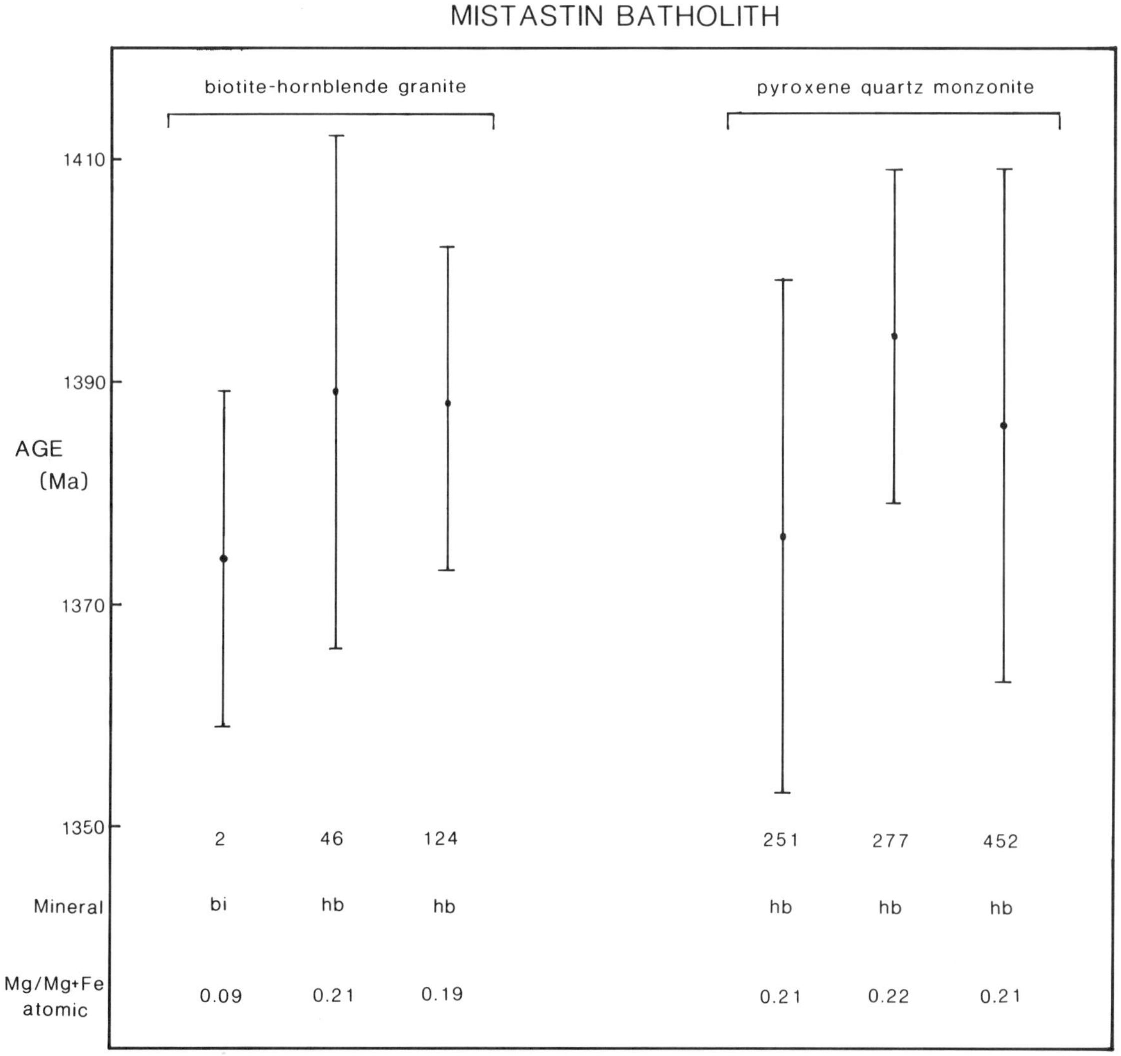

Figure 2. Plot of K-Ar isotopic ages from this report and the previous one in the Yathkyed Lake area.

GSC 87-76 Whole rock, K-Ar age
1445 ± 16 Ma

K = 2.56 %, radiogenic
Ar = 2204 × 10^{-7} cc/gm
K-Ar 3574 atmos. Ar = 0.5 %

From a mafic diorite dyke
(75 E/8) Northeast shore of MacInnis Lake, approximately 750 m north of N-S Bay, N.W.T., 61°23′10″N, 110°07′40″W. Sample GFA-80-14, collected and interpreted by S.S. Ghandi.

The sample is from a mafic diorite dyke that cuts tightly folded sedimentary rocks of the early Proterozoic Nonacho Group (McGlynn, 1971; Gandhi and Prasad, 1980). The dyke was originally mapped as 'granodiorite' during early exploration of the area (Stephen, 1956). Field check in 1980 revealed however that it is a mafic diorite or felsic gabbro dyke, similar to some of the 'Sparrow' dykes in the Nonacho Lake area described by McGlynn et al. (1974). The dyke is approximately 20 m wide, and has irregular intrusive contacts and a general northwesterly trend but its outcrops are limited to a few tens of metres.

The sample is medium to coarse gabbroic in texture, and contains clinopyroxene as the main mafic mineral. Hornblende rims some of the pyroxene crystals. Interstitial minerals include biotite, oxides, quartz, alkali feldspar and granophyre. Major oxide percentages in the rock are: SiO_2: 56.4, Al_2O_3: 14.5, FeO: 5.7, Fe_2O_3: 2.6, MgO: 4.11, CaO: 6.09, Na_2O: 2.8 and K_2O: 3.18.

The K-Ar date on the sample is comparable to similarly dated 3 'sparrow' dykes viz. 1480, 1390 and 1550 Ma, but $^{40}Ar/^{39}Ar$ studies on one of them and three other dykes of the same group, reported by McGlynn et al. (1974), gave an estimated age of 1700 Ma.

REFERENCES

Gandhi, S.S. and Prasad, N.
1980: Geology and uranium occurrences of the MacInnis Lake area, District of Mackenzie; *in* Current Research, Part B, Geological Survey of Canada, Paper 80-1B, p. 107-127.

McGlynn, J.C.
1971: Stratigraphy, sedimentology and correlation of the Nonacho Group, District of Mackenzie, in Report of Activities, Part A, Geological Survey of Canada, Paper 71-1A, p. 140-142.

McGlynn, J.C., Hanson, G.N., Irving, E. and Park, J.K.
1974: Paleomagnetism and age of Nonacho Group sandstones and associated Sparrow dykes, District of Mackenzie; Canadian Journal of Earth Sciences, v. 11, no. 1, p. 30-42.

Stephen, H.C.
1956: Geological report on the WALT claim group, MacInnis Lake, Taltson lake map area, District of Mackenzie, Northwest Territories; Newkirk Mining Corporation Limited; Department of Indian and Northern Affairs, Document no. N 017080, 9 p.

GSC 87-77 Whole rock, K-Ar age
1115 ± 13 Ma

K = .426 %, radiogenic
Ar = 255.4 × 10^{-7} cc/gm
K-Ar 3575 atmos. Ar = 2.2 %

From a gabbro separate
(75 L/10) At the base of a steep north-facing 50 m high escarpment of diabase, capping the ridge of Pethei Peninsula, above the talus slope. N.W.T. 62°43′20″N, 110°51′00″W, elevation: 335 m. Map-unit 15, GSC Map 1122A, Christie Bay. Sample GFA-83-149, collected and interpreted by S.S. Gandhi.

The sample represents the lower part of a sill that is up to 190 m thick and caps the cuesta of the Pethei Peninsula over a 75 km length. A geological cross-section through the sample locality is given by Roscoe et al. (1985, cross-section AB). The rock is a coarse grained gabbro rich in pyroxene, and is virtually unaltered.

The K-Ar date on this sample is very close to that obtained on a sample from the central part of a thin diabase sill belonging to the same intrusive suite (GSC 87-78: 1085 ± 34 Ma). The latter sample has higher K content (2.06 %) and is more altered than is the case with this sample. These two factors thus do not seem to cause significant variation in K-Ar age on rocks from the same suite. On the other hand, both the dates are more than 100 million years younger than the two dates obtained from the chilled margins of the thin sill (GSC 87-79 and 87-80). This implies that the coarser parts of these cogenetic intrusive sheets retain radiogenic argon less effectively than the chilled margins.

GSC 87-78 Whole rock, K-Ar age
1085 ± 34 Ma

K = 2.06 %, radiogenic
Ar = 1192 × 10^{-7} cc/gm
K-Ar 3572 atmos. Ar = 2.5 %

From a diabase sill
(75 L/15) Central part of a 1.3 m thick diabase sill.
See GSC 87-79 for location and interpretation.

GSC 87-79 Whole rock, K-Ar age
1249 ± 14 Ma

K = .88 %, radiogenic
Ar = 616 × 10^{-7} cc/gm
K-Ar 3571 atmos. Ar = 6.4 %

From a diabase sill
(75 L/15) Western corner of Viren Island, McLeod Bay, East Arm of Great Slave Lake,

N.W.T. shore outcrop; map-unit 15, GSC Map 1122A, Christie Bay; 62°47′05″N, 110°57′30″W. Samples GFA-'83-326, 327 and 328 were collected at the same locality and represent the upper chilled margin (GSC 87-79), the central part (GSC 87-78) and the lower chilled margin (GSC 87-80) respectively, of a 1.3 m thick diabase sill.

The diabase sill sampled is apparently an off-shoot of a 5 m thick sill located approximately 1.5 m above it at the sample site (Roscoe et al., 1987, p. 19 and 68). These sills dip gently to the south, and are related to thicker and extensive diabase sills that intruded the strata of the Great slave Supergroup (Stockwell et al., 1968; Hoffman, 1968). The strata were deposited 1895 to 1865 Ma ago. They are exposed on steep side of cuestas capped by the diabase sills. The sills are cut by diabase dykes of the northwest-trending Mackenzie swarm (Fuhrig and Jones, 1969; Fahrig and west, 1986).

The chilled margin samples in thin section show scattered phenocrysts of plagioclase and pyroxene, 0.5 to 1 mm in length, in fine grained matrix containing abundant slender feldspar microlites and interstitial mafic silicates and ozides. The upper chilled margin is altered, and shows pervasive chloritic alteration through the matrix, partially chloritized pyroxene phenocrysts, and patchy cloudiness in some plagioclase crystals due to sericitization. The lower chilled margin sample in contrast is quite fresh, and shows layers of light and dark grey colour, due to difference in abundance of oxides in the matrix but there is not significant textural variation. The sample of the central part of the sill has medium grained diabasic texture, with plagioclase laths ranging from 1 to 2 mm in length, euhedral to subhedral pyroxene and abundant well formed oxide crystals. The rock is moderately altered, as seen from partially chloritized pyroxene and sericitized feldspar, and the presence of chlorite-sericite-rich patches up to a few millimetres in diameter.

The differences in age of the three samples from the same thin sill at the same location is large, amounting to a spread of 160 million years. K-Ar dates on the chilled margins of intrusions are generally regarded as the most reliable, hence the Viren Island sill may be regarded as 1250 Ma old (Roscoe et al., 1987, p. 18). This age is within range of 1200 to 1300 Ma K-Ar dates for the northwest- trending, vertical dykes of the Mackenzie swarm (Fahrig and Jones, 1969; Fahrig and West, 1986; Fahrig, 1987). The dates on the dykes and the sill corroborate paleomagnetic evidence that the two sets of intrusions belong to a single, approximately 1250 Ma of Mackenzie igneous event (Fahrig and Jones, ibid). Although the sills are cut by younger dykes, they themselves must have been fed by dykes from which the magma spread laterally along nearly horizontal strata. The magmatic activity thus most likely proceeded in pulses.

The difference in age between the upper and lower chilled margins, of 30 million years is within the experimental errors for rocks containing relatively small amounts of K and Ar. On the other hand, relatively fresh lower margin sample would be expected to yield an older date but reverse is the case. This probably reflects some accumulation of argon near the root during a slow cooling of the sill, mostly from the central part of the sill. From this standpoint the date from the lower chilled margin is closer to the true age of the intrusion. The central part of the sill in comparison has obviously lost a significant amount of radiogenic argon. This is probably due to its coarser, more pervious texture compared to the margins, and also greater alteration reflected by abundance of chlorite and sericite. It gives a date that is 160 and 135 Ma younger than the two dates from the chilled margins. This is equal to practically the entire time span assigned to the Mackenzie igneous event, viz. 1350 to 1200 Ma, based on K-Ar dates. It is much more likely that the dates give an unduly long time span for the event that was probably short lived, perhaps a few million years, as pointed out by Fahrig (1967) based on two remarkable facts about the Mackenzie igneous rocks: (i) all of their paleomagnetic poles lie within a rather small area and (ii) all have the same magnetic polarity. It is well known that the emplacement of mafic dykes and sills occur in extensional regimes where movement of magma is rapid.

REFERENCES

Bowring, S.A., Van Schmus, W.R., and Hoffman, P.F.
1984: U-Pb zircon ages from Athapuscow Aulacogen, East Arm of Great Slave Lake, N.W.T., Canada; Canadian Journal of Earth Sciences, v. 21, no. 11, p. 1315-1324.

Fahrig, W.F.
1987: The tectonic setting of continental mafic dyke swarms: Failed arm and early passive margin; *in* Mafic Dyke Swarms, ed. Halls, H.C., and Fahrig, W.F., Geological Association of Canada, Special Paper 34.

Fahrig, W.F., and Jones, D.L.
1969: Paleomagnetic evidence for the extent of Mackenzie igneous events; Canadian Journal of Earth Sciences, v. 6, p. 679-688.

Fahrig, W.F., and West, T.D.
1986: Diabase dyke swarms of the Canadian Shield; Geological Survey of Canada, Map 1627A (Scale 1:4 873 900 approx.).

Roscoe, S.M., Gandhi, S.S., Charbonneau, B.W., Maurice, Y.T., and Gibb, R.A.
1987: Mineral resource assessment of the area in the East-Arm (Great Slave Lake) and Artillery Lake region, N.W.T., proposed as a National Park (NTS 75 J, K, L, N, O) Geological Survey of Canada Open File 1434, 92 p.

Stockwell, C.H., Brown, I.C., Barnes, F.Q., and Wright, G.M.
1968: Christie Bay, District of Mackenzie; Geological Survey of Canada Map 1122A (Scale 1:253 440).

GSC 87-80 Whole Rock, K-Ar age
1219 ± 14 Ma

K = 0.819%, radiogenic
Ar = 554.2×10^{-7} cc/gm
K-Ar 3573 atmos. Ar = 5.0%

From a diabase sill
(75 L/15) Lower chilled margin of a 1.3 m thick diabase sill.
See GSC 87-79 for location and interpretation.

GSC 87-81 Muscovite, K-Ar age
2543 ± 21 Ma

K = 8.3%, radiogenic
Ar = 18000 × 10^{-7} cc/gm
K-Ar 3581 atmos. Ar = 0.3%

From a pegmatite separate
(75 O/4) Northwest shore of Artillery Lake. District of Mackenzie. N.W.T., 63°11′00″N, 107°54′07″W. Map-unit 11 Map 1216A, unit 6, GSC Memoir 350. Sample GFA-83-568, collected and interpreted by S.S. Gandhi.

This sample and another one located 55 m to the south-southwest of it (see GSC 87-82) are from muscovite-quartz-feldspar pegmatites in a zone of metasedimentary rocks within the granitic terrain on the north side of Artillery Lake (Gandhi, 1984, 1985). The metasediments are greywacke- type and include some biotite-rich, amphibolitic and quartzose layers and lenses. They trend northwesterly, and have steep layering and foliation. They contain numerous veins, lenses and irregular bodies of pegmatite. All these rocks are unconformably overlain by little disturbed early Proterozoic dolomite of the Artillery Lake formation (Gandhi, ibid.). The two pegmatite samples are located at the unconformity, which is characterized by dolomitization of the basement rocks to a varying degree and the presence of veins of chert and dolomite as well as of quartz-carbonate veins carrying galena and sphalerite (Showing A-7, Roscoe et al., 1987, p. 78-79). One of the pegmatite samples (GFA-83-568, GSC 87-81) is cut by veinlets of dolomite, which in thin section show sharp irregular boundaries across the coarse crystals. Fragments of pegmatite minerals occur as inclusions in the veins. The minerals are in some places partially replaced by dolomite.

The K-Ar dates of the 2543 and 2498 Ma (GSC 87-81, 82) on the two samples are very close with the given margin of inherent uncertainty. They show that the pegmatites were formed at the end of Archean time in the Slave Province, and were not affected by younger Hudsonian events along the Thelon Front which is close to the Artillery Lake on the southeast side. Furthermore, their isotopic equilibrium was apparently little affected by paleoweathering and alteration at the unconformity beneath the little disturbed Artillery Lake formation. The samples are close to low temperature hydrothermal veins carrying galena and sphalerite, which post-date the Artillery Lake formation (Gandhi, 1984, 1985). This hydrothermal event also does not seem to have affected the K-Ar isotopic equilibrium. Lead isotopic data on galena from these veins and several other veins in the Artillery Lake area, show an addition of radiogenic lead most probably derived from the pegmatites and related granites, and suggest a time of lead mixing approximately 1890 to 1990 Ma ago (Roscoe et al., 1987).

REFERENCES

Gandhi, S.S.
1984: Galena-sphalerite-chalcopyrite veins in Aphebian dolomite and Archean basement at Artillery Lake, Northwest Territories, *in* Current Research, Part B, Geological Sruvey of Canada, Paper 84-1B, p. 33-40.

1985: Geology of the Artillery Lake Pb-Zn-Cu district, District of Mackenzie; *in* Current Research, Part A, Geological Survey of Canada, Paper 85-1A, p. 354-363.

Roscoe, S.M., Gandhi, S.S., Charbonneau, B.W., Maurice, Y.T., and Gibb, R.A.
1987: Mineral resource assessment of the area in the East Arm (Great Slave Lake) and Artillery Lake region, N.W.T., proposed as a National park (NTS 75, J, K, L, N, O); Geological Survey of Canada, Open File 1434, 92 p.

GSC 87-82 Muscovite, K-Ar age
2498 ± 21 Ma

K = 8.8%, radiogenic
Ar = 18470 × 10^{-7} cc/gm
K-Ar 3582 atmos. Ar = 0.1%

From a pegmatite separate
(75 O/4) Same location as for GSC 87-81, but 55 m south-southwest. Sample GFA-83-571.
For interpretation see GSC 87-81.

GSC 87-83 Ferrorichterite, K-Ar age
2071 ± 47 Ma

K = 1.24%, radiogenic
Ar = 1.871 × 10^{-4} cc/gm
K-Ar 3409 atmos. Ar = 1.4%

From a granite separate
(85 I) South shore of southwestern Blachford Lake, District of Mackenzie, 62°10′08″N, 112°41′26″W. Sample 77DM-163, collected and interpreted by A. Davidson.

These three ages of amphibole from peralkaline granite of the Grace Lake unit, Blachford Lake Intrusive Suite, are in agreement with those obtained previously from amphiboles of the same unit (GSC 76-188, 2057 ± 56 Ma, and 78-128, 2133 ± 112 Ma; *see* K-Ar Isotopic Ages, Report 14, p. 35-36). They are also in accord with ages obtained by other methods (Rb-Sr whole rock and U-Pb zircon) for older adjacent and related units of this intrusive suite, e.g., 2186 ± 5 Ma (zircon, Whiteman Lake quartz syenite, GSC) and 2175 ± 5 Ma (zircon, Hearne channel granite, Bowring et al, 1984) and with the age of Zr mineralization and related alteration in the adjacent Thor Lake syenite (2094 ± 10 Ma; Bowring et al., 1984).

For reference see GSC 87-79.

GSC 87-84 Ferrorichterite, K-Ar age
2182 ± 47 Ma

K = 1.32%, radiogenic
Ar = 2.177 × 10^{-4} cc/gm
K-Ar 3411 atmos. Ar = 0.8%

From a granite separate
(85 I/2) South shore of southwestern part of Blachford Lake, District of Mackenzie, 62°05′37″N, 112°33′29″W. Sample 77-DM-188, collected and interpreted by A. Davidson.
See GSC 87-83 for description.

GSC 87-85 Riebeckite, K-Ar age
2124 ± 51 Ma

K = 1.14 %, radiogenic
Ar = 1.795 × 10^{-4} cc/gm
K-Ar 3413 atmos. Ar = 0.6 %

From a granite separate
(85 I/2) Peninsula on north shore of Blac Ford Lake, District of Mackenzie, 62°09′49″N, 112°38′57″W. Sample JHI-005, collected by John Hill and interpreted by A. Davidson.
See GSC 87-83 for description and interpretation.

GSC 87-86 Biotite, K-Ar
1816 ± 24 Ma

K = 6.87 %, radiogenic
Ar = 8.364 × 10^{-4} cc/gm
K-Ar 3430 atmos. Ar = 0.5 %

From a paragneiss separate
(86 E/3) Southeast side of Leith Ridge, west of Hottah Lake Exploration Grid 'c' (Noranda-AGIP Joint venture 1978-1980); Line 36 NE, 10 + 50 NW. District of Mackenzie, N.W.T. 65°04′27″N, 119°29′W. Collected and interpreted by S.S. Gandhi.

The paragneiss sample is part of the Holly Lake metamorphic suite exposed along the southeast side of the Leith Ridge, in Hottah Terrane west of the Great Bear magmatic zone (Hildebrand et al., 1984). The suite includes rocks derived from abundant psammitic and pelitic sediments, and mafic to intermediate volcanics. These are metamorphosed to amphibolite facies assemblages, and show well developed foliation trending north to northwest and dipping steeply to vertically. The supracrustal rocks are intruded by numerous plutons of dioritic to granitic composition. The plutons are highly deformed and have yielded U-Pb zircon dates in the range of 1914 to 1900 Ma (Hildebrand and Roots, 1985). They are unconformably overlain by volcanic sequence related to the Great Bear magmatic activity ca 1875 Ma in age. Granitic intrusions related to the Great Bear batholith occur along the Leith Ridge.

The paragneiss sample is medium grained metapelite containing quartz (~ 55 %), feldspar (~ 20 %), biotite (~ 20 %), muscovite (~2 %), oxides, and traces of apatite and zircon. The felsic and mafic minerals are concentrated along folia 1 to 2 mm thick. The folia are crenulated in v-shaped, steeply plunging folds on centimetre scale. Muscovite is concentrated in 2 cm long lensoid aggregates. Quartz forms a number of coarser aggregates up to 5 mm wide. The paragneiss at this locality is unconformably overlain by flatlying lower Paleozoic sediments of the Ronning Group (Seaton, 1984).

The K-Ar date of 1816 ± 24 Ma on biotite concentrate is rather young in view of the zircon dates mentioned above, and it probably reflects cooling and uplift after the intrusion of granites related to the Great Bear batholith. The age on muscovite is anomalously young, and may be attributed to inadequate sample material or other unknown factor.

REFERENCES

Hildebrand, R.S., Annesley, I.R., Bardoux, M.V., Davis, W.J., Heon, D., Reichenback, I.B., and Van Nostrand, T.
1984: Geology of the early Proterozoic rocks in parts of the Leith Peninsula map area, District of Mackenzie; *in* Current Research, Part A, Geological Survey of Canada, Paper 84-1A, p. 217-221.

Hildebrand, R.S., and Roots, C.F.
1985: Geology of the Rivière Grandin map area (Hottah Terrane and western Great Bear Magmatic zone). District of Mackenzie; *in* Current Research, Part A, Geological Survey of Canada, paper 85-1A, p. 373-383.

Seaton, J.B.
1984: Bear Structural Province; *in* Mineral Industry Report 1980-81, Northwest Territories; EGS 1984-5; ed. J.A. Brophy and C.E. Ellis; Department of India ı and Northern Affairs, Ottawa; p. 298-303.

GSC 87-87 Muscovite, K-Ar age
1421 ± 20 Ma

K = 7.45 %, radiogenic
Ar = 6.260 × 10^{-4} cc/gm
K-Ar 3431 atmos. Ar = 0.82

(86 E/3) From a paragneiss separate
See GSC 87-86 for location and interpretation.

GSC 87-88 Biotite, K-Ar age
1789 ± 23 Ma

K = 8.31 %, radiogenic
Ar = 9.881 × 10^{-4} cc/gm
K-Ar 3464 atmos. Ar = 4.0 %

From a lamprophyre separate
(86 G) Bedrock Lake map sheet (86 G). District of Mackenzie, N.W.T., 65°53′N, 115°58′W. Sample P-5, collected by M. St-Onge.

GSC 87-89 Biotite, K-Ar age
91 ± 2 Ma

K = 6.91 %, radiogenic
Ar = 251.8 × 10^{-7} cc/gm
K-Ar 3451 atmospheric Ar = 7.4 %
Concentrate: dark brown biotite with approximately 9 % chlorite alteration

From plagioclase-biotite-quartz porphyry dyke
(105 I) 24 km southeast of major confluence of South Nahanni river, 42 km south of peak 7719 feet, Nahanni map area, southwestern District of Mackenzie, 62°33′24″N, 128°33′54″W; see Anderson (1982, 1983). Sample GGA-79-44E-1, collected by S.P. Gordey and interpreted by R.G. Anderson.

The sample is from a biotite-plagioclase-quartz porphyry dyke which contains 12 % quartz, 9 % plagioclase and 9 % pale to dark brown biotite phenocrysts in a groundmass of plagioclase (27 %), quartz (24 %) and alkali feldspar (19 %). The dyke intrudes the southern South Nahanni pluton.

For interpretation see GSC 87-129.

GSC 87-90 Biotite, K-Ar age
92 ± 2 Ma

K = 7.30 %, radiogenic
Ar = 266.3 cc/gm
K-Ar 3461 atmos. Ar = 9.9 %
Concentrate: light brown biotite with approximately 7 % chlorite contamination and (or) alteration

From granodiorite
(105 I) 14.5 km south-southeast of Mount Wilson, 12 km east of Clea Property, Nahanni map area, southeastern Yukon, 62°45′30″N, 129°39′30″W; see Anderson (1982, 1983). Sample GGAA-80-21a-2, collected and interpreted by R.G. Anderson.

The sample was collected from white weathering, massive, equigranular, medium-grained biotite granodiorite which contains plagioclase (36 %), quartz (34 %), alkali feldspar (19 %), reddish brown biotite (11 %), "secondary" muscovite (0.5 percent) and trace accessory minerals apatite and zircon. The sample is typical of the equigranular phase of the Pelly River pluton.

For interpretation see GSC 87-129.

GSC 87-91 Biotite, K-Ar age
117 ± 3 Ma

K = 3.91 %, radiogenic
Ar = 183.7 × 10^{-7} cc/gm
K-Ar 3465 atmos. Ar = 8.6 %
Concentrate: dark brown impure biotite concentrate with about 6 % chlorite and 5 % hornblende contamination.

From granite
(105 I) 17 km SSE of O'Grady Lakes, 16 km southwest of peak 7719 feet, Nahanni map area, southwestern District of Mackenzie, 62°50′12″N, 128°56′12″W; see Anderson (1982, 1983). Sample GGAB-81-31a-2, collected by Joerg Beekman and interpreted by R.G. Anderson.

The sample was collected from a mottled pink and white weathering, massive, crowded megacrystic biotite-hornblende granite which contains alkali feldspar (43 %), quartz (22 %), plagioclase (20 %), hornblende (11 %), dark brown biotite (2 %) and accessory minerals (1.5 %) magnetite, sphene, apatite, allanite and zircon. Biotite and hornblende are altered to trace amounts (less than 0.2 %) of chlorite and epidote; trace calcite and sericite alteration of plagioclase also occurs. The sample is typical of the crowded megacrystic phase of the O'Grady batholith (see also GSC 87-92). Note very low K content for a biotite which is an average of two separate determinations. The isotopic age is probably meaningless.

For interpretation see GSC 87-129.

GSC 87-92 Hornblende, K-Ar age
78 ± 4 Ma

K = 0.97 %, radiogenic
Ar = 29.86 × 10^{-7} cc/gm
K-Ar 3466 atmos. Ar = 21.3 %
Concentrate: clean, unaltered, pleochroic brown to dark green hornblende, with no visible contamination.

(105 I) Details as for GSC 87-91.
For interpretation see GSC 87-129

GSC 87-93 Biotite, K-Ar age
90 ± 2 Ma

K = 8.00 %, radiogenic
Ar = 287.3 × 10^{-7} cc/gm
K-Ar 3468 atmos. Ar = 20.4 %
Concentrate: relatively clean, unaltered, light brown biotite with a trace of chlorite contamination.

From granodiorite
(105 I) 15 km northwest of southeast corner of Nahanni map area, near headwaters of Flat River, Nahanni map area, southwestern District of Mackenzie, 62°05′54″N, 128°09′24″W; see Anderson (1982, 1983). Sample GGAA-80-102b-1, collected and interpreted by R.G. Anderson.

The sample was collected from white weathering, massive, megacrystic biotite granodiorite containing plagioclase (36 %), quartz (33 %), alkali feldspar (17 %), biotite (12 %), muscovite (0.6 %) and accessory minerals (0.4 %) zircon and tourmaline. Minor (less than 0.2 %), alteration of plagioclase to epidote and calcite and of biotite to chlorite and muscovite occurs. The sample is typical of the megacrystic phase which makes up the homogeneous Shelf Lake pluton (see also GSC 87-94).
For interpretation see GSC 87-129.

GSC 87-94 Muscovite K-Ar age
94 ± 5 Ma

K = 5.3 %, radiogenic
Ar = 198.8 × 10^{-7} cc/gm
K-Ar 3469 atmos. Ar = 21.5 %
Concentrate: clean, fresh, mainly clear muscovite, with no visible contamination.

(105 I) Details as for GSC 87-93.

For interpretation see GSC 87-129.

GSC 87-95 Biotite, K-Ar age
80 ± 4 Ma

K = 7.54 %, radiogenic
Ar = 240.3 × 10^{-7} cc/gm
K-Ar 3470 (2) atmos. Ar = 22.6 %
Concentrate: clean, unaltered, light brown biotite with no detectable chlorite contamination.

(105 I) From granodiorite
32 km south-southeast of peak 7719 feet, 20 km east of major confluence of South Nahanni River, Nahanni map area, southwestern District of Mackenzie, 62°38′48″N, 128°36′54″W; see Anderson (1982, 1983). Sample ANSN-81-46-2, collected and interpreted by R.G. Anderson.

The sample was collected from white-weathering, fresh, massive, inclusion-bearing, equigranular, medium-grained hornblende-biotite granodiorite which contains plagioclase (34 %), quartz (33 %), alkali feldspar (18 %), and trace accessory minerals magnetite, apatite, zircon. Trace alteration of plagioclase to sericite and alteration of hornblende and biotite to chlorite occurs. The sample is typical of the central South Nahanni pluton (see also GSC 87-96).

For interpretation see GSC 87-129.

GSC 87-96 Hornblende, K-Ar age
84 ± 6 Ma

K = 0.63 %, radiogenic
Ar = 21.11 × 10^{-7} cc/gm
K-Ar 3471(2) atmos. Ar = 25.2 %
Concentrate: clean, unaltered light brown to dark green hornblende with no visible contamination.

(105 I) Deatils as for GSC 87-95.

For interpretation see GSC 87-129.

GSC 87-97 Biotite, K-Ar age
91 ± 2 Ma

K = 6.57 %, radiogenic
Ar = 237.8 × 10^{-7} cc/gm
K-r 3473 atmos. Ar = 12.4 %
Concentrate: brown biotite with approximately 10 % chlorite alteration.

(105 I) From aplitic monzogranite
17 km south-southeast of O'Grady Lakes, 16 km southwest of peak 7719 feet, Nahanni map area, southwestern District of Mackenzie, 62°50′12″N, 128°56′12″W; see Anderson (1982, 1983). Sample GGAB-80-31b-1, collected by J. Beekman and interpreted by R.G. Anderson.

The sample was collected from pink weathering, anastomosing aplitic monzogranite intrusion in the crowded megacrystic phase of the O'Grady batholith. The aplite consists of plagioclase (41 %), quartz (32 %), alkali feldspar (25 %), brown biotite (1 %), and accessory minerals (1 %) magnetite, sphene, zircon, and allanite. Trace alteration of biotite to chlorite and plagioclase to sericite occurs. Sample is typical of the late aplitic granite dykes which intrude the O'Grady batholith.

For interpretation see GSC 87-129.

GSC 87-98 Biotite, K-Ar age
88 ± 2 Ma

K = 7.97 %, radiogenic
Ar = 273.7 × 10^{-7} cc/gm
K-Ar 3474(2) atmos. Ar = 22.3 %
Concentrate: relatively clean, light reddish brown biotite with a trace of chlorite alteration.

(105 I) From granodiorite
16 km south-southeast of Mount Wilson, 12 km east of Clea Property, Nahanni map area, southwestern District of Mackenzie, 62°44′45″N, 129°38′30″W; see Anderson (1982, 1983). Sample ANPR-81-32-5, collected and interpreted by R.G. Anderson.

The sample was collected from white weathering, fresh, foliated, magacrystic biotite granodiorite which contains quartz (36 %), plagioclase (34 %), reddish brown biotite (14 %) and trace accessory minerals apatite, zircon, and "secondary" muscovite. Trace alteration of biotite to chlorite and plagioclase to saussurite and sericite occurs. The sample is typical of the megacrystic phase of the Pelly River pluton.

For interpretation see GSC 87-129.

GSC 87-99 Biotite, K-Ar age
91 ± 2 Ma

K = 7.55 %, radiogenic
Ar = 272.7 × 10^{-7} cc/gm
K-Ar 3475 atmos. Ar = 15.6 %
Concentrate: light brown biotite with approximately 3 % chlorite alteration.

(105 I) From granodiorite
30 km south of peak 7719 feet, 18 km east-southeast of major confluence of South Nahanni River, Nahanni map area, southwestern District of Mackenzie, 62°39′30″N, 128°39′30″W; District of Mackenzie, 62°39′30″N, 128°39′30″W; see Anderson (1982, 1983). Sample ANSN-81-42-3, collected and interpreted by R.G. Anderson.

The sample was collected from white weathering, fresh, massive, inclusion-bearing, equigranular, medium-grained hornblende-biotite granodiorite which contains plagioclase (43 %), alkali feldspar (12 %), hornblende (3 %) and trace accessory minerals (0.2 %) magnetite, apatite, zircon and allanite. Rare porphyritic andesite dykes intrude the unit within 10-20 m of site. The sample is typical of the central South Nahanni pluton (see also GSC 87-100).

For interpretation see GSC 87-129.

GSC 87-100 Hornblende, K-Ar age
90 ± 7 Ma

K = 0.55 %, radiogenic
Ar = 19.21 × 10^{-7} cc/gm

K-Ar 3476 atmos. Ar = 33.4 %
Concentrate: clean, unaltered, light brown to green hornblende, with no visible contamination.

(105 I) Details as for GSC 87-99.

For interpretation see GSC 87-129.

GSC 87-101 Biotite, K-Ar age
92 ± 2 Ma

K = 7.84 %, radiogenic
Ar = 290.3 × 10^{-7} cc/gm
K-Ar 3477 atmos. Ar = 15.0 %
Concentrate: clean, unaltered, brownish orange biotite with no detectable chlorite.

From granodiorite
(105 I) 16 km south-southeast of Mount Wilson, 11 km east-southeast of Clea Property, Nahanni map area, southwestern District of Mackenzie, 62°44′54″N, 129°40′00″W; see Anderson (1982, 1983). Sample ANPR-81-22-8, collected and interpreted by R.G. Anderson.

The sample was collected from white-weathering, fresh, faintly foliated, equigranular, medium-grained biotite granodiorite which contains plagioclase (37 %), quartz (25 %), alkali feldspar (19 %), reddish brown biotite (17 %) and trace accessory minerals (2 %) apatite and zircon. The sample is typical of the equigranular phase of the Pelly River pluton.

For interpretation see GSC 87-129.

GSC 87-102 Biotite, K-Ar age
92 ± 2 Ma

K = 7.01 %, radiogenic
Ar = 256.9 × 10^{-7} cc/gm
K-Ar 3478 atmos. Ar = 13.4 %
Concentrate: light brownish orange biotite with approximately 10 % chlorite alteration.

From granodiorite
(105 I) 14.5 km south-southeast of Mount Wilson, 12 km east of Clea Property, Nahanni map area, southwestern District of Mackenzie, 62°45′30″N, 129°39′48″W; see Anderson (1982, 1983). Sample ANPR-81-25-1, collected and interpreted by R.G. Anderson.

The sample was collected from white-weathering, fresh, massive equigranular, medium-grained biotite granodiorite which contains plagioclase (41 %), quartz (30 %), alkali feldspar (14 %), reddish brown biotite (15 %) and trace accessory minerals (1 %) apatite, zircon, "secondary" muscovite, monazite and allanite. The sample is typical of the equigranular phase of the Pelly River pluton near the contact with the megacrystic phase.

For interpretation see GSC 87-129.

GSC 87-103 Biotite, K-Ar age
91 ± 2 Ma

K = 7.50 %, radiogenic
Ar = 273 × 10^{-7} cc/gm
K-Ar 3481 atmos. Ar = 10.5 %
Concentrate: light brown biotite with approximately 6 % chlorite alteration.

From granodiorite
(105 I) 15 km south-southeast of Mount Wilson, 12 km east of Clea Property, Nahanni map area, southwestern District of Mackenzie, 62°45′42″N, 129°38′00″W; see Anderson (1982, 1983). Sample ANPR-81-27-7, collected and interpreted by R.G. Anderson.

The sample was collected from white-weathering, fresh massive, sparingly megacrystic, medium-grained biotite granodiorite which contains plagioclase (33 %), quartz (34 %), alkali feldspar (19 % of which 9 % are megacrysts), reddish brown biotite (15 %) and trace accessory minerals "secondary" muscovite, apatite, and zircon. The sample is typical of the equigranular phase of the Pelly River pluton.

For interpretation see GSC 87-129.

GSC 87-104 Biotite, K-Ar age
90 ± 2 Ma

K = 7.34 %, radiogenic
Ar = 264.5 × 10^{-7} cc/gm
K-Ar 3486 atmos. Ar = 13.6 %
Concentrate: light brown biotite with approximately 2 % chlorite alteration.

From hornblende-biotite-quartz-feldspar porphyry dyke
(105 I) 26 km south-southeast of peak 7719 feet, 21 km east-northeast of major confluence of South Nahanni River, Nahanni map area, southwestern District of Mackenzie, 62°42′12″N, 128°36′06″W; see Anderson (1982, 1983). Sample ANSN-81-58-1, collected and interpreted by R.G. Anderson.

The sample was collected from a porphyritic dyke which intrudes the central South Nahanni pluton and which contains phenocrysts of plagioclase (30 %), dark brown biotite (8 %), quartz (5 %), hornblende (3 %) and accessory minerals (0.5 %) magnetite, alkali feldspar, sphene, apatite, zircon, and allanite in a groundmass (54 %) of similar composition (see also GSC 87-105). The sample is typical of the porphyritic dykes which commonly intrude the central South Nahanni pluton.

For interpretation see GSC 87-129.

GSC 87-105 Hornblende, K-Ar age
87 ± 5 Ma

K = 0.73 %, radiogenic
Ar = 25.20 × 10^{-7} cc/gm

K-Ar 3487 atmos. Ar = 19.9 %
Concentrate: clean, unaltered, light brown to dark green hornblende with no visible contamination.

(105 I) Details as for GSC 87-104.

For interpretation see GSC 87-129.

GSC 87-106 Biotite, K-Ar age
93 ± 2 Ma
K = 7.47 %, radiogenic
Ar = 278.3 × 10^{-7} cc/gm
K-Ar 3488 atmos. Ar = 11.8 %
Concentrate: clean, unaltered, brown biotite with a trace of chlorite contamination.

From biotite-hornblende-quartz-plagioclase porphyry dyke
(105 I) 33 km south of peak 7719 feet, 20 km east of major confluence of South Nahanni River, Nahanni map area, southwestern District of Mackenzie, 62°38′24″N, 128°37′24″W; see Anderson (1982, 1983). Sample ANSN-81-43-2, collected and interpreted by R.G. Anderson.

The sample was collected from a porphyry dyke which intrudes typical granodiorite of the central South Nahanni pluton. The dyke contains phenocrysts of plagioclase (18 %), hornblende (7 %), dark brown biotite (4 %), quartz (2 %) and trace accessory minerals (1 %) alkali feldspar, magnetite, sphene, apatite, zircon, and allanite in a fine-grained groundmass (68 %) of nearly identical composition (see also GSC 87-107). The sample is typical of the porphyritic dykes which commonly intrude the central South Nahanni pluton.

For interpretation see GSC 87-129.

GSC 87-107 Hornblende, K-Ar age
106 ± 6 Ma
K = 0.87 %, radiogenic
Ar = 36.92 × 10^{-7} cc/gm
K-Ar 3489 atmos. Ar = 85.4 %
Concentrate: clean, unaltered light brown to dark green hornblende, with a trace of biotite contamination.

(105 I) Details as for GSC 87-106.

The very low radiogenic argon content of this sample renders the calculated age unreliable and it should be interpreted with caution. No concentrate remained for re-extraction.

For additional interpretation see GSC 87-129.

GSC 87-108 Biotite, K-Ar age
92 ± 2 Ma
K = 7.58 %, radiogenic
Ar = 279.0 × 10^{-7} cc/gm
K-Ar 3490 atmos. Ar = 8.5 %
Concentrate: relative clean, unaltered, light brown biotite with a trace of chlorite contamination.

From granodiorite.
(105 I) 31 km southeast of peak 7719 feet, 22 km east-northeast of major confluence of South Nahanni River, Nahanni map area, southwestern District of Mackenzie, 62°39′48″N, 128°34′45″W; see Anderson (1982, 1983). Sample ANSN-81-54-2, collected and interpreted by R.G. Anderson.

The sample is a white weathering, massive, inclusion-bearing, equigranular, medium-grained hornblende-biotite granodiorite which contains plagioclase (42 %), quartz (25 %), alkali feldspar (19 %), brown biotite (12 %), hornblende (2 %) and accessory minerals (0.2 %) apatite, zircon, allanite, tourmaline and magnetite. Trace alteraton of hornblende and biotite to chlorite and epidote occurs. The sample is typical of the central South Nahanni pluton (see also GSC 87-109).

For interpretation see GSC 87-129.

GSC 87-109 Hornblende, K-Ar age
89 ± 7 Ma
K = 0.58 %, radiogenic
Ar = 20.59 × 10^{-7} cc/gm
K-Ar 3491 atmos. Ar = 19.5 %
Concentrate: clean, unaltered light brown to dark green hornblende with no visible contamination.

(105 I) Details as for GSC 87-108

For interpretation see GSC 87-129.

GSC 87-110 Biotite, K-Ar age
92 ± 2 Ma
K = 7.28 %, radiogenic
Ar = 266.8 × 10^{-7} cc/gm
K-Ar 3492 atmos. Ar = 14.4 %
Concentrate: clean, unaltered light brown biotite with no detectable chlorite.

From granodiorite
(105 I) 30 km south of peak 7719 feet, 19 km east-northeast of major confluence of South Nahanni River, Nahanni map area, southwestern District of Mackenzie, 62°39′42″N, 128°38′24″W; see Anderson (1982, 1983). Sample ANSN-81-38-1, collected and interpreted by R.G. Anderson.

The sample was collected form white weathering, fresh, massive, inclusion-bearing, equigranular, medium-grained hornblende-biotite granodiorite which contains plagioclase (42 %), quartz (28 %), brown biotite (15 %), alkali feldspar (12 %), hornblende (3 %) and trace amounts of accessory minerals (magnetite, apatite, zircon). Trace (less than 0.1 %) amounts of chlorite after hornblende and biotite. The sample is typical of the central South Nahanni pluton (see also GSC 87-111).

For interpretation see GSC 87-129.

GSC 87-111 Hornblende, K-Ar age
93 ± 6 Ma

K = 0.71 %, radiogenic
Ar = 26.45 × 10^{-7} cc/gm
K-Ar 3493 atmos. Ar = 4.3 %
Concentrate: clean, unaltered, light brown to light green hornblende with no visible contamination.

(105 I) Details as for GSC 87-110

For interpretation see GSC 87-129.

GSC 87-112 Biotite, K-Ar age
91 ± 2 Ma

K = 7.28 %, radiogenic
Ar = 2.641 × 10^{-7} cc/gm
K-Ar 3494 atmos. Ar = 7.3 %
Concentrate: very fine-grained, clean, unaltered brownish orange biotite with no detectable chlorite.

From biotite lamprophyre dyke.

(105 I) 15 km south-southeast of Mount Wilson, 12 km east-southeast of Clea Property, Nahanni map area, southwestern District of Mackenzie, 62°45′15″N, 129°38′42″W; see Anderson (1982, 1983). Sample ANPR-81-30-2, collected and interpreted by R.G. Anderson.

The sample was collected from a dark green weathering, massive, tough, biotite porphyry lamprophyre dyke which intrudes the equigranular phase of the Pelly River pluton. The dyke contains reddish brown biotite (31 %), clinopyroxene (17 %), and hornblende (0.3 %), a groundmass (51 %) of plagioclase, alkali feldspar, quartz and apatite.

For interpretation see GSC 87-129.

GSC 87-113 Biotite, K-Ar age
95 ± 1 Ma

K = 7.294 %, radiogenic
Ar = 277.69 × 10^{-7} cc/gm
K-Ar 3595 atmos. Ar = 2.58 %

From monzogranite
(105 I) 0.1 km west of summit of Mount Wilson, Nahanni map area, southwestern District of Mackenzie, 62°53′12″N, 129°42′30″W; see Anderson (1982, 1983). Sample ANMW-82-328-1, collected and interpreted by R.G. Anderson.

The sample was collected from white weathering, fresh, massive, homogeneous, seriate, medium-grained, biotite monzogranite which consists of quartz (50 %), plagioclase (29 %), alkali feldspar (15 %), biotite (9 %), and "secondary" muscovite (0.6 %). The sample is from the structurally highest part of the Mount Wilson pluton and is typical of the seriate variety of the Mount Wilson pluton.

For interpretation see GSC 87-129.

GSC 87-114 Hornblende, K-Ar age
101 ± 2 Ma

K = 0.554 %, radiogenic
Ar = 22.26 × 10^{-7} cc/gm
K-Ar 3598 atmos. Ar = 18.98 %

From granite
(105 I) 15.8 km southeast of O'Grady Lakes, 18.4 km southwest of peak 7719 feet, Nahanni map area, southwestern District of Mackenzie, 62°49′00″N, 129°02′30″W; see Anderson (1982, 1983). Sample ANOG-82-130-1, collected and interpreted by R.G. Anderson.

The sample was collected from white weathering, fresh massive, inclusion-bearing, equigranular biotite-hornblende granite which contains alkali feldspar (41 %), plagioclase (18 %), quartz (18 %), hornblende (19 %), dark brown biotite (2 %), clinopyroxene (1 %) and accessory minerals (2 %) sphene, apatiti, tourmaline, magnetite, and allanite. Trace alteration of hornblende and biotite to epidote and chlorite and plagioclase to sericite occur. The sample is typical of the equigranular marginal phase of the O'Grady batholith.

For interpretation see GSC 87-129.

GSC 87-115 Biotite, K-Ar age
94 ± 1 Ma

K = 6.68 %, radiogenic
Ar = 250.01 × 10^{-7} cc/gm
K-Ar 3600 atmos. Ar = 12.15 %

From aplitic granite
(105 I) 12.3 km southeast of O'Grady Lakes, 15.8 km west-southwest of peak 7719 feet, Nahanni map area, southwestern District of Mackenzie, 62°54′00″N, 128°57′00″W; see Anderson (1982, 1983). Sample ANOG-82-166-2, collected and interpreted by R.G. Anderson.

The sample was collected from an irregular pink- weathering, fine- to medium-grained aplitic granite dyke which intrudes the crowded megacrystic phase of the O'Grady batholith. It contains alkali feldspar (43 %), quartz (40 %), plagioclase (15 %), brown biotite (1 %), and trace (0.2 % each) hornblende and clinopyroxene. Minor (1 %) sericite alteration of plagioclase occurs. The sample is typical of the late aplitic granite dykes which intrude the O'Grady batholith.

For interpretation see GSC 87-129.

GSC 87-116 Biotite, K-Ar age
98 ± 2 Ma

K = 7.784 %, radiogenic
Ar = 303.66 × 10^{-7} cc/gm
K-Ar 3605 atmos. Ar = 9.73 %

From granodiorite.
(105 I) 28 km southeast of O'Grady Lakes, 7 km south-southeast of peak 7719 feet, Nahanni map area, southwestern District of Mackenzie, 62°52′24″N,

128°38′30″W; see Anderson (1982, 1983). Sample ANOG-82-232-1, collected and interpreted by R.G. Anderson.

The sample was collected form white weathering, fresh, massive, inclusion-bearing, medium-grained hornblende-biotite granodiorite which contains plagioclase (34 %), quartz (22 %), alkali feldspar (17 %), pale to dark brown biotite (14 %), hornblende (11 %) and accessory minerals (1 %) magnetite, tourmaline and apatite. The sample is typical of the O'Grady batholith's marginal equigranular phase (see also GSC 87-117).

For interpretation see GSC 87-129.

GSC 87-117 Hornblende, K-Ar age
95 ± 2 Ma

K = 0.557 %, radiogenic
Ar = 21.16 × 10^{-7} cc/gm
K-Ar 3606 atmos. Ar = 36.36 %

(105 I) Details as for GSC 87-116.

For interpretation see GSC 87-129.

GSC 87-118 Biotite, K-Ar age
96 ± 1 Ma

K = 7.301 %, radiogenic
Ar = 279.54 × 10^{-7} cc/gm
K-Ar 3608 atmos. Ar = 4.50 %

From monzogranite
(105 I) 19.3 km south-southeast of peak 7719 feet, 28 km northwcst of major confluence of south Nahanni River, Nahanni map area, southwestern District of Mackenzie, 62°46′48″N, 128°38′54″W; see Anderson (1982, 1983). Sample ANNN-82-177-3, collected and interpreted by R.G. Anderson.

The sample was collected from white-weathering, fresh, massive, inclusion-bearing hornblende-biotite monzogranite which contains quartz (43 %), plagioclase (28 %), alkali feldspar (17 %), brown biotite (11 %) and minor hornblende and magnetite (1 %). The sample is typical of the northern South Nahanni pluton.

For interpretation see GSC 87-129.

GSC 87-119 Biotite, K-Ar age
96 ± 1 Ma

K = 7.583 %, radiogenic
Ar = 289.55 × 10^{-7} cc/gm
K-Ar 3612 atmos. Ar = 6.80 %

From biotite porphyry dyke
(105 I) 46.4 km south of peak 7719 feet, 21.9 km east-southeast of major confluence of South Nahanni River, Nahanni map area, southwestern District of Mackenzie, 62°33′24″N, 128°34′18″W; see Anderson (1982, 1983). Sample ANSS-82-293-1, collected and interpreted by R.G. Anderson.

The sample is from a biotite porphyry dyke containing 16 % brown biotite in a fine-grained groundmass. The sample is typical of rare porphyry dykes which intrude the southern South Nahanni pluton.

For interpretation see GSC 87-129.

GSC 87-120 Biotite, K-Ar age
96 ± 1 Ma

K = 7.592 %, radiogenic
Ar = 289.41 × 10^{-7} cc/gm
K-Ar 3614 atmos. Ar = 7.27 %

From coarsely megacrystic monzogranite
(105 I) 44.6 km south of peak 7719 feet, 24 km east-southeast of major confluence of South Nahanni River, Nahanni map area, southwestern District of Mackenzie, 62°33′54″N, 128°36′00″W; see Anderson (1982, 1983). Sample ANSS-82-298-1, collected and interpreted by R.G. Anderson.

The sample is typical of the white-weathering, weakly foliated, inclusion-poor, coarsely megacrystic monzogranite and contains quartz (35 %), plagioclase (31 %), alkali feldspar (26 % including up to 7 % megacrysts), brown biotite (8 %) and hornblende (0.5 %). Alteration of biotite to chlorite and plagioclase to sericite amounts to 0.5 % each. Multiple generations of aplite dykes intrude the southern South Nahanni pluton, although not where the sample was collected.

For interpretation see GSC 87-129.

GSC 87-121 Biotite, K-Ar age
93 ± 1 Ma

K = 7.889 %, radiogenic
Ar = 292.82 × 10^{-7} cc/gm
K-Ar 3616 atmos. Ar = 5.81 %

From coarsely megacrystic granite
(105 I) 10.5 km south-southeast of confluence of Little and South Nahanni Rivers, 6.3 km east of Drill Lake, Nahanni map area, southwestern District of Mackenzie, 62°23′12″N, 128°35′30″W; see Anderson (1982, 1983). Sample ANLD-82-274-1, collected and interpreted by R.G. Anderson.

The sample was collected from white weathering, fresh, massive, homogeneous, medium- to very coarse-grained, coarsely megacrystic biotite granite near the margin of the LENED pluton. The rock consists of quartz (38 %), alkali feldspar (23 %), plagioclase (23 %), reddish brown biotite (13 %), ''secondary'' muscovite (1 %) and accessory minerals (less than 1 %) apatite, monazite, and zircon. Minor chlorite alteration of biotite (1 %) and fine-grained white mica alteration of plagioclase (2 %) occurs. The sample is typical of the coarsely megacrystic phase of the LENED pluton.

For interpretation see GSC 87-129.

GSC 87-122 Biotite K-Ar age
87 ± 2 Ma

K = 6.7 %, radiogenic
Ar = 232 × 10^{-7} cc/gm
K-Ar 3618 atmos. Ar = 14.0 %

From aplitic monzogranite sill
(105 I) 17 km southeast of confluence of Little and South Nahanni Rivers, 12.5 km east-southeast of Drill Lake, Nahanni map area, southwestern District of Mackenzie, 62°20′56″N, 128°29′48″W; see Anderson (1982, 1983). Sample ANRU-82-241-3, collected and interpreted by R.G. Anderson.

The sample was collected from the centre of a 6-8 m wide, steeply northeast-dipping sill which intrudes calc-silicate hornfels along the northeast margin of the RUDI pluton. The rock consists of quartz (41 %), alkali feldspar (26 %), plagioclase (26 %), "primary" muscovite (4 %), reddish brown biotite (1 %), and accessory minerals (less than 1 %) garnet and tourmaline. Chlorite alteration products (0.3 %) of biotite and fine-grained white mica alteration products (0.6 %) of plagioclase occur. The lithology is typical of satellitic or marginal peraluminous phases of two-mica plutons associated with tungsten skarns (see also GSC 87- 123).

For interpretation see GSC 87-129.

GSC 87-123 Muscovite K-Ar age
84 ± 1 Ma

K = 5.748 %, radiogenic
Ar = 192.04 × 10^{-7} cc/gm
K-Ar 3619 atmos Ar = 10.18 %

(105 I) Details as for GSC 87-122.

For interpretation see GSC 87-129.

GSC 87-124 Biotite, K-Ar age
94 ± 1 Ma

K = 7.659 %, radiogenic
Ar = 285.92 × 10^{-7} cc/gm
K-Ar 3621 atmos. Ar = 7.79 %

From megacrystic hornblende-biotite monzogranite
(105 I) 44.6 km south of peak 7719 feet, 24 km east-southeast of major confluence of South Nahanni River, Nahanni map area, southwestern District of Mackenzie, 62°34′18″N, 128°35′06″W; see Anderson (1982, 1983). Sample ANSS-82-300-1, collected and interpreted by R.G. Anderson.

The coarsely megacrystic hornblende-biotite monzogranite is white weathering, fresh, faintly foliated, and contains quartz (38 %), alkali feldspar (35 % which includes up to 5 % coarse megacrysts), plagioclase (21 %) brown biotite (4 %) and hornblende (0.5 %). Alteration of biotite to chlorite amounts to 0.8 %. The sample is typical of the homogeneous, sparingly hornblende-bearing, coarsely megacrystic monzogranite of the southern South Nahanni pluton (see also GSC 87-125).

For interpretation see GSC 87-129.

GSC 87-125 Hornblende, K-Ar age
93 ± 1 Ma

K = 0.835 %, radiogenic
Ar = 31.01 × 10^{-7} cc/gm
K-Ar 3622 atmos. Ar = 24.60 %

(105 I) Details as for GSC 87-124

For interpretation see GSC 87-129.

GSC 87-126 Biotite, K-Ar age
92 ± 1 Ma

K = 7.193 %, radiogenic
Ar = 264.84 × 10^{-7} cc/gm
K-Ar 3624 atmos. Ar = 7.99 %

From granite
(105 I) 15.8 km southeast of O'Grady Lakes, 18.4 km west-southwest of peak 7719 feet, Nahanni map area, southwestern District of Mackenzie, 62°52′12″N, 128°59′00″W; see Anderson (1982, 1983). Sample ANOG-82-138-1, collected and interpreted by R.G. Anderson.

The sample was collected from white weathering, fresh, homogeneous, massive, medium- to coarse-grained crowded megacrystic biotite-hornblende granite which contains alkali feldspar (50 %), quartz (22 %), plagioclase (15 %), hornblende (9 %), dark brown biotite (1 %) and accessory minerals (1 %) magnetite, apatite and sphene. Minor (2 %) sericitic alteration of plagioclase occurs. The sample is typical of the crowded megacrystic phase of the O'Grady batholith (see also GSC 87-127).

For interpretation see GSC 87-129.

GSC 87-127 Hornblende, K-Ar age
95 ± 1 Ma

K = 0.965 %, radiogenic
Ar = 36.72 × 10^{-7} cc/gm,
K-Ar 3625 atmos. Ar = 8.93 %

(105 I) Details as for GSC 87-126.

For interpretation see GSC 87-129.

GSC 87-128 Biotite, K-Ar age
94 ± 1 Ma

K = 7.378 %, radiogenic
Ar = 276.37 × 10^{-7} cc/gm
K-Ar 3627 atmos. Ar = 3.76 %

From monzogranite
(105 I) 6 km south-southeast of south end of Shelf Lake, 2.5 km east-southeast of Zenchuk Lake, Nahanni map area, southwestern District of Mackenzie,

62°04′58″N, 128°07′00″W; see Anderson (1982, 1983). Sample ANFL-82-312-6, collected and interpreted by R.G. Anderson.

The sample was collected from a white weathering, massive, sparsely inclusion-bearing, seriate to megacrystic (hornblende-) biotite monzogranite near the contact with calc-silicate hornfels. The sample consists of quartz (31 %), alkali feldspar (31 %), plagioclase (30 %), brown biotite (6 %), hornblende (with clinopyroxene cores; 1.5 %) and accessory minerals (1 %) apatite, sphene, zircon and allanite. Chlorite alteration products of biotite and calcite products of plagioclase are rare (less than 0.1 %). The sample is typical of the homogeneous, sparingly hornblende-bearing megacrystic Shelf Lake pluton.

For interpretation see GSC 87-129.

GSC 87-129 Biotite, K-Ar age
91 ± 1 Ma

K-Ar 3695 — K = 6.54 %, radiogenic Ar = 237.3 × 10^{-7} cc/gm atmos. Ar = 7.9 %

(105 I) — From biotite lamprophyre 6.5 m southeast of south end of Drill Lake, 13 km south of confluence of Little and South Nahanni Rivers, central Nahanni map area, southwestern District of Mackenzie, 62°21′46″N, 128°37′20″W; see Anderson (1982, 1983). Sample ANLZ-82-278-1, collected and interpreted by R.G. Anderson.

The sample was collected from Union Carbide Exploration Corporation drill core (DDH 80-L-72-F-118.8 m) intersection of a biotite lamprophyre dyke which clearly crosscuts garnet skarn in calc-silicate hornfels of Cambro- Ordovican limestone. The dark grey, massive equigranular, fine-grained homogeneous lamprophyre consists of reddish- brown biotite (19 %), opaque minerals (1 %), hornblende (30 %) in a fine-grained groundmass (50 %). The sample is typical of ''post-ore'' lamprophyre rarely intersected in drill core.

INTERPRETATION OF SELWYN PLUTONIC SUITE K-Ar AGES

Field mapping (Anderson, 1982, 1983) complemented by petrography, geochemistry and mineralogical and isotopic studies (Anderson et al., 1983) defined the Selwyn plutonic suite in terms of two ''end-member'' plutonic types distributed in northwest-trending belts and a medial ''transitional'' group of plutons (Figure 3). Granodiorite and granite plutons with widespread hornblende (Emerald Lake, O'Grady, northern South Nahanni and central South Nahanni plutons) formed the northeastern belt. Monzogranite and lesser granodiorite plutons which contain biotite and ''primary'' or ''secondary'' muscovite (''two-mica'' plutons) but lack hornblende made up the southwestern belt (MACTUNG, CLEA, Pelly River, LENED, RUDI, CAC, and CANTUNG plutons). ''Transitional'' plutons, mainly biotite-bearing but which also contain rare, scattered hornblende, clinopyroxene and ''secondary'' muscovite, occupy a medial position (Mt. Wilson, southern South Nahanni, and Shelf Lake plutons). The suite intruded miogeoclinal limestone, chert, shale and coarse clastics of platform and basinal aspect and of Proterozoic to Late Triassic age. Slaty cleavage developed in Jura-Cretaceous deformation is annealed in narrow, andalusite-bearing aureoles around the high level plutons. There are no stratigraphic constraints on the youngest age of intrusion.

World class size (MACTUNG and CANTUNG) and other important (CLEA and LENED-RUDI-CAC) tungsten-base metal (Cu ± Zn ± Mo ± Sn) mineral deposits (Dawson and Dick, 1978; Dick, 1979, 1980; Dick and Hodgson, 1982; Godwin et al., 1980; Atkinson and Baker, 1986; Glover and Burson, 1986) are associated with the two-mica plutons whereas base metal (Cu ± Mo ± gold) occurrences are associated with the hornblende-bearing plutons.

The oldest of a range of isotopic ages (80-102 Ma) determined in earlier studies of a few of the plutons in the suite were interpreted to indicate an intrusive interval between 89-96 Ma for the suite (see Table 2; Baadsgaard et al., 1961; Wanless et al., 1970, 1974; Archibald et al., 1978, Godwin et al., 1980; Archibald, 1980, 1981; Smit et al, 1985; partly summarized in Anderson, 1982, 1983). The present study, which added 45 new K-Ar isotopic ages (summarized in Table 1 and Figure 4), had a five-fold objective: (1) to confirm and expand the geochronometric data base for the complete plutonic suite; (2) to investigate the suspected consanguinity of constituent phases mapped in various plutons; (3) to compare isotopic ages between tungsten skarn-associated two-mica and barren hornblende- bearing plutons; (4) to date late dykes to provide a minimum age for the plutons in the absence of tight stratigraphic age constraints; (5) to provide some age constraints on the formation of tungsten skarn associated with the two-mica plutons.

Newly determined mid-Cretaceous K-Ar ages range from 78-117 Ma (Table 1). Two of the oldest isotopic ages (GSC 87-91) (O'Grady batholith) and 87-107 (central South Nahanni pluton) are unreliable. All reliable Rb-Sr isochron, U-Pb, and earlier K-Ar isotopic ages from Table 2 average 91.3 ± 4.7 Ma (1 s.d.) for the 44 determinations on plutonic and satellitic plutonic phases. The 78 ± 4 Ma isotopic age for hornblende for one sample of the O'Grady batholith's crowded megacrystic phases is anomalously young and was not included in the average. Determinations for most plutons are internally concordant except for CLEA and central South Nahanni plutons and O'Grady batholith (Table 2). Mineral pairs are generally concordant (except for CLEA pluton (KTP-Bi, KTP-Mu) and O'Grady batholith (G.S.C. 67-65 and 67-66 and GSC 87-126 and 87-127 in Table 2). However, O'Grady batholith (G.S.C. 67-65 and 67-66 and GSC 87-126 and 87-127 in Table 2). However, biotite (in two-mica pairs) or hornblende (in hornblende-biotite pairs) do not commonly retain the oldest ages as might be expected from their blocking temperatures (e.g., CLEA pluton (KTP-Bi, KTP-Mu), southern South Nahanni pluton (GSC 87-124 and 87-125), Shelf Lake pluton (GSC 87-93 and 87-94), O'Grady batholith (GSC 67-65 and 67-66 and GSC 87-92, 87-116 and 87-117) and central South Nahanni pluton (GSC 87-163, 87-96, 87-99, 87-100, 87-108, 87-109)). Isotopic

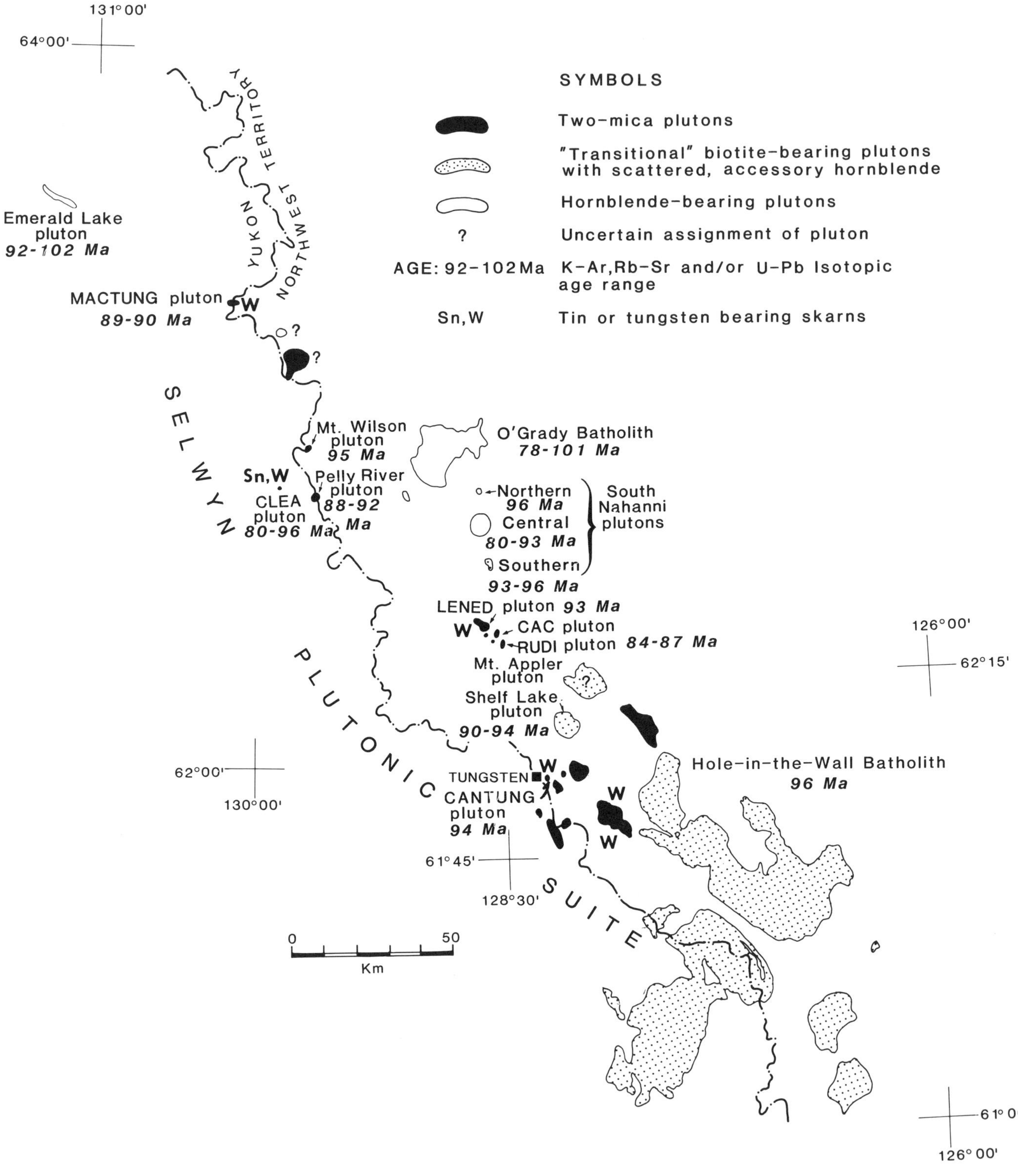

Figure 3. Distribution of some of the plutons in the Selwyn plutonic suite and their K-Ar isotopic ages. See Table 2 for sources of data and uncertainties for individual determinations.

Table 1. Compilation of new GSC K-Ar Isotopic ages Selwyn Plutonic Suite -- southeastern Yukon and southwestern District of Mackenzie

GSC No.	Field Number	Location	Lithology	Mineral Dated	Age (Ma)
		Two-Mica Plutons			
		Mactung Pluton			
GSC 87-130	ANMT-81 8-1	1 km northeast of Mt. Allan, west end of "Cirque Lake", N.W.T.	granodiorite	Bi	89 ± 2
GSC 87-131	ANMT-81 5-6	2 km northeast of Mt. Allan, 17 km southeast of Keele Peak, N.W.T.	monzogranite	Bi	90 ± 2
GSC 87-132	"	"	"	Mu	90 ± 11
GSC 87-133	ANMT-82 331-1	2 km northeast of Mt. Allan, 16 km southeast of Keele Peak, N.W.T.	aplitic monzogranite	Mu	90 ± 1
		Pelly River Pluton			
GSC 87-90	GGAA-80 21a-2	14.5 km south-southeast of Mount Wilson, 12 km east of Clea Property, Y.T.	granodiorite	Bi	92 ± 2
GSC 87-98	ANPR-81 32-5	16 km south-south-east of Mount Wilson, 12 km east of Clea Property, N.W.T.	granodiorite	Bi	88 ± 2
GSC 87-101	ANPR-81 22-8	16 km south-south-east of Mount Wilson, 11 km east-southeast of Clea Property, N.W.T.	granodiorite	Bi	92 ± 2
GSC 87-102	ANPR-81 25-1	14.5 km south-southeast of Mount Wilson, 12 km east of Clea Property, N.W.T.	granodiorite	Bi	92 ± 2
GSC 87-103	ANPR-81 27-7	15 km south-south-east of Mount Wilson, 12 km east of Clea Property, N.W.T.	granodiorite	Bi	91 ± 2
GSC 87-112	ANPR-81 30-2	15 km south-south-east of Mount Wilson, 12 km east-southeast of Clea Property, N.W.T.	biotite lamprophyre	Bi	91 ± 2
		Mount Wilson Pluton			
GSC 87-113	ANMW-82 328-1	0.1 km west of summit of Mount Wilson, N.W.T.	monzogranite	Bi	95 ± 1
		Lened Pluton			
GSC 87-121	ANLD-82 274-1	10.5 km south-southeast of con-fluence of Little and South Nahanni Rivers, 6 km east of Drill Lake, N.W.T.	granite	Bi	93 ± 1
GSC 87-129	ANLZ-82 278-1	6.5 km southeast of south end of Drill Lake, 13 km south of confluence of Little and South Nahanni Rivers, N.W.T.	biotite lamprophyre	Bi	91 ± 1
		Rudi Pluton			
GSC 87-122	ANRU-82 241-3	17 km southeast of confluence of Little and South Nahanni rivers, 12.5km east-southeast of Drill Lake, N.W.T.	aplitic monzogranite	Bi	87 ± 2
GSC 87-123	"	"	"	Mu	84 ± 1
		"Transitional" plutons			
		Southern South Nahanni Pluton			
GSC 87-120	ANSS-82 298-1	45 km south of peak 7719 feet, 24 km east-south-east of major con-fluence of South Nahanni River, N.W.T.	monzogranite	Bi	96 ± 1
GSC 87-124	ANSS-82 300-1	Same as for K-Ar Lab No. GSC 87-120	monzogranite	Bi	94 ± 1
GSC 87-125	"	"	"	Hb	93 ± 1
GSC 87-89	GGA-79 44E-1	42 km south of peak 7719 feet, 24 km southeast of major conflu-ence of South Nahanni River, N.W.T.	porphyry dyke	Bi	91 ± 2
GSC 87-119	ANSS-82 293-1	46 km south of peak 7719 feet, 22 km east-southeast of major confluence of South Nahanni River, N.W.T.	porphyry dyke	Bi	96 ± 1
		Shelf Lake Pluton			
GSC 87-93	GGAA-80 102b-1	15 km northwest of southeast corner of Nahanni map area, near headwaters of Flat River, N.W.T.	granodiorite	Bi	90 ± 2
GSC 87-94	"	"	"	Mu	94 ± 5
GSC 87-128	ANFL-82 312-6	6 km south-south-east of south end of Shelf Lake, 2.5 km east-southeast of Zenchuk Lake, N.W.T.	monzogranite	Bi	94 ± 1
		Hornblende-bearing plutons			
		O'Grady Batholith			
GSC 87-91	GGAB-80 31a-2	16 km southwest of peak 7719 feet, N.W.T.	granite	Bi	117 ± 3*
GSC 87-92	"	"	"	Hb	78 ± 4
GSC 87-126	ANOG-82 138-1	16 km southwest of O'Grady Lakes, 18 km west-southwest of peak feet, N.W.T.	granite	Bi	92 ± 1
GSC 87-127	"	"	"	Hb	95 ± 1
GSC 87-114	ANOG-82 130-1	16 km southwest of O'Grady Lakes, 18 km southwest of peak 7719 feet, N.W.T.	granite	Hb	101 ± 2

Table 1. (cont'd)

GSC No.	Field Number	Location	Lithology	Mineral Dated	Age (Ma)
GSC 87-116	ANOG-82 232-1	28 km southeast of O'Grady Lakes, 7 km south-south-east of peak 7719 feet, N.W.T.	granodiorite	Bi	98 ± 2
GSC 87-117	"	"	"	Hb	95 ± 2
GSC 87-97	GGAB-80 31b-1	17 km south-south-east of O'Grady Lakes, 16 km southwest of peak 7719 feet, N.W.T.	aplitic monzogranite	Bi	91 ± 2
GSC 87-115	ANOG-82 166-2	12 km southeast of O'Grady Lakes, 16 km west-southwest of peak 7719 feet, N.W.T.	aplitic granite	Bi	94 ± 1
		Northern South Nahanni Pluton			
GSC 87-118	ANNN-82 177-3	19.3 km south-southeast of peak 7719 feet, 28 km northwest of major confluence of South Nahanni River, N.W.T.	monzogranite	Bi	96 ± 1
		Central South Nahanni Pluton			
GSC 87-95	ANSN-81 46-2	32 km south-south-east of peak 7719 feet, 20 km east of major conflu-ence of South Nahanni River, N.W.T.	granodiorite	Bi	80 ± 4
GSC 87-96	"	"	"	Hb	84 ± 6
GSC 87-99	ANSN-81 42-3	30 km south of peak 7719 feet, 18 km east-south-east of major con-fluence of South Nahanni River, N.W.T.	granodiorite	Bi	91 ± 2
GSC 87-100	"	"	"	Hb	90 ± 7
GSC 87-108	ANSN-81 54-2	31 km southeast of peak 7719 feet, 22 km east-northeast of major confluence of South Nahanni River, N.W.T.	granodiorite	Bi	92 ± 2
GSC 87-109	"	"	"	Hb	89 ± 7
GSC 87-110	ANSN-81 38-1	30 km south of peak 7719 feet, 19 km east-north east of major con-fluence of South Nahanni River, N.W.T.	granodiorite	Bi	92 ± 2
GSC 87-111	"	"	"	Hb	93 ± 6
GSC 87-104	ANSN-81 58-1	26 km south-south-east of peak 7719 feet, 21 km east-northeast of major confluence of South Nahanni River, N.W.T.	porphyry dyke	Bi	90 ± 2
GSC 87-105	"	"	"	Hb	87 ± 5
GSC 87-106	ANSN-81 43-2	33 km south of peak 7719 feet, 20 km east of major confluence of South Nahanni River, N.W.T.	porphyry dyke	Bi	93 ± 2
GSC 87-107	"	"	"	Hb	106 ± 6*

ages of the Selwyn plutonic suite's three varieties and of constituent phases in individual plutons are indistinguishable. Archibald's (1980, 1981) unpublished K-Ar ages for northern and southern South Nahanni, Mt. Applar, LENED, CAC, RUDI, and Shelf Lake plutons are similar to the overall range of the Selwyn plutonic suite and to the isotopic ages reported here for a particular pluton.

Aplite, porphyry and lamprophyre dykes were sampled to provide minimum ages for intrusion of the Selwyn plutonic suite (specifically Pelly River, LENED, southern South Nahanni, and central South Nahanni plutons and O'Grady batholith) and to ascertain if the plutons were the locus for later, Tertiary magmatism (e.g., Pigage and Anderson, 1985; Jackson et al., 1986). The mid-Cretaceous isotopic ages for the 9 samples vary little (range = 87-96 Ma; average = 91.6 ± 2.6 Ma (1 s.d.)) and are clearly coeval with the cooling ages for their host plutons.

Biotite lamprophyre collected in drill core crosscuts garnet skarn associated with tungsten mineralization at Union Carbide Exploration corporation's LENED camp (GSC 87-129). The mid-Cretaceous K-Ar isotopic age (91 ± 1 Ma) for the dyke's biotite is identical to that for other lamprophyre and some porphyry dykes sampled from other plutons and indicates that the skarn mineralization is no younger than the mid-Cretaceous two-mica plutons which host it. The dyke's isotopic age is concordant with or slightly younger than the 92-95 Ma ages determined by Archibald et al. (1978) for biotite- and amphibole-bearing skarn at Cantung. The isotopic age is coeval with mid- Cretaceous ages determined for sericitic alteration of the LENED-CAC-RUDI plutons locally associated with subparallel quartz-tourmaline veins (Archibald, 1980).

Other isotopic systems (Rb-Sr and U-Pb (zircon)) for the apparently simple, post-tectonic plutons of the Selwyn plutonic suite are surprisingly poorly behaved and unpredictable (e.g., Anderson et al., 1983). K-Ar isotopic ages provide only estimates for the cooling of an individual pluton. The oldest, reliable, hornblende K-Ar isotopic ages for plutons and ages for hornblende or biotite in crosscutting dykes are likely the best estimates for the minimum age for the Selwyn plutonic suite and suggest a youngest age of 95-101 Ma for the intrusion of the suite which is in accord with reliable Rb-Sr isochron (e.g. Godwin et al., 1978) and U-Pb (e.g., Smit et al., 1985) isotopic ages (Table 2).

REFERENCES

Anderson, R.G.

1982: Geology of the Mactung pluton in Niddery Lake map area and some of the plutons in Nahanni map area, Yukon Territory and District of Mackenzie; *in* Current Research, Part A, Geological Survey of Canada, Paper 82-1A, p. 299-304.

1983: Selwyn plutonic suite and its relationship to tungsten skarn mineralization, southeastern Yukon and District of Mackenzie; *in* Current Research, Part B, Geological Survey of Canada, Paper 83-1B, p. 151-163.

Anderson, R.G., Armstrong, R.L., Parrish, R. and Bowman, J.R.

1983: Potential of SE Selwyn plutonic suite for W-skarn deposits; (abstract), Geological Association of Canada, Program withAbstracts, v. 8, p. A2.

Archibald, D.A.
1980: Description and K-Ar geochronology of the intrusive rocks, Lened property, N.W.T.; unpublished Union Carbide Company report.

1981: Preliminary report on the K-Ar geochronology and petrography of intrusive rocks, Selwyn Mountains, N.W.T. and Y.T.; unpublished Union Carbide Company report, 31 p.

Archibald, D.E., Clark, A.H., Farrar, E., and Zaw, U Khin
1978: Potassium-argon ages of intrusion and scheelite mineralization, Cantung, Tungsten, Northwest Territories; Canadian Journal of Earth Sciences, v. 15, p. 1205-1207.

Atkinson, D. and Baker, D.J.
1986: Recent developments in the geologic understanding of MacTung; *in* J.A. Morin, editor, Mineral Deposits of Northern Cordillera, Geology Division, Canadian Institute of Mining and Metallurgy, Special Volume 37, p. 234-244.

Baadsgaard, H., Folinsbee, R.E., and Lipson, J.
1961: Potassium-argon dates from Cordilleran granites; Geological Society of America Bulletin, v. 72, p. 689-702.

Dawson, K.M. and Dick, L.A.
1978: Regional metallogeny of the northerncordillera: tungsten and base metal-bearing skarns in southeastern Yukon and southwestern Mackenzie; *in* Current Research, Part A, Geological Survey of Canada, Paper 78-1A, p. 287-292.

Dick, L.A.
1979: Tungsten and base metal skarns in the northern Cordillera; *in* Current Research, Part A, Geological Survey of Canada, Paper 79-1A, p. 259-266.

1980: A comparative study of the geology, mineralogy and conditions of formation of contact metasomatic mineral deposits in the northeastern Canadian Cordillera; unpubl. Ph.D. thesis, Queen's University, 471 p.

Dick, L.A. and Hodgson, C.J.
1982: The MacTung W-Cu (Zn) contact metasomatic and related deposits of the northeastern Canadian Cordillera; Economic Geology, v. 77, no. 4, p. 845-867.

Glover, J.K. and Burson, M.J.
1986: Geology of the Lened tungsten skarn, Logan Mountains, Northwest Territories; *in* J. Morin, editor, Mineral Deposits of Northern Cordillera, Geology Division, Canadian Institute of Mining and Metallurgy, Special Volume 37, p. 255-265.

Godwin, C.I., Armstrong, R.L., and Tompson, K.M.
1980: K-Ar and Rb-Sr dating and the genesis of tungsten at the Clea tungsten skarn property, Selwyn Mountains, Yukon Territory; Canadian Institute of Mining and Metallurgy Bulletin, v. 73, no. 821, p. 90-93

Jackson, L.E., Jackson, S.P. Gordey, Armstrong, R.L. and Harakal, J.E.
1986: Bimodal Paleogene volcanics near Tintina Fault, east-central Yukon and their possible relationship to placer gold; *in* J.A. Morin and D.S. Emond (editors), Indian and Northern Affairs Canada, Yukon Geology, v. 1, p. 139-147.

Pigage, L.C. and Anderson, R.G.
1985: The Anvil plutonic suite, Faro, Yukon Territory; Canadian Journal of Earth Sciences, v. 22, p. 1204-1216.

Smit, H., Armstrong, R.L., and van der Heyden, P.
1985: Petrology, chemistry and radiogenic isotope (K-Ar, Rb-Sr and U-Pb) study of the Emerald Lake pluton, eastern Yukon Territory; *in* Current Research, Part B, Geological Survey of Canada, paper 85-1B, p. 347-359.

Wanless, R.K., Stevens, R.D., Lachance, G.R., and Delabio, R.N.
1970: Age determinations and geological studies. K-Ar isotopic ages, report 9; Geological Survey of Canada, Paper 69-2A, p. 27-28.

1974: Age determinations and geological studies, K-Ar isotopic ages, Report 12; Geological Survey of Canada, Paper 74-2, p. 12-18.

GSC 87-130 Biotite, K-Ar age
89 ± 2 Ma

K = 7.04 %, radiogenic
Ar = 248.5 × 10^{-7} cc/gm
K-Ar 3480 atmos. Ar = 13.8 %
Concentrate: light brown biotite with approximately 9 % chlorite alteration.

From granodiorite
(105 O) 1 km northeast of Mount Allan, west end of ''Cirque Lake'', 18 km southeast Keele Peak, Niddery Lake map area, southwestern District of Mackenzie, 63°17′30″N, 130°07′54″W; see Anderson (1982, 1983). Sample ANMT-81-8-1, collected and interpreted by R.G. Anderson.

The sample was collected from white weathering, fresh, massive, equigranular, comparatively more mafic, biotite granodiorite which contains quartz (38 %), plagioclase (34 %), alkali felsdpar (17 %), reddish-brown biotite (11 %), ''secondary'' muscovite (0.2 %) and trace zircon. The sample is typical of the more mafic variety of the Mactung pluton's equigranular phase.

For interpretation see GSC 87-129.

GSC 87-131 Biotite, K-Ar age
90 ± 2 Ma

K = 6.97 %, radiogenic
Ar = 248.6 × 10^{-7} cc/gm
K-Ar 3484 atmos. Ar = 13.4 %
Concentrate: light brown biotite with approximately 9 % chlorite alteration.

From monzogranite
(105 O) 2 km northeast of Mount Allan, 17 km southeast of Keele Peak, Nidderly Lake map area, southwestern District of Mackenzie, 63°18′00″N, 130°07′54″W; see Anderson (1982, 1983). Sample ANMT-81-5-6, collected and interpreted by R.G. Anderson.

The sample was collected from white weathering, massive, equigranular to seriate biotite monzogranite which contains quartz (36 %), plagioclase (31 %), alkali feldspar (30 %), reddish-brown biotite (2 %), ''secondary'' muscovite (0.3 %) and trace garnet and spinel as accessory phases. The sample is typical of the peraluminous marginal phase of the Mactung pluton (see also GSC 87-132).

For interpretation see GSC 87-129.

Table 2. Compilation of reliable isotopic age determinations for Selwyn Plutonic Suite

Unit[1]	GSC No.	Sample Number	Age (Ma)	Mineral	Comments
		Two-Mica plutons			
		Mactung Pluton			
equigranular phase	GSC 73-74[1] (Wanless et al., 1974, p. 26)	FJ68-320-2	89 ± 4	Bi	
maficequi-granular phase	GSC 87-130	ANMT-81-8-1	89 ± 2	Bi	
peraluminous equigranular phase	GSC 87-131	ANMT-81-5-6	90 ± 2	Bi	
peraluminous satellitic dyke	GSC 87-133	ANMT-82-331-1	90 ± 1	Mu	
		Itsi Batholith			
	AK 125 (Baadsgaard et al., 1961, p. 459)	same	98 ± 5	Bi	U.of Alberta K-Ar determination
		CLEA Pluton			
megacrystic phase	KTP-Bi (Godwin et al. 1980, p. 92)	same	80 ± 3	Bi	U.B.C. K-Ar determination
same as above	KTP-Mu[2] (Godwin et al. 1980,p. 92)	same	87 ± 3	Mu	same as above
megacrystic phase	KTP-WR[2] KTP-Bi KTP-Mu (Godwin et al. 1980, p. 92)	same	96 ± 3		U.B.C. Rb-Sr whole-rock-mineral isochron
		Mount Wilson Pluton			
megacrystic phase	GSC 87-113	ANMW-82-328-1	95 ± 1	Bi	
		Pelly River Pluton			
equigranular phase	GSC 87-90	GGAA-80-21a-2	92 ± 2	Bi	
equigranular phase	GSC 87-101	ANPR-81-22-8	92 ± 2	Bi	
equigranular phase	GSC 87-102	ANPR-81-25-1	92 ± 2	Bi	
equigranular phase	GSC 87-103	ANPR-82-27-7	91 ± 2	Bi	
megacrystic phase	GSC 87-98	ANPR-81-32-5	88 ± 2	Bi	
biotite lamprophyre	GSC 87-112	ANPR-81-30-2	91 ± 2	Bi	
		Lened Pluton			
coarsely megacrystic phase	GSC 87-121	ANLD-82-274-1	93 ± 1	Bi	
biotite lamprophyre dyke	GSC 87-129	ANLZ-82-278-1	91 ± 1	Bi	dyke crosscuts skarn
		Rudi Pluton			
peraluminous satellitic dyke	GSC 87-122	ANRU-82-241-3	87 ± 2	Bi	
same as above	GSC 87-123	"	84 ± 1	Mu	

[1] after Anderson (1982, 1983)

Table 2. (cont'd)

Unit[1]	GSC No.	Sample Number	Age (Ma)	Mineral	Comments
		Cantung Pluton			
	AHC-11 (Archibald et al., 1978, p. 1207)	same	94 ± 3	Bi	Queen's Univ. K-Ar determination
		Transitional Plutons			
		Southern South Nahanni Pluton			
coarsely mega-crystic phase	GSC 87-120	ANSS-82-298-1	96 ± 1	Bi	
coarsely mega-crystic phase	GSC 87-124	ANSS-82-300-1	94 ± 1	Bi	
same as above	GSC 87-125	"	93 ± 1	Hb	
porphyry dyke	GSC 87-89	GGA-79-44E-1	91 ± 2	Bi	
porphyry dyke	GSC 87-119	ANSS-82-293-1	96 ± 1	Bi	
		Shelf Lake Pluton			
megacrystic phase	GSC 87-93	GGAA-80-102b-1	90 ± 2	Bi	
same as above	GSC 87-94	"	94 ± 5	Mu	
megacrystic phase	GSC 87-128	ANFL-82-312-6	94 ± 1	Bi	
		Hole-In-The-Wall Batholith			
	AK-107 (Baadsgaard et al., 1961, p. 459)	same	96 ± 5	Bi	U. of Alberta K-Ar determination
		Hornblende-Bearing Plutons			
		Emerald Lake Pluton			
Bluetrachy-ticphase	X-5 (Smit et al., 1985)	X-5	92 ± 3	Bi	U.B.C. K-Ar determination
same as above	same as above	"	92 ± 3	Hb	same as above
samea s above	X-5(-Bi) (Smit et al., 1985)	"	156 ± 30	WR,Hb, Plag, K-spar	U.B.C. Rb-Sr whole-rock-mineral isochron age
same as above	X-5 (Smit et al., 1985)	"	93.5 ± 0.5 95.6 ± 0.8	zircon	U.B.C. U-Pb determination
mainphase	X-2 (Smit et al., 1985)	X-2	102 ± 6	WR,Hb, Plag, K-spar	U.B.C. Rb-Sr whole-rock-mineral isochron age
biotitephase	X-6 (Smit et al., 1985)	X-6	83 ± 2	WR, Bi, Plag, K-spar	same as above
same as above	X-6(-Bi) (Smit et al., 1985)	"	76 ± 5	WR, Plag K-spar	same as above
		O'Grady Batholith			
equigranular phase	GSC 87-114	ANOG-82-130-1	101 ± 2	Hb	
equigranular phase	GSC 87-116	ANOG-82-232-1	98 ± 2	Bi	
same as above	GSC 87-117	"	95 ± 2	Hb	
crowded mega-crystic phase	GSC 67-65 (Wanless et al., 1970, p. 37)	BU66-27-5	80 ± 5	Hb	
same as above	GSC 67-66 (Wanless et al., 1970, p. 37)	"	87 ± 4	Bi	

[1] after Anderson (1982, 1983)

Table 2. (cont'd)

Unit[1]	GSC No.	Sample Number	Age (Ma)	Mineral	Comments
crowded mega-crystic phase	GSC 87-92	GGAB-81-31a-2	78 ± 4	Hb	
crowded mega-crystic phase	GSC 87-126	ANOG-82-138-1	92 ± 1	Bi	
same as above	GSC 87-127	"	95 ± 1	Hb	
aplite intrusions	GSC 87-97	GGAB-80-31b-1	91 ± 2	Bi	
aplite intrusions	GSC 87-115	ANOG-82-166-2	94 ± 1	Bi	
		Northern South Nahanni Pluton			
equigranular phase	GSC 87-118	ANNN-82-177-3	96 ± 1	Bi	
		Central South Nahanni Pluton			
equigranular phase	GSC 87-95	ANSN-81-46-2	80 ± 4	Bi	
same as above	GSC 87-96	"	84 ± 6	Hb	
equigranular phase	GSC 87-99	ANSN-81-42-3	91 ± 2	Bi	
same as above	GSC 87-100	"	90 ± 7	Hb	
equigranular phase	GSC 87-108	ANSN-81-54-2	92 ± 2	Bi	
same as above	GSC 87-109	"	89 ± 7	Hb	
equigranular phase	GSC 87-110	ANSN-81-38-1	92 ± 2	Bi	
same as above	GSC 87-111	"	93 ± 6	Hb	
porphyry dyke	GSC 87-104	ANSN-81-58-1	90 ± 2	Bi	
same as above	GSC 87-105	"	87 ± 5	Hb	
porphyry dyke	GSC 87-106	ANSN-81-43-2	93 ± 2	Bi	

[1] after Anderson (1982, 1983)

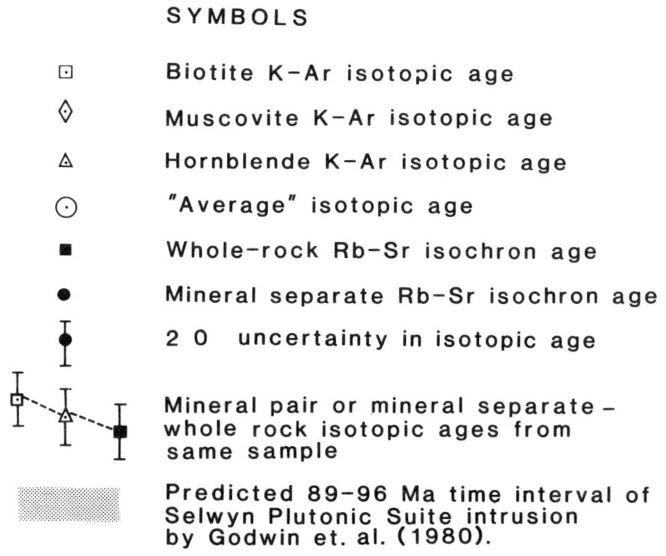

Figure 4. Plot of K-Ar isotopic ages (and 2 s.d. uncertainties) with respect to time scale of Armstrong (1978, 1982). Shaded interval (87-96 Ma) is period of intrusion of Selwyn plutonic suite predicted from earlier studies (Godwin et al., 1978). Determinations for K-Ar Lab No. 3618 (87 ± 2 Ma for biotite from a two-mica granite sill associated with the RUDI pluton) and K-Ar Lab No. 3695 (91 ± 1 Ma for a biotite lamprophyre dyke associated with tungsten skarn mineralization at the LENED property) are not included in the diagram.

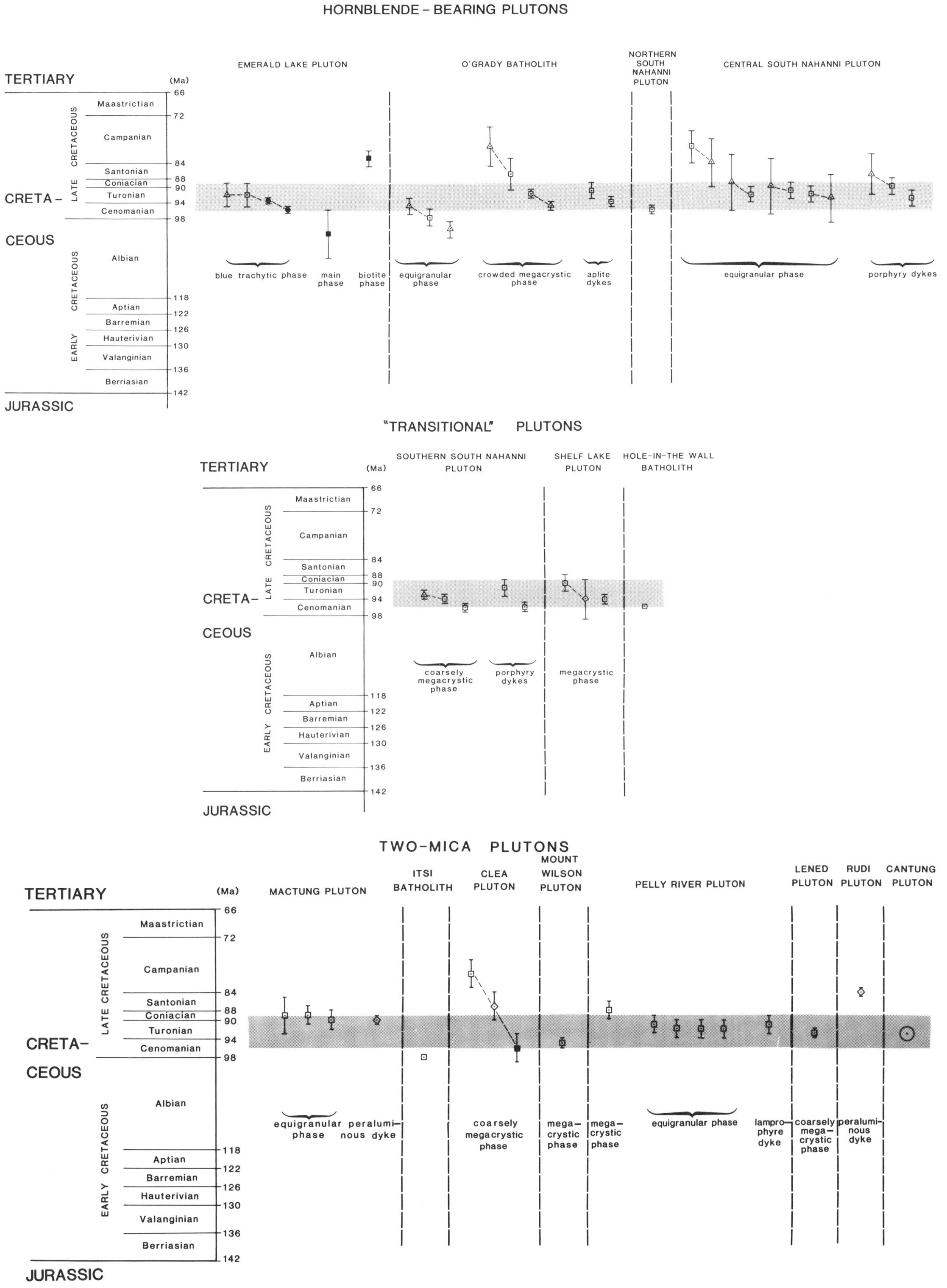
HORNBLENDE - BEARING PLUTONS
EMERALD LAKE PLUTON
O'GRADY BATHOLITH
NORTHERN SOUTH NAHANNI PLUTON
CENTRAL SOUTH NAHANNI PLUTON
TERTIARY
(Ma)
66
Maastrictian
72
Campanian
84
Santonian
88
Coniacian
90
Turonian
94
Cenomanian
98
Albian
118
Aptian
122
Barremian
126
Hauterivian
130
Valanginian
136
Berriasian
142
LATE CRETACEOUS
EARLY CRETACEOUS
CRETA-
CEOUS
JURASSIC
blue trachytic phase
main phase
biotite phase
equigranular phase
crowded megacrystic phase
aplite dykes
equigranular phase
porphyry dykes
"TRANSITIONAL" PLUTONS
SOUTHERN SOUTH NAHANNI PLUTON
SHELF LAKE PLUTON
HOLE-IN-THE WALL BATHOLITH
coarsely megacrystic phase
porphyry dykes
megacrystic phase
TWO-MICA PLUTONS
MACTUNG PLUTON
ITSI BATHOLITH
CLEA PLUTON
MOUNT WILSON PLUTON
PELLY RIVER PLUTON
LENED PLUTON
RUDI PLUTON
CANTUNG PLUTON
equigranular phase
peraluminous dyke
coarsely megacrystic phase
mega-crystic phase
mega-crystic phase
equigranular phase
lamprophyre dyke
coarsely mega-crystic phase
peraluminous dyke

GSC 87-132 Muscovite K-Ar age
90 ± 11 Ma

K = 5.81 %, radiogenic
Ar = 207.7 × 10^{-7} cc/gm
K-Ar 3485 atmos. Ar = 98.4 %
Concentrate: clean, unaltered, mainly clear muscovite with no visible contamination.

(105 O) Details as for GSC 87-131.

Because of the very low radiogenic argon content of the extracted gas from this sample, the calculated age should be regarded only as a confirmation of GSC 87-131 and should not stand alone.

For additional interpretation see GSC 87-129.

GSC 87-133 Muscovite K-Ar age
90 ± 1 Ma

K = 8.445 %, radiogenic
Ar = 304.19 × 10^{-7} cc/gm
K-Ar 3593 atmos. Ar = 17.19 %

From aplitic monzogranite dyke
(105 O) 2 km northeast of Mount Allan, 16 km southeast of Keele Peak, Niddery Lake map area, southwestern District of Mackenzie, 63°18′06″N, 130°08′24″W; see Anderson (1982, 1983). Sample ANMT-82-331-1, collected and interpreted by R.G. Anderson.

The sample was collected from a white weathering, fresh, aplitic monzogranite dyke which contains quartz (38 %), alkali feldspar (29 %), plagioclase (24 %), ''primary'' muscovite (4 %), tourmaline (4 %) and trace euhedral garnet. The dyke intrudes country rock along the southeastern margin of the mactung pluton and is an apophysis of the pluton's peraluminous marginal phase. The sample is typical of satellitic or marginal peraluminous phases characteristic of two-mica plutons associated with tungsten skarns.

For interpretion see GSC 87-129.

DISTRICT OF FRANKLIN
(GSC 87-134 TO GSC 87-148)

GSC 87-134 Muscovite, K-Ar age
1640 ± 23 Ma

K = 8.53 %, radiogenic
Ar = 8.864 × 10^{-4} cc/gm
K-Ar 3441 atmos. Ar = 0.7 %

From a granite pegmatite separate.
(27 B) North Flank of Dewar Lakes dome, Baffin Island, District of Franklin, 68°33′N, 70°55′W. Sample 1981 HSA 57-1, collected and interpreted by J.R. Henderson

Samples 81HSA57-1 and 81HSA57-2 were collected 20 m apart from the same granite pegmatite dyke. Large book muscovite was hand picked in the field for dating. The dyke is not deformed and intrudes the basal formations of the Aphebian Piling Group. Similar dykes; nearby intrude the underlying Archean granitic gneiss. The ages of 1640 ± 23 and 1617 ± 22 Ma may be interpreted to signify the time at which muscovite began to retain radiogenic argon (i.e. when the ambient temperature fell below 350°C). However these same samples yield muscovite Rb-Sr ages of 1592 and 1633 Ma, suggesting that the muscovite may contain excess argon. On the other hand, two samples of large book muscovite from a similar dyke cutting the basal Penrhyn Group on the *south* flank of the Dewar Lakes dome yielded Rb-Sr ages of 1722 and 1712 Ma, which suggests that the younger Rb-Sr ages from the other dyke may be in error. These ages may be compared further with a U-Pb zircon age of 1895 + 47/−21 Ma separated from granulite grade megacrystic granitic gneiss from the east coast of Baffin Island (Henderson, 1985) which yielded biotite K-Ar ages of 1727, 1688 and 1685 Ma.

To summarize K-Ar and Rb-Sr ages of large book muscovite from post-kinematic dykes in the central part of Foxe Fold Belt, Baffin Island show apparent closure temperatures of 500°C at 1722 to 1712 Ma and 350°C at 1640 to 1617 Ma. These temperatures and times may be compared with a U-Pb zircon age of 1895 Ma from late-kinematic granulite grade granite intruding Piling Group 175 km east. Biotite K-Ar ages from the granite gave ages of 1727, 1688, and 1685 Ma. Together these mineral ages appear to indicate that subsequent to regional dynamothermal metamorphism, the high grade rocks in Foxe Fold Belt cooled from 700-800° to less than 300°C in about 170 Ma, whereas in the lower grade region in central Baffin Island post kinematic cooling from 500° to 300°C took about 280 Ma.

REFERENCE

Henderson, J.R.
1985: Geology, Ekalugad Fiord-Home Bay, District of Franklin, Northwest Territories, Geological Survey of Canada, Map 1606, scale 1:250,000.

GSC 87-135 Muscovite, K-Ar age
1617 ± 22 Ma

K = 8.85 %, radiogenic
Ar = 9.002 × 10^{-4} cc/gm
K-Ar 3442 atmos. Ar = 1.7 %

From a granite pegmatite separate.
(27 B) North flank of Dewar Lakes dome Baffin Island, District of Franklin, 68°38′N, 70°55′W. Sample 81H5A57-2, collected and interpreted by J.R. Henderson.
See GSC 87-134 for description and interpretation.

GSC 87-136 Muscovite, K-Ar age
1659 ± 23 Ma

K = 8.74%, radiogenic
Ar = 9.243 × 10^{-4} cc/gm
K-Ar 3443 atmos. Ar = 1.4%

From a pegmatite separate.
(27 C) Pegmatite dike on top of Eskimo Hill, 2 km south of SE margin of Barnes Icecap and Generator Lake 69°32′N, 71°55′W. Sample 1981HSA-87B-1, collected and interpreted by J.R. Henderson.

Samples 81HSA-87B-1 and 81HSA-87B-2 are samples of large book muscovite collected from the same undeformed tourmaline-bearing granite pegmatite intruding metagabbro southeast of Barnes Icecap (Henderson, 1985). The ages may be interpreted to signify the time at which muscovite began to retain radiogenic argon (i.e. when the ambient temperature fell below 350°C). These K-Ar ages of 1659 ± 23 and 1609 ± 22 Ma may be compared with muscovite Rb-Sr ages of 1756 and 1738 Ma from the same material. If the Rb-Sr ages reflect the time at which the ambient temperature fell below 450°C, it suggests a very slow cooling history for the north margin of the Foxe Fold Belt. A previously reported (Henderson, 1985) K-Ar age of 1805 Ma appears to be too old; possibly due to excess argon.

REFERENCE

Henderson, J.R.
1985: Geology McBeth Fiord-Cape Henry Kater, District of Franklin, Northwest Territories, Geological Survey of Canada, Map 1605, scale 1:250,000.

GSC 87-137 Muscovite, K-Ar age
1609 ± 22 Ma

K = 9.0%, radiogenic
Ar = 9.086 × 10^{-4} cc/gm
K-Ar 3444 atmos. Ar = 1.7%

From a granite pegmatite separate.
(27 C) Top of 'Eskimo Hill', 2 km south of south-east margin of Barnes Icecap and Generator Lake, Baffin Island, District of Franklin, 69°32′N, 71°55′W. Sample 1981HSA87B-2, collected and interpreted by J.R. Henderson.
See GSC 87-136 for description and interpretation.

GSC 87-138 Hornblende, K-Ar age
1740 ± 47 Ma

K = 1.08%, radiogenic
Ar = 1.229 × 10^{-4} cc/gm
K-Ar 3436 atmos. Ar = 2.3%

From a mafic orthopyroxene tonalite gneiss separate.
(48 E) Southern Devon Island: 23 km north of Lemieux Point, 17 km east of Croker Bay. District of Franklin, N.W.T., 74°43′N, 82°33′W. Collected and interpreted by T. Frisch. Sample FS-78-25. See GSC 87-139 for interpretation.

GSC 87-139 Biotite, K-Ar age
1769 ± 24 Ma

K = 7.53%, radiogenic
Ar = 8.796 × 10^{-4} cc/gm
K-Ar 3437 atmos. Ar = 0.6%

From a granitic gneiss separate.
(48 E) Southern Devon Island: 23 km north of Lemieux Point and 17 km east of Croker Bay. District of Franklin, N.W.T., 74°43′N, 82°33′W. Collected and interpreted by T. Frisch. Sample FS-78-26.

The hornblende and biotite dated come from mafic orthopyroxene tonalite gneiss and pink granite gneiss, respectively, which are interlayered and highly flattened. The gneisses, thought to have been igneous rocks prior to granulite facies metamorphism, belong to the northernmost part of the Churchill Structural Province (Frisch, 1983, p. 9). The hornblende K-Ar age agrees with the lower U-Pb concordia intercept age of zircon from the same rock, 1723 Ma; the upper intercept age is 2518 + 56/ −33 Ma (unpublished data). Several hornblende and mica K-Ar ages of 1.7 ± 0.1 Ga have been obtained from the Precambrian Shield around northern Baffin Bay, including an age of 1760 Ma on biotite from gneiss in southeast Ellesmere Island (GSC 61-49 in Lowdon et al., 1963; Larsen and Dawes, 1974). The interlayered gneisses are therefore late Archean-early Proterozoic rocks that apparently underwent uplift and cooling ca. 1.7 Ga.

REFERENCES

Frisch, T.
1983: Reconnaissance geology of the Precambrian Shield of Ellesmere, Devon and Coburn islands, Arctic Archipelago: a preliminary account; Geology Survey of Canada, Paper 82-10, 11 p.

Larsen, O. and Dawes, P.R.
1974: K/Ar and Rb/Sr age determinations on Precambrian crystalline rocks in the Inglefield Land-Inglefield Bredning region, Thule district, western North Greenland; Gronlands Geologiske Undersogelse, Rapport 66 p. 5-8.

Lowdon, J.A., Stockwell, C.H., Tipper, H.W. and Wanless, R.K.
1963: Age determinations and geological studies (including isotopic ages — Report 3); Geological Survey of Canada, Paper 62-17, 140 p.

GSC 87-140 Whole rock, K-Ar age
1870 ± 18 Ma

K = .395%, radiogenic
Ar = 503.8 × 10^{-7} cc/gm
K-Ar 3525 atmos. Ar = 3.9%

	From a diabase separate
(48 E)	West side of Johnson Bay, Dundas Harbour, Devon Island, NWT, District of Franklin. 74°31′30″N, 82°24′. Map unit Hd, GSC Map 1574A. Sample FS-78-51, collected and interpreted by T. Frisch. See GSC 87-142 for interpretation.

GSC 87-141	Whole rock, K-Ar age **204 ± 3 Ma**
	K = 4.2%, radiogenic Ar = 352.2 × 10^{-7} cc/gm
K-Ar 3526	atmos. Ar = 5.6%
	From a diabase separate.
(48 E)	West side of Johnson Bay, Dundas Harbour, Devon Island, NWT, District of Franklin. 74°31′30″N, 82°24′W, Map Unit Hd, GSC Map 1574A. Sample FS-78-50, collected and interpreted by T. Frisch. See GSC 87-142 for interpretation.

GSC 87-142	Whole rock, K-Ar age **837 ± 11 Ma**
	K = .589%, radiogenic Ar = 243.9 × 10^{-7} cc/gm
K-Ar 3524	atmos. Ar = 6.2%
	From a diabase separate
(48 F)	Island at head of Burnett Inlet, Devon Island, NWT. District of Franklin, 74°36′N, 86°12′W, Map Unit Hd, GSC Map 1574A. Sample FS-78-53A, collected and interpreted by T. Frisch.

GSC 87-140 to 87-147 are all determinations on whole-rock samples of unmetamorphosed diabase dykes cutting late Archean-early Proterozoic granulite facies gneisses and predating deposition of Cambrian and younger Paleozoic sediments on Devon Island. The dykes form significant swarms and trend mainly east-west (in rough conformity with the general gneissic foliation); northwest-striking dykes are locally common in southern Devon Island and probably represent extensions of Franklin dykes exposed on Bylot Island (Jackson and Davidson, 1975). Most of the dyke rocks dated come from chilled margins and consist of porphyritic to glomeroporphyritic diabase, with phenocrysts of plagioclase, clinopyroxene and/or olivine in an aphanitic groundmass. Dyke interiors, which were also dated in two instances, have subophitic texture. Partial alteration of the silicate minerals is generally apparent in thin section but nowhere severe. No dyke on Devon Island had been dated prior to this work.

GSC 87-142/87-143 and GSC 87-140/87-141 refer to interior/margin sample pairs from two dykes on the south coast of Devon Island. As the dykes are seen to overlain by lower Paleozoic strata, the ages obtained from the chilled margins are clearly too young. GSC 87-140, on the other hand, is too old, as it approaches U-Pb, and exceeds K-Ar, ages given by metamorphosed basement rocks. The age (GSC 87-142), 837 ± 11 Ma, of the other dyke interior, although plausible, must be treated with circumspection in view of the anomalously low age, 97 Ma, yielded by the chilled margin. There is no obvious reason, either in lithology of field setting, for the anomalous ages obtained.

GSC 87-144, 1336 ± 15 Ma, is considerably older than the typical ages of diabase dykes in the region. Christie (1962) reported that, near Cape Combermere, on the east coast of Ellesmere Island, diabase dykes appear to be overlain unconformably by unmetamorphosed Thule Group rocks, which are now known to be 1.1-1.2 Ga old. In northwest Greenland, an unmetamorphosed, pre-Thule Group diabase dyke has been dated at 1563 ± 60 Ma (Dawes et al., 1973). Further isotopic data are needed before this single 1.3 Ga date from Canada can be accepted.

The remaining three determinations were made on E-W or WNW-trending dykes in northern Devon Island. The ages obtained, 625-790 Ma, are similar to many others from diabase dykes in neighbouring parts of the Precambrian shield (Ellesmere, Baffin and Somerset islands and Greenland), which are assigned to, or coeval with, the Franklin swarm (Fahrig and West, 1986).

REFERENCES

Christie, R.L.
1962: Geology, southeast Ellesmere Island, District of Franklin (map with marginal notes); Geological Survey of Canada, Map 12-1962.

Dawes, P.R., Rex, D.C. and Jepsen, H.F.
1973: K/Ar whole rock ages of dolerites from the Thule district, western North Greenland; Gronlands Geologiske Undersogelse, Rapport 55, p. 61-66.

Fahrig, W.F. and West, T.D.
1986: Diabase dyke swarms of the Canadian Shield; Geological Survey of Canada, Map 1627A.

Jackson, G.D. and Davidson, A.
1975: Geology of the Bylot Island map-area; Geological Survey of Canada, Paper 74-29, 12 p.

GSC 87-143	Whole rock, K-Ar age **96.6 ± 1.5 Ma**
	K = 2.96%, radiogenic Ar = 114.2 × 10^{-7} cc/gm
K-Ar 3527	atmos. Ar = 8.7%
	From a diabase separate
(48 F)	Island at head of Burnett Inlet, Devon Island. 74°36′N, 86°12′W, Map-unit Hd, GSC Map 1574A. Sample FS-78-53, collected and interpreted by T. Frisch. See GSC 87-142 for interpretation.

GSC 87-144	Whole rock, K-Ar age **1336 ± 15 Ma**
	K = .619%, radiogenic Ar = 476.3 × 10^{-7} cc/gm
K-Ar 3528	atmos. Ar = 1.9%
	From a diabase separate
(48 F)	West side of lake, 5 km north of head of Powell Inlet, NWT, District of Franklin,

74°43′30″, 85°31′, Map-unit Hd, GSC Map 1574A. Sample No. FS-78-58, collected by T. Frisch.
See GSC 87-142 for interpretation.

GSC 87-145 Whole Rock, K-Ar age
625.4 ± 8.3 Ma

K = .417%, radiogenic
Ar = 121.2 × 10^{-7} cc/gm
K-Ar 3529 atmos. Ar = 11.1%

From a diabase separate
(48 G) 1.5 km south of lake, 12 km SSW of Cape Newman Smith, Devon Island, NWT, district of Franklin. 75°32′N, 85°13′W. Map unit Hd, GSC Map 1574A. Sample FS-80-9A, collected and interpreted by T. Frisch.
See GSC 87-142 for interpretation.

GSC 87-146 Whole rock, K-Ar age
725.8 ± 9.4 Ma

K = .432%, radiogenic
Ar = 150 × 10^{-7} cc/gm
K-Ar 3530 atmos. Ar = 4.7%
From a diabase separate
(48 G) Five km SE of the head of Truelove Inlet, Devon Island. N.W.T., District of Franklin. 75°36′30″N, 84°22′W, Map-unit Hd, GSC Map 1574A. Sample FS-80-68, collected and interpreted by T. Frisch.
See GSC 87-142 for interpretation.

GSC 87-147 Whole rock, K-Ar age
790 ± 10 Ma

K = 0.176%, radiogenic
Ar = 67.76 × 10^{-7} cc/gm
K-Ar 3531 atmos. Ar = 16.0%

From a diabase separate
(48 G) South shore of easternmost inlet off Sverdrup Inlet, Devon Island. District of Franklin, N.W.T., 75°24′30″N, 85°47′00″W. Map unit Hd, GSC map 157A, FS-80-80, collected and interpreted by T. Frisch. See GSC 87-142 for interpretation.

GSC 87-148 Muscovite, K-Ar age
364 ± 7 Ma

K = 5.52%, radiogenic
Ar = 8.653 × 10^{-5} cc/gm
K-Ar 3453 atmos. Ar = 88%

From a granitic gneiss separate.
(340 E) East coast of Ayles Fiord, Ellesmere Island, District of Franklin, N.W.T.; 82°47′50″N, 79°49′W. Sample FS-75-194, collected and interpreted by T. Frisch.

The sample comes from a body of poorly foliated and highly cataclastic granitic gneiss underlying greenschist- grade schist and phyllite and is pictured in Frisch (1974, Fig. 12A). The gneiss is regarded as belonging to the crystalline basement (succession I) of Pearya, a composite exotic terrane in northernmost Ellesmere Island (Trettin, 1987). U-Pb and Rb-Sr work has established that the basement is at least Neohelikian (1.0-1.1 Ga) in age. The muscovite K-Ar age agrees with that obtained from pegmatite in gneiss 2 km to the west, 354 ± 15 Ma (GSC67-50 in Wanless et al., 1970). Both samples are believed to be Precambrian rocks that were strongly affected by a later tectonothermal event, such as heating during Devonian magmatism (Trettin et al., 1987).

REFERENCES

Frisch, T.
1974: Metamorphic and plutonic rocks of northernmost Ellesmere Island, Canadian Arctic Archipelago; Geological Survey of Canada, Bulletin 229, 87 p.

Trettin, H.P.
1987: Pearya: a composite terrane with Caledonian affinities in northern Ellesmere Island; Canadian Journal of Earth Sciences, v. 24, p. 224-245.

Trettin, H.P., Parrish, R. and Loveridge, W.D.
1987: U-Pb age determinations on Proterozoic to Devonian rocks from northern Ellesmere Island, Arctic Canada; Canadian Journal of Earth Sciences, v. 24, p. 246-256.

Wanless, R.K., Stevens, R.D., Lachance, G.R., Delabio, R.N.
1970: Age determinations and geological studies, K-Ar isotopic ages, Report 9; Geological Survey of Canada, paper 69-2A, 78 p.

YUKON TERRITORY
(GSC 87-149 TO GSC 87-175)

GSC 87-149 Muscovite, K-Ar age
109 ± 2 Ma

K = 4.00%, radiogenic
Ar = 1.747 × 10^{-5} cc/gm
K-Ar 3367 atmos. Ar = 36.3%

From a tungsten- and molybdenum-bearing quartz veinlet stockwork.
(105 B) Logtung tungsten-molybdenum deposit, 23 km west of Swift River, Yukon, near the headwaters of Logjam Creek, 60°00′18″N, 131°35′47″W. Sample SYA80-109, collected and interpreted by W.D. Sinclair.

The muscovite appears to be hydrothermal in origin. It is fine grained, translucent apple-green and occurs in small clusters associated with quartz, feldspar, fluorite and pyrite along fractures cutting quartz porphyry.

The Logtung deposit a tungsten-molybdenum porphyry deposit with some associated skarn-type mineralization. Scheelite and molybdenite occur mainly in at least four superimposed quartz vein systems associated with quartz porphyry dykes on the north flank of a granitic stock. Minor scheelite is associated with a thin sporadic skarn zone that occurs adjacent to the granitic stock (Noble et al., 1983). The muscovite dated in this report is from late-stage fractures in the mineralized quartz porphyry dyke rock. These fractures cut earlier-formed quartz-molybdenite veinlets and appear to post-date the tungsten-molybdenum mineralization directly associated with the porphyry dykes. A more widespread system of polymetallic, sheeted quartz veins that is centred on the quartz porphyry dykes likely post-dates the muscovite-bearing fractures.

The age of the muscovite reported herein (109 ± 2 Ma) is likely a close approximation of the age of tungsten-molybdenum mineralization associated with the quartz porphyry dykes. However, Stewart and Evensen (1983) suggested that the age of the granitic stock and associated quartz porphyry dykes, based on Rb/Sr isotopic analysis, is 118 ± 2 Ma. If both of these ages are valid, then crystallization of the granitic rocks and formation of the mineralized and hydrothermally-altered zones must have occurred over a period of 8 to 10 Ma.

REFERENCES

Noble, S.R., Spooner, E.T.C., and Harris, F.R.
1983: The Logtung W (scheelite)-Mo deposit, S. Yukon: an example of large tonnage, low grade, porphyry-style, stockwork and sheeted vein mineralization; Geological Association of Canada, Program with Abstracts, v. 8, p. A51.

Stewart, J.P. and Evensen, N.M.
1983: The Logtung W. (scheelite)-Mo deposit, S. Yukon: petrology and geochemistry of spatially associated felsic igneous rocks; Geological Association of Canada, Program with Abstracts, v. 8, p. A65.

GSC 87-150	Hornblende, K-Ar age **235 ± 12 Ma**
	K = 0.81 %, radiogenic Ar = 7.902×10^{-6} cc/gm
K-Ar 3399	atmos. Ar = 20.4 %
(105 B)	From a quartz monzonite separate 3 km west of Hidden Lake, Yukon 60°13′N, 131°21′W. Map unit 15b, GSC map 10-1960. Sample TOA80-9-4, collected and interpreted by G. Abbott.

Sample TOA80-9-4 is coarse grained hornblende monzogranite from the Ram Stock, a foliated and in places, mylonitized batholith which is elongated parallel to the regional metamorphic fabric and tectonic grain. The specimen analysed is intensely fractured, but unfoliated. Thin sections show that plagioclase is intensely saussuritized. Most hornblende is euhedral and fresh, but some is altered to chlorite. The intense deformation of the entire batholith and weak alteration of the analysed specimen suggests that the 235 Ma age is only a minimum.

The samples are from a variety of intrusions which cut or are part of the Yukon-Tanana (Kootenay) Terrane in the Swift River area of southern Yukon. There, the YTT is mainly a middle to late Paleozoic assemblage of tectonically interleaved shale, sandstone, chert, carbonate, volcanic, and ultramafic rocks. The age determinations are thought to represent three separate intrusive events (late Paleozoic or older, Middle Jurassic, and late Early Cretaceous).

The Yukon-Tanana (Kootenay) Terrane is thought to have been obducted onto ancestral North America during Mesozoic arc-continent collision (Tempelman-Kluit, 1979). The Ram Stock represents a Paleozoic intrusive event of uncertain origin that predates that collision. The Jurassic intrusions resemble I-type intrusions, and may reflect partial melting of subducted or thickened oceanic basement prior to collision. The Mid-Cretaceous intrusions resemble S-type intrusions and probably reflect partial melting of thickened continental crust following collision. No chemical or isotopic analyses of the intrusions, which would support or refute the last two hypotheses, are known to the writer. The ages of the two Mesozoic suites correspond to two widespread plutonic episodes outlined by Armstrong (in press), and Anderson (1985). The Jurassic suite is confined to accreted terranes whereas the mid-Cretaceous group is also widespread in ancestral North America.

REFERENCES

Abbott, J.G.
1981: Geology of the Seagull tin district, *in* Yukon Geology and Exploration 1979-80; Exploration and Geological Services Div., D.I.A.N.D., Whitehorse, Yukon, p. 32-44.

Anderson, R.G.
1985: An overview of some Mesozoic and Tertiary plutonic suites and their associated mineralization in the northern Canadian Cordillera; *in* R.P. Taylor and D.F. Strong eds., Granite Related Mineral Deposits, Geology, Petrogenesis, and Tectonic Setting, Extended abstracts of papers presented at C.I.M. Conference on Granite Related Mineral Deposits, Sept. 15-17, C.I.M., Geology Division, p. 1-20.

Armstrong, R.L.
in press: Mesozoic and early Cenozoic magmatic evolution of the Canadian Cordillera; *in* Rodgers Symposium Volume.

Noble, S.R., Spooner, S.R., and Harris, F.R.
1984: The Logtung large tonnage, low-grade W (Scheelite)-Mo porphyry deposit, south-central Yukon Territory; *in* Econ. Geol. V. 79, p. 848-868.

Poole, W.H., Roddick, J.A., and Green, L.H.
1960: Geology, Wolf Lake map area, Yukon Territory; Geol. Surv. Can., map 10-1960.

Stewart, J.P. and Evenson, N.M.
1983: The Logtung W (scheelite)-Mo deposit, S. Yukon,: petrology and geochemistry of spacially associated felsic igneous rock (abs); *in* Geol. Assoc. can., Program with Abstracts, V. 8, p. 65

Tempelman-Kluit, D.J.
1979: Transported cataclasite, ophiolite and granodiorite in Yukon: Evidence of arc-continent collision; Geol. Surv. Can., Bull. 79-11, 27 p.

Wanless, R.K., Stevens, R.D., Lachance, G.R., and Delabio, R.N.
1970: Age determinations and geological studies; Geol. Surv. Can., Paper 69-2A.

Wanless, R.K., Stevens, R.D., Lachance, G.R., and Delabio, R.N.
1972: Age determinations and geological studies; Geol. Surv. Can., Paper 71-2.

GSC 87-151	Hornblende, K-Ar age **181 ± 14 Ma**
	K = 0.53%, radiogenic Ar = 3.922 × 10^{-6} cc/gm
K-Ar 3400	atmos. Ar = 47.9%
	From a monzonite separate
(105 B)	Saddle at head of Logjam Creek, 4.0 km northeast of Logtung tungsten-molybdenum deposit, Wolf Lake map area, Yukon, 60°02′30″N, 131°32′30″W. Sample TOA80-22-6, collected and interpreted by G. Abbott.

Sample TOA80-22-6 (GSC 87-151) is medium grained, hornblende-biotite monzonite from a composite stock that also contains phases of peridotite, pyroxenite, and diorite. The Middle Jurassic age of is like that obtained by Wanless and others (1969, p. 6.7) from the Simpson Peak and Nome Lake Batholiths, which also cut the Yukon-Tanana Terrane, 30 km to the south in Jennings River map area (104-0).

GSC 87-152	Biotite, K-Ar age **108 ± 2 Ma**
	K = 7.55%, radiogenic Ar = 3.266 × 10^{-5} cc/gm
K-Ar 3407	atmos. Ar = 17.1%
	From a granite separate
(105 B)	South face of cirque, 2.5 km south of the west end of Crescent Lake, Yukon, 60°10′N, 131°15′W. Sample TOA80-10-4, collected and interpreted by G. Abbott.

Sample TOA80-10-4 (GSC 87-152) is from a small stock of fine grained, equigranular biotite monzogranite. The sample is a 1.5 m interval (47.8 m-49.3 m) of diamond drill core from a drill hole that intersected the stock. The composition and age (Wanless and others, 1972, p. 32) of the intrusion resemble the nearby Seagull Batholith. Cassiterite is reported from a small breccia near the margin of the stock (STQ prospect). The Seagull Batholith also contains several tin prospects.

GSC 87-153	Biotite, K-Ar age **127 ± 3 Ma**
	K = 7.04%, radiogenic Ar = 3.600 × 10^{-5} cc/gm
K-Ar 3420	atmos. Ar = 10.2%
	From a diorite separate.
(105 B)	23 km west of Swift River, Yukon, 60°02′N, 131°36.5′W. Map unit 14, GSC map 10-1960. Sample TOA80-25, collected and interpreted by G. Abbott.

Sample TOA80-25 (GSC 87-153) is fine-grained, equigranular, hornblende biotite diorite from a vertical northwest-trending dike as wide as 1 km. The dike is cut by a monzogranite stock located 1.5 km south of the sample location. That intrusion, which hosts the Logtung porphyry molybdenum-tungsten deposit, has returned an Rb/Sr age of 118 ma ± 2 Ma (Stewart, 1983, Noble, 1984). The dike is unlike the Cretaceous stock, but does resemble the Jurassic suite of intrusions represented by sample TOA80-22-6, and the 127 ma age probably reflects reheating by the younger intrusion.

GSC 87-154	Muscovite, K-Ar age **50.78 ± .79 Ma**
	K = 8.82%, radiogenic Ar = 176.6 × 10^{-7} cc/gm
K-Ar 3679	atmos Ar = 7.7%
	From a quartz-wolframite vein separate
(105 B)	Fiddler tungsten prospect, 11.2 km northeast of Rancheria, southern Yukon, 60°08′09″N, 130°26′03″W. Sample SYA84-32, collected and interpreted by W.D. Sinclair

The rock is from greisen alteration in phyllite adjacent to a wolframite-bearing quartz vein. This alteration consists of approximately 50% quartz and 50% muscovite with minor amounts of fluorite and trace amounts of cassiterite. The muscovite is medium-grained and relatively clear with a slight dusting of very fine grained opaque minerals. No visible chlorite is present.

Quartz veins exposed on the property are up to nearly one metre wide and appear to strike northeasterly and dip 30 degrees southeast. They contain coarse, bladed wolframite and trace amounts of molybdenite, cassiterite, chalcopyrite and galena.

The quartz veins and adjacent alteration are probably associated with an unexposed granitic pluton. Although the closest granitic rocks exposed are those of the Cassiar batholith, more than 3 km to the west, diamond drilling on the property has suggested that the Fiddler prospect, along with numerous silver-lead-zinc veins and replacement zones in the Rancheria area, may be associated with small and in many cases unexposed, granitic plutons of early Tertiary age, rather than the mid-Cretaceous Cassiar batholith. The date of 50.8 ± 0.8 Ma determined on hydrothermal muscovite from the Fiddler prospect supports this contention.

GSC 87-155	Biotite, K-Ar age **87.3 ± 2.0 Ma**
K-Ar 3162 (2)	See GSC 81-35 for analysis
	From molybdenite-bearing quartz-veinlet stockwork
( )	Red Mountain molybdenum deposit, 75 km east-northeast of Whitehorse, Yukon, 60°59′16″N, 134°44′30″W. Sample SYA79-56 from Amoco Canada Petroleum Company Limited drill hole

78-4 at 430-444 m. Collected and interpreted by W.D. Sinclair.
For description see GSC 81-37 (*in* Stevens et al., 1982).

This date was obtained by re-analysis of reserve biotite concentrate of GSC 81-35 which had previously yielded an age of 95.6 ± 2.8 Ma. The reason for the discrepancy is not apparent. However, the older date is considered less reliable and is superceded by the new date of 87.3 ± 2.0 Ma.

REFERENCE

Stevens, R.D., Delabio, R.N., and Lachance, G.R.
1982: Age determinations and geological studies, K-Ar isotopic ages, Report 16; Geological Survey of Canada, Paper 82-2, 56 p.

GSC 87-156 Phlogopite, K-Ar age
79.0 ± 1.8 Ma

K-Ar 3354 (2) See GSC 81-36 for analysis

( ) From quartz monzonite porphyry
Red Mountain molybdenum deposit, 75 km east-northeast of Whitehorse, Yukon, 60°59′17″N, 134°44′30″W. Sample SYA80-33 from Amoco Canada Petroleum Company Limited drill hole 79-7 at a depth of 692-699.5 m. Collected and interpreted by W.D. Sinclair.
For description see GSC 81-36 (*in* Stevens et al., 1982).

This date is the result of re-analysis of reserve phlogopite of GSC 81-36 on which a data of 87.4 ± 1.9 Ma had been determined previously. The reason for the discrepancy is not known. However, the older date is considered less reliable and is superceded by the younger date of 79.0 ± 1.8 Ma.

REFERENCE

Stevens, R.D., Delabio, R.N. and Lachance, G.R.
1982: Age determinations and geological studies, K-Ar isotopic ages, Report 16; Geological Survey of Canada, Paper 82-2, 56 p.

GSC 87-157 Muscovite, K-Ar age
98.2 ± 1.5 Ma

K = 7.71%, radiogenic
Ar = 302.4 × 10^{-7} cc/gm
K-Ar 3587 atmos. Ar = 7.7%

From a muscovite-biotite granite separate.
(105 F) Risby tungsten deposit, 1 km northwest of Twin Mountain, 13 km west-southwest of Ross River, Quiet Lake map-area, Yukon Territory, 61°51′20″N, 133°23′10″W. Map unit Kqm, GSC Open File 486. Sample SYA83-42 from Hudson Bay Exploration and Development Limited diamond drill hole 41 at a depth of 290.5 m, collected and interpreted by W.D. Sinclair.

The sample is medium-grained, equigranular granite consisting of 35 % quartz, 36 % K-feldspar, 25 % oligoclase 3 % muscovite and 1 % biotite. Muscovite ranges from relatively clear, medium-sized grains to fine-grained clusters replacing biotite. See GSC 87-158 for interpretation.

GSC 87-158 Muscovite, K-Ar age
102.7 ± 1.6 Ma

K = 8.17%, radiogenic
Ar = 335.7 × 10^{-7} cc/gm
K-Ar 3588 atmos. Ar = 8.3%

From a muscovite-biotite granite separate
(105 F) Same location as GSC 87-157. Sample SYA83-43 from Hudson Bay Exploration and Development Limited diamond drill hole 45 at a depth of 326 m, collected and interpreted by W.D. Sinclair.

This is a medium-grained, equigranular granite consisting of approximately 35% quartz, 35% K-feldspar, 25% oligoclase, 3% muscovite and 2% biotite. Trace amounts of sulphide minerals are present along fractures. Muscovite is relatively fresh and clear and ranges from medium-sized grains 1 to 2 mm long to fine-grained clusters replacing biotite.

Both samples (GSC 87-157, 87-158) are from a granitic pluton that has intruded metasedimentary rocks of Proterozoic and/or Paleozoic age. Carbonate units close to the intrusion have been altered to skarn and mineralized with scheelite and minor amounts of molybdenite and chalcopyrite. Both ages agree closely and indicate a mid-Cretaceous age for the intrusion and associated skarn. The samples are similar to other granites in the Quiet Lake map-area which have yielded comparable ages (e.g. GSC 78-102 to 109, Wanless et al, 1979).

REFERENCE

Wanless, R.K., Stevens, R.D., Lachance, G.R., and Delabio, R.N.
1979: Age determinations and geological studies, K-Ar isotopic ages, Report 14; Geological Survey of Canada, Paper 79-2, 67 p.

GSC 87-159 Hornblende, K-Ar age
344 ± 19 Ma

K = 0.69%, radiogenic
Ar = 1.016 × 10^{-5} cc/gm
K-Ar 3402 atmos. Ar = 37.9%

From a quartz diorite separate
(105 G) Finlayson Lake map-area, Yukon, 61°13′N, 130°24′W. Sample TOE80-28-3, collected and interpreted by P. Erdmer.

This is a Mississippian granitoid intruded during the Devono-Mississippian pulse (380-340 Ma) recognized elsewhere in the Yukon-Tanana terrane (see Aleinikoff et al., GSA Bull. *97*, 626-637, 1986). Its age is like that of other augen gneiss in Yukon and in Alaska, and the age is considered reliable. The rock is exposed in a thrust slice in Money Klippe, a few kilometres NE of Tintina Fault (see Tempelman-Kluit 1979, and Erdmer, 1985).

REFERENCES

Aleinikoff, I.N., Dusel-Bacon, C., and Foster, H.L.
1986: Geochronology of augen gneiss and related rocks, Yukon-Tanana terrane, east-central Alaska. GSA Bull., *97*, 626-637.

Erdmer, P.
1985: An examination of the cataclastic fabrics and structures of parts of Nisutlin, Anvil, and Simpson allochthons, central Yukon: test of the arc-continent collision model. Journal of Structural Geology, I, 57-72.

Tempelman-Kluit, D.J.
1979: Transported cataclasite, ophiolite, and granodiorite in Yukon: evidence of arc-continent collision. GSC Paper 79-14.

GSC 87-160 Biotite, K-Ar age
91.6 ± 2.1 Ma

K = 6.29%, radiogenic
Ar = 2.297×10^{-5} cc/gm
K-Ar 3419 atmos. Ar = 20.3%

From a quartz monzonite (monzogranite) separate
(105 J) 11 km west of Traffic mountain, Yukon, 62°07′N, 130°37.5′W. Map unit 13, GSC map 12-1961. Sample TOA80-1-1, collected and interpreted by G. Abbott.

The sample is from a small stock of medium grained, equigranular biotite monzogranite that is associated with the Pike copper-silver occurence. Pyrite, arsenopyrite, chalcopyrite, galena, and sphalerite with associated silver are in a quartz vein stockwork and form disseminations in and along the margins of the intrusion. The age of 92 Ma indicates that the stock belongs to the Selwyn Plutonic Suite. These intrusions are widespread in east-central Yukon, but no others are known to contain mineral occurences like the Pike.

GSC 87-161 Hornblende, K-Ar
108.8 ± 1.7 Ma
K = .701%, radiogenic
Ar = 30.57×10^{-7} cc/gm
K-Ar 3668 atmos. Ar = 21.6%

From a crystalline tuff separate
(105 J) Southwestern Sheldon Lake Map-area 62°32′30″N, 131°46′31″W. Yukon. Sample GGAA-83-9-3, collected by R.G. Anderson and interpreted by S.P. Gordey.

This sample is a quartz-feldspar-hornblende-biotite crystal lithic tuff from the South Fork Volcanics, in east-central Yukon. The stratigraphic position within the pile is not established. Combined with concordant hornblende and biotite dates (94.6 Ma) from quartz-feldspar-hornblende- biotite crystal tuff (GSC 87-162, 87-163) from about 2.8 km to the southeast a range for volcanism of at least 14 Ma is indicated. These results agree with some previously published dates from the same formation (Wood and Armstrong, 1982).

REFERENCE

Wood, D.H. and Armstrong, R.L.
1982: Geology, chemistry and geochronometry of the Cretaceous South Fork Volcanics, Yukon Territory, *in* Current Research, Part A, Geological Survey of Canada, Paper 82-1A, p. 309-316.

GSC 87-162 Biotite, K-Ar age
94.61 ± 1.5 Ma

K = 5.92%, radiogenic
Ar = 223.3×10^{-7} cc/gm
K-Ar 3669 atmos. Ar = 9.1%

From dacite separate
(105 J) Southwestern Sheldon Lake map-area. Yukon, 62°31′21″N, 131°44′36″W. Sample GGAA-83-18-1, collected by R.G. Anderson and interpreted by S.P. Gordey.

This sample is a quartz-feldspar-hornblende-biotite crystal tuff from the South Fork Volcanics, in east-central Yukon. The stratigraphic position within the pile is not established. Hornblende from the same sample yields the same age (GSC 87-163). Combined with a hornblende date (108.8 Ma) from quartz-feldspar-hornblende-biotite crystal lithic tuff (GSC 87-161) from about 2.8 km to the northwest a range for volcanism of at least 14 Ma is indicated. These results agree with some previously published dates from the same formation (Wood and Armstrong, 1982).

REFERENCE

Wood, D.H. and Armstrong, R.L.
1982: Geology, chemistry and geochronometry of the Cretaceous South Fork Volcanics, Yukon Territory, *in* Current Research, Part A, Geological Survey of Canada, paper 82-1A, p. 309-316.

GSC 87-163 Hornblende, K-Ar age
94.58 ± 1.5 Ma

K = .751%, radiogenic
Ar = 28.34×10^{-7} cc/gm
K-Ar 3670 atmos. Ar = 29.6%

From a dacite separate
(105 J) See GSC 87-162 for location.

This sample is a quartz-feldspar-hornblende-biotite crystal tuff form the South Fork Volcanics, in east-central Yukon. The stratigraphic position within the pile is not established. Biotite from the sample yields the same age (GSC 87-162). Combined with a hornblende date (108.8 Ma) from quartz-feldspar-hornblende-biotite crystal lithic tuff (GSC 87-161) from about 2.8 km to the northwest a range for volcanism of at least 14 Ma is indicated. These results agree with some previously published dates from the same formation (Wood and Armstrong, 1982).

REFERENCE

Wood, D.H. and Armstrong, R.L.
1982: Geology, chemistry and geochronometry of the Cretaceous South Fork Volcanics, Yukon Territory, *in* Current Research, Part A, Geological Survey of Canada, Paper 82-1A, p. 309-316.

GSC 87-164 Whole rock, K-Ar age
85.3 ± 2.1 Ma

K = 3.46%, radiogenic
Ar = 1.175 × 10^{-5} cc/gm
K-Ar 3472 atmos. Ar = 3.3%

From a rhyolite separate
(105 M) On the south side of Minto Creek, 13 km north-northwest of Mayo, Yukon Territory, 63°42′N, 135°58′N. map unit 14a, GSC Map 890A. Sample SYA81-141, collected and interpreted by J.A. Morin.

The sample is from a rhyolite flow that unconformably overlies quartzite and schist of Proterozoic and/or Paleozoic age. The rock is fine-grained and contains 5% plagioclase phenocrysts, 2% quartz phenocrysts and 2% biotite phenocrysts. Biotite phenocrysts are 1 to 2 mm long and are slightly chloritized.

The age of the flow corresponds closely with ages of compositionally similar intrusions in the Mayo area such as the quartz-feldspar porphyry dykes on Mount Haldane (e.g. GSC 80-74, Stevens et al., 1982).

REFERENCE

Stevens, R.D., Delabio, R.N., and Lachance, G.R.
1982: Age determinations and geological studies, K-Ar isotopic ages, Report 15; Geological Survey of Canada, Paper 81-2, 56 p.

GSC 87-165 Whole rock, K-Ar age
90.7 ± 1.4 Ma

K = 3.82%, radiogenic
Ar = 138.4 × 10^{-7} cc/gm
K-Ar 3680 atmos. Ar = 4.0%

From a biotite hornfels separate
(105 M) Kalzas tungsten-tin-molybdenum deposit, 4 km northwest of Big Kalzas Lake and 70 km southeast of Mayo, Yukon, 63°16′03″N, 134°41′50″W. Sample SYA84-103 from Union Carbide drill hole K2 at 318 m, collected and interpreted by W.D. Sinclair.

The Kalzas deposit consists of large (up to 1/2 m wide), subparallel quartz veins that are connected by quartz veinlet stockworks. The quartz veins contain coarse, euhedral wolframite crystals and lesser cassiterite, molybdenite, scheelite, pyrite, galena and native bismuth. Tourmaline occurs in the quartz veinlets and as pervasive alteration in the associated clastic sedimentary host rocks.

Although igneous rocks are not exposed in the area of the deposit, the worldwide association of quartz-wolframite veins with granitic intrusions, and the presence of biotite hornfels in drill holes suggest that a hidden pluton underlies the Kalzas deposit. The date of 91 ± 1 Ma obtained on biotite hornfels is consistent with the mid-Cretaceous age of granitic plutons in the Mayo area generally and thus supports this contention.

GSC 87-166 Biotite, K-Ar age
94 ± 2 Ma

K-Ar 3198 (2) See GSC 81-41 for analysis

From a biotite quartz monzonite
( ) Bulldozer trench near top of ridge at headwaters of Skate Creek, 10 km west of Hansen Lakes, Yukon, 64°03′01″N, 135°33′44″W. Sample SYA79-86, collected and interpreted by W.D. Sinclair. For description, see GSC 81-41 (*in* Stevens et al., 1982).

The date reported herein is the result of re-analysis of reserve biotite concentrate of GSC 81-41 which had previously yielded an age of 98.8 ± 3.7 Ma. The reason for the discrepancy in the two ages is not known. However, the older date is considered less reliable and is superceded by the younger 93.8 ± 2 Ma date.

REFERENCE

Stevens, R.D., Delabio, R.N. and Lachance, G.R.
1982: Age determinations and geological studies, K-Ar isotopic ages, Report 16; Geological Survey of Canada, Paper 82-2, 56 p.

GSC 87-167 Muscovite, K-Ar age
63.03 ± 1.4 Ma

K = 8.3%, radiogenic
Ar = 206.9 × 10^{-7} cc/gm
K-Ar 3697 atmos. Ar = 15.5%

From a granite separate
(115 H) Hatch property, 2.5 km west of Sekulmum Lake on the south side of Thatchell Creek, Aishihik Lake map-area, Yukon, 61°33′10″N, 137°38′46″W. Map unit Tgal, GSC Map 17-1973. Sample SYA84-21 from Hudson Bay Exploration and Development Limited drill hole 84-2 at a depth of 81.7 m, collected and interpreted by W.D. Sinclair.

The sample is a leucocratic, medium-grained, equigranular granite with muscovite-rich bands and pods. It consists of 30% quartz, 20% K-feldspar, 20% albite, 30% muscovite and trace amounts of disseminated carbonate and opaque minerals. The muscovite is clear with trace amounts of included carbonate and opaque minerals.

The sample is from the apical zone of a small, unexposed pluton that has intruded biotite schist/hornfels of Proterozoic and/or Paleozoic age. Molybdenite occurs as disseminated grains in the granite and in quartz veinlet stockworks in hornfels. Gold and base metal sulphides occur in quartz veins that cut both granite and hornfels and crosscut the molybdenite bearing quartz veinlet stockworks. The intrusion belongs to the Nisling Range alaskite, although the age reported here (63 Ma) is slightly older than the range of 50-60 Ma reported by Tempelman-Kluit and Wanless (1975). The age of the intrusion represents a maximum age for the associated molybdenum and gold mineralization.

REFERENCE

Templeman-Kluit, D.J. and Wanless, R.K.
1975: Potassium-argon age determinations of metamorphic and plutonic rocks in the Yukon Crystalline Terrane; Canadian Journal of Earth Sciences, v. 12, p. 1895-1909.

GSC 87-168 Feldspar, K-Ar age
122.9 ± 1.9 Ma

K = 6.87 %, radiogenic
Ar = 339.4 × 10^{-7} cc/gm
K-Ar 3538 atmos. Ar = 2.3 %

From a gold-silver vein
(115 I) Heustis Vein, 10 km southwest of Mount Nansen and 45 km west of Carmacks, Carmacks map-area, Yukon Territory, 62°03′N, 137°09′W. Sample SYA82-130, collected by J.A. Morin and interpreted by W.D. Sinclair.

The sample consists of about 30 % coarse-grained adularia, 50 % fine-grained quartz and 20 % clay minerals. X- ray diffraction of the adularia gave a sharp, well-defined pattern for orthoclase with no other feldspar phases indicated.

The Huestis Vein is a set of narrow, steeply-dipping quartz veins in metamorphic rocks of Precambrian and/or Paleozoic age. Pyrite, arsenopyrite, sphalerite, galena and freibergite are the principal sulphide minerals of the veins which have been mined for their gold and silver content (Saager and Bianconi, 1971). The adularia-bearing sample probably represents part of the upper alteration assemblage of the veins.

The Huestis and other gold-silver veins in the Mount Nansen area are thought to be genetically related to associated Late Cretaceous-early Tertiary volcanic and intrusive rocks (Biaconi and Saager, 1971). The Early Cretaceous date reported here is thus either anomalous or indicates that gold-silver mineralization is older than associated volcanic and intrusive rocks. considering anomalously young ages encountered in dating adularia elsewhere (e.g. Halliday and Mitchell, 1976), this date must be interpreted with caution. The possibility that the adularia contains excess radiogenic argon to give an anomalously old age should be considered.

REFERENCES

Halliday, A.N. and Mitchell, J.G.
1976: Structural, K-Ar and 40Ar-39Ar age studies of adularia K-feldspar from the Lizard Complex, England; Earth and Planetary Science Letters, v. 29, p. 227-237.

Saager, R. and Bianconi, F.
1971: The Mount Nansen gold-silver deposit, Yukon Territory, Canada; Mineralium Deposita, v. 6, p. 209-244.

GSC 87-169 Biotite, K-Ar age
75.1 ± 1.9 Ma

K = 3.36 %, radiogenic
Ar = 1.001 × 10^{-5} cc/gm
K-Ar 3405 atmos. Ar = 28.1 %

From quartz monzonite separate
(115 N) 10 km due west of Mt. Mort, Yukon, 63°10′N, 140°35′15″W. Sample T080-2-1, collected by D. Tempelman-Kluit.

GSC 87-170 Hornblende, K-Ar age
86.5 ± 1.9 Ma

K = 0.76 %, radiogenic
Ar = 2.617 × 10^{-6} cc/gm
K-Ar 3406 atmos. Ar = 70.5 %

From quartz monzonite separate
(115 N) See GSC 87-169 for location.

GSC 87-171 Whole rock, K-Ar age
51.4 ± 3.9 Ma

K = 0.67 %, radiogenic
Ar = 1.358 × 10^{-6} cc/gm
K-Ar 3483 atmos. Ar = 30.8 %

From a tuff separate.
(115 N) On the west bank of Sixty Mile River, 8 km downstream from the confluence of Matson Creek, Yukon, 63°39′40″, 140°08′20″W. Sample 81-LGW-5, collected and interpreted by G.W. Lowey.

The tuff is approximately 1 m thick and horizontally laminated. It varies from a vitric tuff to a crystal tuff and is recrystallized, although glass shards are still recognizable. The coarser-grained laminae are graded.

The tuff is interbedded with clastic sedimentary rocks that are considered Eocene and/or younger in age by Tempelman-Kluit (1974), and the whole rock date tends to confirm this interpretation. However, tuffs are notorious for giving inaccurate K-Ar ages due to the loss of Ar from the glass shards. It is possible then, that the tuff is of the same age as volcansim at Haystack Mountain (48 km east) and that the tuff was derived from explosive eruption of Haystack Mountain (see GSC 87-172 for a description and interpretation of volcanic rocks in the Indian River area). This interpretation agrees with palynological dates (Upper Cretaceous, possibly Maastrichtian), for the sedimentary rocks in the Sixty Mile River area (Lowey and Hills, in prep.).

REFERENCES

Lowey, G.W. and Hills, L.V.
(in prep.): New dates (palynological) for clastic sedimentary rocks in the Indian River and Sixty Mile River areas, west-central Yukon, and the Tantalus Formation, south-central Yukon.

Tempelman-Kluit, D.J.
1974: Reconnaissance geology of Aishihik Lake, Snag and part of Stewart River map-areas, west-central Yukon, Geological Survey of Canada, Paper 73-41.

GSC 87-172 Whole rock, K-Ar age
65.2 ± 1.7 Ma

K = 3.54 %, radiogenic
Ar = 9.135 × 10^{-6} cc/gm
K-Ar 3503 atmos. Ar = 8.8 %

From an andesite porphyry separate.

(115 O) From andesite on the west bank of McKinnon Creek, 6.2 km north of Haystack Mountain, Indian River area, Yukon, 63°42′30″N, 139°08′30″W. Sample 81-LGW-1, collected and interpreted by G.W. Lowey.

The andesite is porphyritic, consisting of 15 % plagioclase (An_{30-40}) phenocrysts, 10 % amphibole (hornblende) phenocrysts and 5 % biotite phenocrysts, in a very-finely-crystalline, dark-green groundmass of plagioclase, magnetite and minor apatite. The plagioclase phenocrysts are oscillatory zoned and amphibole and biotite phenocrysts have reaction rims of magnetite. The groundmass has a felted texture.

Volcanic rocks in the Indian River area are predominantly andesite with minor dacite. They contain abundant quartzite, monzonite and schist-gneiss xenoliths; the groundmass is felted to trachytic in texture; and phenocrysts of plagioclase and amphibole are usually subparallel to one another. These rocks occur as dikes, sills and flows that intrude and overlie clastic sedimentary rocks of Lower Cretaceous (Albian) age (Lowey and Hills, in prep.). Haystack Mountain is probably a volcanic centre and represents the erosional remnant of a volcanic neck.

The sample is from a subaerial volcanic flow and the whole rock date is interpreted as a minimum age for volcanism at Haystack Mountain. Bostock (1942) correlated these volcanics with the Carmacks Group and the results of this study confirm his interpretation.

REFERENCES

Bostock, H.S.

1942: Ogilvie, Yukon Territory; Geological Survey of Canada, Map 7 IIA.

Lowey, G.W. and Hills, L.V.

(in prep.): New dates (palynological) for clastic sedimentary rocks in the Indian River and Sixty Mile River areas, west-central Yukon, and the Tantalus Formation, south-central Yukon.

GSC 87-173 Whole rock, K-Ar age
64.8 ± 1.7 Ma

K = 3.19 %, radiogenic
Ar = 8.181 × 10^{-6} cc/gm
K-Ar 3510 atmos. Ar = 6.2 %

From an andesite separate.

(115 O) On the north flank of Haystack Mountain, 1 km from the summit at an elevation of approximately 900 m above mean sea level, Indian River area, Yukon, 63°39′50″N, 139°10′30″W. Sample 81-LGW-2, collected and interpreted by G.W. Lowey.

The andesite is porphyritic to glomeroporphyritic, consisting of 15 % plagioclase (An_{30-40}) phenocrysts, 20 % amphibole (hornblende) phenocrysts and 5 % pyroxene (augite) phenocrysts, in a finely-crystalline, light-grey-green groundmass of plagioclase, magnetite and minor apatite. The plagioclase phenocrysts are oscillatory zoned and amphibole and pyroxene phenocrysts have reaction rims of magnetite. The ground-mass has a felted texture.

The sample is from a subaerial volcanic flow and the whole rock date is interpreted as a minimum age for volcanism at Haystack Mountain. The volcanics are part of the Carmacks Group (see GSC 87-172 for further description and interpretation).

GSC 87-174 Whole Rock, K-Ar age
69.0 ± 1.7 Ma

K = 3.23 %, radiogenic
Ar = 8.830 × 10^{-6} cc/gm
K-Ar 3511 atmos. Ar = 11.5 %

From andesite on the extreme east

(115 O) bank of McKinnon Creek, 5.3 km north of Haystack Mountain, Indian River area, Yukon, 63°42′10″N, 139°07′50″W. Sample 81-LGW-4 is from Cyprus Anvil Mining Corporation drill hole IR-80-2 at a depth of 145 m (approximately 518 m above mean sea level), collected and interpreted by G.W. Lowey.

The andesite is porphyritic, consisting of 20 % plagioclase (An_{35-45}) pheoncrysts, 15 % amphibole (hornblende) phenocrysts and 5 % biotite phenocrysts, in a finely-crystalline, dark-grey-green ground-mass of plagioclase, magnetite and minor apatite. The plagioclase phenocrysts are oscillatory zoned and amphibole phenocrysts have magnetite reaction rims. The groundmass has a felted texture.

The sample is from a sill and/or dike and the whole rock date is interpreted as a maximum age for volcanism at Haystack Mountain. These volcanics are part of the Carmacks Group (see GSC 87-172 for further description and interpretation).

GSC 87-175 Muscovite, K-Ar age
154 ± 3 Ma

K = 6.86 %, radiogenic
Ar = 4.287 × 10^{-5} cc/gm
K-Ar 3408 atmos. Ar = 12.7 %

From a gneiss separate

(115 P) North end of White Mountains, McQuesten map-area Yukon, 63°17′N, 137°09′W. Sample TOE80-12-3, collected and interpreted by P. Erdmer.

The rock is from unit 5 of Bostock (1963), a mylonitic granodiorite orthogneiss with nearly horizontal foliation in southwestern McQuesten map-area (interpreted as Simpson Allochthon by Erdmer, 1985. It has a strongly developed flaser fabric, not unlike Mississippian-age orthogneiss of Finlayson Lake map-area (see report GSC 87-159).

The Late Jurassic age is that of the metamorphic overprint recorded in most of the Yukon-Tanana terrane (Yukon Crystalline Terrane of Tempelman-Kluit & Wanless 1975), and in the regional ''pink quartz-monzonite'' suite. The protolith age is unknown.

REFERENCES

Bostock, H.S.
1963: Geology, McQuesten, Yukon Territory. GSC map 1143A.

Erdmer, P.
1985: An examination of the cataclastic fabrics and structures of parts of Nisutlin, Anvil, and Simpson allochthons, central Yukon: test of the arc-continent collision model. Journal of Structural Geology, 1, 57-72.

Templeman-Kluit, D.J. and Wanless, R.K.
1975: Potassium-argon age determinations of metamorphic and plutonic rocks in the Yukon Crystalline Terrane; Canadian Journal of Earth Sciences, v. 12, p. 1895-1909.

BRITISH COLUMBIA
(GSC 87-176 TO GSC 87-245)

GSC 87-176 Hornblende, K-Ar age
69.8 ± 1.9 Ma

K = .356 %, radiogenic
Ar = 9.851 × 10^{-7} cc/gm
K-Ar 3698 atmos. Ar = 33.9 %

From an amphibolite separate
(82) West shore of Slocan Lake across from New Denver, B.C. UTM 11U, 4705E, 55382N. Sample PCA-458-83, collected and interpreted by R. Parrish.

Hornblende comes from a foliated amphibolite metavolcanic rock; the protolith is probably the Kaslo formation or part of the McHardy assemblage. The amphibolite is intruded by leucocratic granite of the Ladybird granite suite which is 58 ± 1 Ma old (U/Pb zircon, monazite). The 70 Ma hornblende age may reflect nearly complete resetting from an older formation age. Zircons from this amphibolite, are 60 Ma old and originated during Paleogene metamorphism. The hornblende K/Ar age, therefore could also be close to the formation age of the hornblende.

GSC 87-177 Hornblende, K-Ar age
57.9 ± 2.4 Ma

K = 1.1 %, radiogenic
Ar = 25.09 × 10^{-7} cc/gm
K-Ar 3699 atmos. Ar = 13.6 %

From a granodioritic gneiss seperate
(82) 1 km west of Castlegar, B.C. UTM 11U, 4507E, 54639N. Sample PCA-58-84-CG, collected by and interpreted by R. Parrish.

The host Castlegar gneiss is a late Cretaceous megacrystic granitic intrusion strongly deformed in Paleogene time during displacement on the Slocan Lake/Valkyr fault system. The 58 Ma age reflects rapid cooling during this extensional tectonic event and is similar to other hornblende ages within the Valhalla complex.

GSC 87-178 Hornblende, K-Ar age
64.4 ± 1.1 Ma

K = .884 %, radiogenic
Ar = 22.51 % × 10^{-7} cc/gm
K-Ar 3682 atmos. Ar = 22.5 %

From a recrystallized mafic dyke cutting Coryell syenite.
(82 F) Outcrop on highway west of Castlegar between Allandale Creek & Syringa Creek, B.C. UTM 11U, 4376E, 54658N. Sample PCA-21-84-D, collected and interpreted by R. Parrish.

Hornblende from this recrystallized mafic dyke has an age of 64 Ma which exceeds the age of the 52 Ma (U/Pb zircon) Coryell syenite which it intrudes. Therefore an explanation involving excess argon in the hornblende, possibly related to degassing during its modest shearing and recrystallization, appears necessary to explain the age.

GSC 87-179 Muscovite, K-Ar age
47.2 ± 1.1 Ma

K = 8.79 %, radiogenic
Ar = 163.3 × 10^{-7} cc/gm
K-Ar 3686 atmos. Ar = 33.8 %

From 58 Ma mylonitic granite.
(82 F) Pedro Creek Logging Rd., B.C. map unit — PEg, UTM 11U 4605E, 54898N Ref. Paper 85-1A Parish et al. Sample PCA-407-83, collected and interpreted by R. Parrish.
See GSC 87-180 for description and interpretation.

GSC 87-180 Muscovite, K-Ar age
48.7 ± 1.1 Ma

K = 8.56 %, radiogenic
Ar = 164.1 × 10^{-7} cc/gm
K-Ar 3688 atmos. Ar = 24.2 %

From a 58 Ma mylonitic granite
(82 F) Springer Creek Logging Rd.; 2 km from Highway. Map unit PEg., UTM 11U, 4676E, 55122N Ref. Current Res. 85-1A. Parrish et al. Sample Springer Creek, interpreted by R.R. Parrish.

Both PCA-407-83 and Springer Creek muscovites are porphyroblasts in greenschist-grade mylonitic Ladybird granite (58 Ma old, U/Pb zircon). The ages of 47-48 Ma reflect the time of cooling below the Ar closure temperature for muscovite (~ 300-350°C) during uplift and tectonic denudation of Valhalla Complex during the Eocene.

GSC 87-181 Hornblende, K-Ar age
421 ± 11 Ma

K = 1.75%, radiogenic
Ar = 3.225 × 10^{-5} cc/gm
K-Ar 3440 atmos. Ar = 9.6%

From a nepheline syenite separate
(82 N) Ice River Complex Yoho National Park, B.C. Sample Ice River. 51°08′, 116°26′. Collected by W.C. Gussow, interpreted by R. Parrish.

This age determination is a repeat analysis of an improved hornblende separate from nepheline syenite of the complex. Previous analysis of this hornblende has yielded even older ages. Lack of agreement between analyses demonstrates disturbance to the K/Ar system. A $^{40}Ar/^{39}Ar$ age spectrum done on this separate yields a partially flat segment at ~ 380 Ma in general agreement with U/Pb zircon and sphene analyses on this sample. Excess argon is the explanation for the anomalously old K/Ar age.

GSC 87-182 Biotite, K-Ar age
48.52 ± .77 Ma

K = 7.53%, radiogenic
Ar = 143.9 × 10^{-7} cc/gm
K-Ar 3653 atmos. Ar = 31.0%

From a granite separate
(104 I) On ridge crest, elev. 5880′ on northwest side of valley of Major Hart Creek; B.C. 58°49′30″N, 128°34′W. Map unit-Kgm. Ref-NE Cry Lake map area, GSC Open File 610. Sample GA 83-41, collected and interpreted by H. Gabrielse.

The sample was collected from a miarolitic, smoky quartz granite which has intruded oceanic rocks of the oceanic Sylvester Allochthon. It was formerly included in a mid-Cretaceous suite of granite now called the Cassiar Suite. The age documents the importance of Eocene plutons within the general area of the mid-Cretaceous Cassiar Batholith. These rocks appear to be important because of their relationship to base-metal and tungsten deposits.

GSC 87-183 Whole rock, K-Ar age
9.27 ± .15 Ma

K = 5.69%, radiogenic
Ar = 20.55 × 10^{-7} cc/gm
K-Ar 3532 atmos. Ar = 36.7%

(91 I/3) From Late Miocene a basalt, separate Specimen MV82-161, from the Prospect Creek area (lat. 50°04′35″N; long. 121°08′50″W), On old logging road 2 km east of Teepee Creek and north of Prospect Creek, below younger volcanics. Sample is a dark grey, very fine grained basaltic dyke.

The K/Ar whole rock age of 9.27 ± .15 Ma is typical of the Chilcotin Group or plateau basalts, of this region. The dyke is probably a feeder to this group. Interpreted by J.W.H. Monger.

GSC 87-184 Whole rock, K-Ar age
49.12 ± .76 Ma

K = 2.06%, radiogenic
Ar = 39.87 × 10^{-7} cc/gm
K-Ar 3533 atmos. Ar = 6.4%

From an andesite separate
(91 I/3) At elevation 5400 feet, 2 km east of Teepee Creek and north of Prospect Creek, B.C., collected and interpreted by J.W.H. Monger.

Specimen MV82-161a is a very fine grained, flow banded andesite or dacite from a volcanic sequence that overlies mid-Cretaceous volcanics of the Spences Bridge Group, west of Spius Creek (lat. 50°04′35″N; long. 121°08′50″W). The K/Ar whole rock age of 49.0 ± 0.8 Ma is typical of that of the Eocene volcanics of this region.

GSC 87-185 Hornblende K-Ar age
49.2 ± 1.0 Ma

K = 0.508%, radiogenic
Ar = 9.893 × 10^{-7} cc/gm
K-Ar 3683 atmos. Ar = 48.0%

From an andesite separate
(92 H/1) Cliff, above rock glacier near head of Placer creek, elevation 6000′; 4.5 km SW of summit of Placer Mountain, B.C. Sample MV84-664, collected and interpreted by J.W.H. Monger.

Specimen MV84-664 is hornblende andesite from Placer Mountain, southwest of Princeton (lat. 49°06′57″N; 120°27′31″W), that lithologically is similar to other, dated, Eocene volcanics of this region (Princeton Group). The K/Ar date of 49.2 ± 1.0 Ma reflects the time of eruption of these rocks.

GSC 87-186 Biotite, K-Ar age
109.0 ± 1.7 Ma

K = 7.8%, radiogenic
Ar = 340.7 × 10^{-7} cc/gm
K-Ar 3685 atmos. Ar = 12.2%

(92 H1) Specimen MV84-663 is a "grey granodiorite" from the Ashnola River area NE side of ridge, elev. 7000′ .9 km SW, SW shore of Trapper Lake ~ 7 km W of Cathedral Lakes provincial park (lat. 49°02′55″N; long. 120°27′03″W), that is from the northern margin of the Remmell Batholith which is exposed mainly in Washington State. The K/Ar on biotite of of 109.0 ± 1.7 Ma is in general accordance with K/Ar dates from the Remmell. There is the possibil-

ity, based on lithological similarities, that these rocks are partly or wholly of early Mesozoic (ca. 200 Ma) intrusive age but K/Ar systems were reset during a mid-Cretaceous magmatic pulse (represented by sample MV84-660 (GSC 87-190)). Interpreted by J.W.H. Monger.

GSC 87-187 Whole rock, K-Ar age
49.66 ± .78 Ma

K = 1.01%, radiogenic
Ar = 19.79 × 10^{-7} cc/gm
K-Ar 3696 atmos. Ar = 14.5%

From a basalt separate
(92 H1) Specimen MV84-661b is a fine grained olivine basalt, from Cathedral Provincial Park (lat. 49°02′49″N; long. 120°12′27″W) Cathedral Ridge; above and W. of Lady Slipper Lake, elev. 8300′, B.C. that lithologically resembles Neogene rocks of the region (Chilcotin Group). However, the K/Ar whole rock age of 49.66 ± 0.78 Ma is typical of the Eocene volcanics of the Kamloops and Princeton groups, and it is concluded that this is a date reflecting time of eruption of a lithologically atypical unit. Interpreted by J.W.H. Monger.

GSC 87-188 Biotite, K-Ar age
47.94 ± .75 Ma

K = 7%, radiogenic
Ar = 132.1 × 10^{-7} cc/gm
K-Ar 3693 atmos. Ar = 13.3%

From a granodiorite separate
(92 H6) Specimen MV84-651 is a medium grained biotite granodiorite of the Mount Outram Stock, which intrudes the Permian to Jurassic Hozameen Group on the east side of the Cascade Belt (lat. 49°18′16″N; long 120°09′54″W) Near summit of Mt. Outram. Elev. 8000′ ~ 21 km, ESE of Hope B.C. The K/Ar age of 47.94 ± .75 Ma probably reflects the time of intrusion. The stock is one of a number of relatively small granodiorite intrusions in the eastern part of the Cascade orogen that yields Eocene K/Ar dates (cf Needle Peak pluton etc.). Interpreted by J.W.H. Monger.

GSC 87-189 Hornblende, K-Ar age
22.4 ± 1.1 Ma

K = .237%, radiogenic
Ar = 2.077 × 10^{-7} cc/gm
K-Ar 3684 atmos. Ar = 79.5%

(92 H7) From a tuff separate
South side of Warburton Peak, elev. 5600′, 5.5 Km WSW of confluence of Podunk Creek and Tulgmeen River, B.C.

Specimen MV84-649 is from fine-grained hornblende and feldspar bearing lithic tuff of the "Podunk Creek" volcanics from the south side of Warburton Peak (lat. 49°19′45″N; long. 120°58′45″W). The K/Ar date of 22.4 ± 1.1 Ma is similar to those obtained by Berman (M.Sc. thesis, U.B.C.) from the Coquihalla volcanics to the north, and is believed to reflect the time of eruption of these rocks. Interpretated by J.W.H. Monger.

GSC 87-190 Biotite, K-Ar age
100.6 ± 1.5 Ma

K = 7.06%, radiogenic
Ar = 283.8 × 10^{-7} cc/gm
K-Ar 3694 atmos. Ar = 12.9%

From a granitoid separate
(92 H7) Elev. 7500′, above Haystack Lakes, E side of Haystack Mt. Cathedral Peaks area, B.C. Specimen MV84-660 is from Cathedral Provincial Park (lat. 49°01′27″N; long. 120°06′24″W) and is massive, miarolitic granite that the writer proposes to call the Cathedral Pluton. It intrudes the Remmell Batholith, noted above. It is on trend with the lithologically similar Verde Creek stock (K/Ar ca. 90 Ma) and mid-Cretaceous volcanics of the Spences Bridge Group. The 100.6 ± 1.5 Ma date is probably the date of intrusion. Interpreted by J.W.H. Monger.

GSC 87-191 Biotite, K-Ar age
6.3 ± 1.5 Ma

K = 7.6%, radiogenic
Ar = 1.885 × 10^{-5} cc/gm
K-Ar 3482 atmos Ar = 16.3%

From a granodiorite separate
(92 I) Elevation 7500′ northwest of lake, 1.1 km northwest from summit of Antimony Mountain, British Columbia, 50°10′13″N, 121°52′08″W. Sample MV81-195-7, collected by J.W.H. Monger.

GSC 87-192 Hornblende, K-Ar age
174 ± 13 Ma

K = 0.53%, radiogenic
Ar = 3.763 × 10^{-6} cc/gm
K-Ar 3495 atmos. Ar = 15.3%

From a granodiorite separate
(92 I) At elevation 3400′, east side of Mowhokam creek, 4.4 km northeast of summit of Jackass Mountain, British Columbia. UTM: 10-612600-5546600. Sample MV81-13G collected by J.W.H. Monger.

GSC 87-193 Biotite, K-Ar age
186.0 ± 4.0 Ma

K = 7.72 %, radiogenic
Ar = 5.879 × 10^{-5} cc/gm
K-Ar 3496 atmos. Ar = 4.9 %

From a granodiorite separate
(92 I) At elevation 5200′ head to tributary to Nicoameen River, east of summit of Zakwaski Mt. B.C., UTM: 10-617500-5556250. Sample MV81-141a, collected by J.W.H. Monger.

GSC 87-194 Hornblende, K-Ar age
186 ± 15 Ma

K = 0.51 %, radiogenic
Ar = 3.884 × 10^{-6} cc/gm
K-Ar 3497 atmos. Ar = 12.5 %

From a granodiorite separate
(92 I) Details as for 87-193.
For location see 87-193.

GSC 87-195 Biotite, K-Ar age
79.5 ± 1.8 Ma

K = 6.74 %, radiogenic
Ar = 2.047 × 10^{-5} cc/gm
K-Ar 3498 atmos Ar = 17.3 %

From a granodiorite separate
(92 I) North of small lake 3.7 km north-northwest form Mehatt Peak, north of Rutledge Glacier, B.C. UTM: 10-571500-5551350. Sample MV81-201-2, collected by J.W.H. Monger.

GSC 87-196 Hornblende, K-Ar age
220 ± 16 Ma

K = 0.54 %, radiogenic
Ar = 4.911 × 10^{-6} cc/gm
K-Ar 3499 atmos. Ar = 8.1 %

From a granodiorite separate
(92 I) Details as for GSC 87-195.
For location see GSC 87-195.

GSC 87-197 Hornblende, K-Ar age
60.4 ± 4.1 Ma

K = 0.67 %, radiogenic
Ar = 1.600 × 10^{-6} cc/gm
K-Ar 3502 atmos. Ar = 29.2 %

From a granodiorite separate
(92 I) At elevation 2000′, north side of Campbell Creek, 5.3 km northwest of summit of Mount Vicars, British Columbia. Sample MVB81-59, collected by D. Brown and interpreted by J.W.H. Monger.

Specimen MVB81-59 (UTM 10U E69895 N560855) is hornblende biotite granodiorite of the Wildhorse Batholith. The K/Ar date on hornblende of 60.4 Ma probably represents almost total resetting of an Early Jurassic magmatic age. For further interpretation See GSC 87-208.

K/Ar dates from the intermontane and Eastern Coast-Cascade Belts, Hope (92H) and Ashcroft (92I) map-areas, Southwestern British Columbia
J.W.H. Monger, Cordilleran and Pacific Margins division.

The specimens are ''spot'' samples collected during regional mapping of Hope and Ashcroft map-areas. Most are from the southern Intermontane Belt, and a few from the eastern part of the Coast-Cascade Belt. The Intermontane Belt specimens sample the early, middle and late Mesozoic and early and late Tertiary magmatic episodes of this region. Because of this complex magmatic history, most K/Ar dates, at least of Mesozoic age, must be treated with caution and compared with data from other isotopic systems. Below, the K/Ar results are grouped into their probable times of primary recrystallization.

GSC 87-198 Biotite, K-Ar age
144.4 ± 2.2 Ma

K = 8 %, radiogenic
Ar = 467.5 × 10^{-7} cc/gm
K-Ar 3638 atmos. Ar = 2.7 %

From a granodiorite separate
(92 I/1) 3/4 km East of Rock Lake on road. 50°02′15″N, 120°02′15″W.

The ''Pennask Batholith'' lies mainly in easternmost Hope map area, but part is in southeastern Ashcroft. As mapped, it consists of a rim of typically hornblende biotite granodiorite, which includes ''type Pennask'', which surrounds on north, east and south sides a core of megacrystic quartz monzonite to monzodiorite. It is starting to appear from this study that the rim is Early Jurassic in age and the core is a separate, magmatically unrelated, Late Jurassic pluton, which is tentatively named ''Osprey Lake Pluton''.

Specimen MVV82-45 is biotite granodiorite from near Pennask Lake (lat. 50°02′15″N; long. 120°02′15″W), the type locality of the Pennask Batholith. The K/Ar biotite date of 159.5 ± 2.4 Ma is probably a reset date, that may reflect intrusion of the nearby Late Jurassic (ca. 160 Ma) Osprey Lake Pluton. Possibly the time of intrusion was ca. 200 Ma as lithologically similar rocks from elsewhere in the ''Pennask Batholith'' yield preliminary U/Pb dates around 200 Ma (see MV84-41), as do rocks of the lithologically similar Wildhorse Batholith to the north. A U/Pb specimen from this area is being processed. Interpreted by J.W.H. Monger.

GSC 87-199 Biotite, K-Ar age
183.3 ± 2.7 Ma

K = 7.25 %, radiogenic
Ar = 543.8 × 10^{-7} cc/gm
K-Ar 3641 atmos. Ar = 3.8 %

(92 I/1) From a granodiorite separate
Specimen MVK82-134B is medium grained biotite-hornblende granodiorite

from the Douglas Lake Stock, 4.5 km W of W end of Douglas Lake (lat. 50°11′03″N; long. 120°17′15″W), which intrudes the Upper Triassic Nicola Group. The slightly discordant dates of 183.3 ± 2.7 Ma on biotite and 177.5 ± 2.8 Ma on hornblende may reflect slight resetting. Other lithologically similar granodiorites, north and south of this stock (Wildhorse and Pennask batholiths) have yielded U/Pb dates of about 200 Ma, which is probably the age of intrusion. Interpreted by J.W.H. Monger.

GSC 87-200 Hornblende, K-Ar age
177.5 ± 2.8 Ma

K = .537%, radiogenic
Ar = 38.95 × 10^{-7} cc/gm
K-Ar 3642 atmos. Ar = 19.4%

From a granodiorite separate
(92 I/1) See GSC 87-199 for location and interpretation.

GSC 87-201 Biotite, K-Ar age
52.22 ± .81 Ma

K = 6.11%, radiogenic
Ar = 125.8 × 10^{-7} cc/gm
K-Ar 3645 atmos. Ar = 7.7%

From a porphyry separate
(92 I/3) Specimen MV82-163a is from a pale grey biotite feldspar porphyry, from the Nicola River valley (lat. 50°10′01″N; long. 121°02′50″W) On Highway 8; .5 km N of where Nuaitch Creek enters Nicola River, B.C., within the main outcrop area of calc-alkaline, continental volcanics of the mid-Cretaceous Spences Bridge Group.

It yields markedly discordant K/Ar dates on hornblende and biotite. The hornblende date of 109 ± 1.9 Ma contrasts with a biotite date of 52.22 ± 0.81 Ma. Lithologically the rock resembles the mid-Cretaceous Spences Bridge Group lithology. Accordingly, the hornblende date may be the crystallization date, whereas the biotite may have been reset by early Tertiary magmatism which is present in the area (e.g. MV82-161a). Sample interpretated by J.W.H. Monger.

GSC 87-202 Hornblende, K-Ar age
109 ± 1.9 Ma

K = .477%, radiogenic
Ar = 20.84 × 10^{-7} cc/gm
K-Ar 3646 atmos. Ar = 15.7%

(92 I/3) See GSC 87-201 for location, description, and interpretation.

GSC 87-203 Whole rock, K-Ar age
82 ± 3.2 Ma

K = 1.23%, radiogenic
Ar = 4.011 × 10^{-6} cc/gm
K-Ar 3462 atmos. Ar = 15.4%

From an andesite separate
(92 I/3) At elevation 930 metres above tributary of Nicoamen River, 6.7 km NE of Mount Lytton and 7.9 km NW of Zakwaski Mountain, British Columbia, UTM 10-616300-5562620. Sample MV81-148, collected by J.W.H. Monger.

GSC 87-204 Biotite, K-Ar age
63.8 ± 1.5 Ma

K = 7.15%, radiogenic
Ar = 1.805 × 10^{-5} cc/gm
K-Ar 3500 atmos. Ar = 18.3%

From granodiorite.
(92 I/4) West side of small lake at elevation 7300′, 1.4 km west of Kiowa Mountain, 3 km NNW of confluence of Kwoiek and North Kwoiek Creeks, B.C., UTM: 10-592110-5557470. Sample MV81-195-12, collected by J.W.H. Monger.

GSC 87-205 Hornblende, K-Ar age
69 ± 10 Ma

K = 0.84%, radiogenic
Ar = 2.303 × 10^{-6} cc/gm
K-Ar 3501 atmos. Ar = 18.1%

From a granodiorite separate
(92 I/4) Details as for GSC 87-204.
See GSC 87-204 for location.

GSC 87-206 Biotite, K-Ar age
48.85 ± .76 Ma

K = 7.46%, radiogenic
Ar = 143.7 × 10^{-7} cc/gm
K-Ar 3643 atmos. Ar = 8.9%

From a granodiorite separate
(92 I/7) Specimen MVK82-133 is from coarse grained hornblende biotite granodiorite from the southern part of the Nicola Batholith (lat. 50°16′56″N; long. 120°33′46″W). The dates of 48.85 ± .76 Ma from biotite and 61.0 ± 1.1 Ma from hornblende probably represent slight resetting from the original date of intrusion, since a preliminary U/Pb date of 65.3 ± 1.5 Ma from this specimen probably reflects time of intrusion. Resetting presumably occurred during rapid uplift of a horst comprising mainly the Nicola Batholith in Eocene (ca. 50 Ma) time.

From the northern part of this batholith, R.L. Armstrong (pers. comm.) has obtained Rb/Sr measurements that suggest an Early Jurassic age (ca. 200 Ma), which is also indicated by the lithological similarity between many granitic rocks in the Nicola Batholith and those of the Wildhorse Batholith, which exhibits a similar spectrum of isotopic ages. More detailed isotopic studies on these intrusions are clearly needed. Interpreted by J.W.H. Monger.

GSC 87-207 Hornblende, K-Ar age
61.0 ± 1.1 Ma

K = .837%, radiogenic
Ar = 20.17 × 10^{-7} cc/gm
K-Ar 3644 atmos. Ar = 33.5%

From granodiorite.
(92 I/7) See GSC 87-206 for sample location and interpretation.

GSC 87-208 Biotite, K-Ar age
161.8 ± 2.4 Ma

K = 7.09%, radiogenic
Ar = 466.5 × 10^{-7} cc/gm
K-Ar 3639 atmos. Ar = 4.5%

From a granodiorite separate
(92 I/8)

A widespread episode of early Mesozoic calc-alkaline and alkaline magmatism is present in the southern Intermontane Belt. It is represented by calc-alkaline plutons such as the Guichon, Wildhorse and Pennask batholiths. All intrude Upper Triassic stratified rocks of the Nicola Group and the Guichon Batholith is overlain nonconformably by Early Jurassic (Sinemurian; ca. 200 Ma) strata of the Ashcroft Formation. Both Guichon and Wildhorse batholiths have previously yielded ca. 200 Ma K/Ar dates, and preliminary U/Pb dates of similar vintage from rocks of the "Pennask Batholith" are just starting to appear, although most K/Ar dates from it are younger.

The Wildhorse Batholith lies southeast of the town of Kamloops, northeastern Ashcroft map-area, and is typically hornblende biotite granodiorite.

Specimen MVM82-20 form 3 km WSW of Dardanelles Lake, B.C. (lat. 50°21′10″N; long. 120°11′34″W) is typical hornblende biotite granodiorite of the Wildhorse Batholith. The K/Ar dates of 161.8 ± 2.4 Ma on biotite and 169.3 ± 2.6 Ma on hornblende are possibly reset from an Early Jurassic magmatic age, as other K/Ar dates from elsewhere in this pluton are around 200 Ma (Preto et al., 1979, CJES, p. 1658) and a U/Pb date is 196 + 6/-1 Ma (MVK82-20). Interpreted by J.W.H. Monger.

GSC 87-209 Hornblende, K-Ar age
169.3 ± 2.6 Ma

K = .589%, radiogenic
Ar = 40.65 × 10^{-7} cc/gm
K-Ar 3640 atmos. Ar = 12.1%

From a granodiorite separate
(92 I/8) See GSC 87-208 for location and interpretation.

GSC 87-210 Whole rock, K-Ar age
47.35 ± .74 Ma

K = 3.28%, radiogenic
Ar = 61.16 × 10^{-7} cc/gm
K-Ar 3536 atmos. Ar = 3.7%

From an andesite separate
(92 I/8) At elevation 4500′, 2.5 km east of Blackwill Lake, east Stump Lake map sheet, B.C., collected by K. Kettles and interpreted by J.W.H. Monger. Specimen MVK82-60C is a fine grained, slightly porphyritic andesite with feldspar phenocrysts, from near Stump Lake, south of Kamloops (lat. 50°24′40″N; long. 120°01′30″W), with a K/Ar date of 47.4 ± 0.7 Ma is typical of those of the Eocene Kamloops Group.

GSC 87-211 Whole rock, K-Ar age
25.5 ± .4 Ma

K = .697%, radiogenic
Ar = 6.958 × 10^{-7} cc/gm
K-Ar 3535 atmos. Ar = 19.4%

From an olivine basalt separate
(92 I/9) Specimen MVD82-82 is from flat-lying olivine basalt near Knutsford (lat. 50°32′37″N; long. 120°18′48″W) that unconformably overlies faulted Eocene and older rocks.

Lithologically the basalt resembles Neogene (ca. 14-2 Ma) basalt of the Chilcotin Group (or "plateau basalt"). Ewing (1981, Ph.D. thesis, U.B.C.) obtained a K/Ar date of 31 Ma from this formation, which to the writer appeared to be too old. The K/Ar whole rock date of 25.5 ± 0.4 Ma is similar to one obtained recently by W.H. Mathews at U.B.C., and so may be a "real age". If so, it is anomalous with respect to other Chilcotin Group ages, but significant, as it may reflect some kind of "back-arc" magmatism, relative to the ca. 23 Ma Coquihalla calc-alkaline volcanics. It also puts a minimum age on the faults involving the underlying Eocene rocks. Interpretated by J.W.H. Monger.

GSC 87-212 Whole rock, K-Ar age
41.42 ± .65 Ma

K = 3.88%, radiogenic
Ar = 63.18 × 10^{-7} cc/gm
K-Ar 3534 atmos. Ar = 7.3%

From an andesite separate
(91 I/13) About 50 metres west and below Highway 12, 5.5 km south along highway from Pavilion, B.C. Collected and interpreted by J.W.H. Monger.

Specimen MV82-177 is from a suite of pink hornblende feldspar andesites occurring in fault bounded blocks along the Fraser River near Pavilion (lat. 50°49′34″N; long. 121°51′37″W) and formerly mapped as part of the mid-Cretaceous Spences Bridge Group. Lithologically it resembles other Eocene volcanics in the region (Kamloops Group) and this correlation is borne out by the K/Ar whole rock age of 41.4 ± 0.7 Ma, although this is slightly younger than other ca. 50 Ma dates from these volcanics.

GSC 87-213 Biotite, K-Ar age
49.1 ± 1.4 Ma

K = 7.24%, radiogenic
Ar = 1.401 × 10^{-5} cc/gm
K-Ar 3404 atmos. Ar = 12.1%

From a gneiss separate
(93 F) Road cut about 7 km south of west end of Sinkut Lake, Nechako River area, British Columbia, 53°50′7″N, 124°01′9″W. Sample 80-TD-110AG, collected by H.W. Tipper.

GSC 87-214 Biotite, K-Ar age
113.2 ± 1.7 Ma

K = 7.42%, radiogenic
Ar = 337 × 10^{-7} cc/gm
K-Ar 3566 atmos. Ar = 2.6%

From a granite separate
(94 E) Toodoggone River map area, Stikine Ranges, Cassiar Mountains; 6 km west of Frog River and 3 km south of Toodoggone-Kechika map boundary. 57°58′N, 127°15′30″W. Sample GAD 81-57, collected by C.J. Dodds and interpreted by H. Gabrielse.

The sample is typical of mid-Cretaceous granite formerly included within a suite of hornblende granodiorite in northwestern Toodoggone (94E) and adjacent Kechika (94K) map-areas. The association of Early Jurassic granodorite with mid-Cretaceous granite appears to be common in the Omineca Belt of the Cassiar and Omineca mountains.

GSC 87-215 Biotite, K-Ar age
43.22 ± .67 Ma

K = 7.02%, radiogenic
Ar = 119.3 × 10^{-7} cc/gm
K-Ar 3584 atmos. Ar = 7.7%

From a lamprophyre separate
(94 E) At elevation 6500 feet on east side of ridge about 1.5 km north of Warner peak, Toodoggone River map area, Sifton Ranges, B.C. 57°41′N, 126°07′W. Sample GAE 82-659-1, collected by H. Gabrielse.

GSC 87-216 Biotite, K-Ar age
170.6 ± 2.6 Ma

K = 7.29%, radiogenic
Ar = 507.1 × 10^{-7} cc/gm
K-Ar 3651 atmos. Ar = 4.4%

From a biotite, muscovite, quartz schist separate
(94 E) Crest of ridge south of Eastern Lake on Ludwig Creek, B.C. 57°58′N, 126°30′W, Sifton Ranges, Toodoggone River map area, NE Corner. Map unit HICm; GSC Open File 483, Sample GAE 83-787, collected by H. Gabrielse.

GSC 87-217 Muscovite, K-Ar age
85.5 ± 1.3 Ma

K = 7.72%, radiogenic
Ar = 262.6 × 10^{-7} cc/gm
K-Ar 3652 atmos. Ar = 11.3%

From a biotite, muscovite, quartz schist separate
(94 E) For location see GSC 87-216.

STIKINE BATHOLITH

GSC 87-218 Hornblende, K-Ar age
210 ± 4 Ma

K = 1.207%, radiogenic
Ar = 104.70 × 10^{-7} cc/gm
K-Ar 3671 atmos. Ar = 5.77%

From clinopyroxene-hornblende porphyry dyke
(94 E/13) Northwestern Toodoggone map area, north central British Columbia, 57°51′55″N, 127°52′06″W; see Anderson (1984). Sample GAAN-83-43-1, collected and interpreted by R.G. Anderson.

The sample was collected from a light grey, 1-1.5 m wide, steeply north-northwesterly-dipping clinopyroxene-horblende porphyry dyke. The sample contains euhedral light to dark greenish brown, commonly zoned hornblende (11%) and clear clinopyroxene (5%) phenocrysts in a groundmass (84%) of plagioclase, opaque minerals acicular hornblende and apatite. The sample is typical of the easterly-trending, ± hornblende-clinopyroxene dykes which intrude and characterize the Late Triassic plutonic suite along the eastern Stikine Arch. The dykes are considered comagmatic with upper part of Middle and Upper Triassic Stuhini Group volcanics.

For interpretation see GSC 87-222.

GSC 87-219 Biotite, K-Ar age
208 ± 4 Ma

K = 7.41 %, radiogenic
Ar = 633.63 × 10^{-7} cc/gm
K-Ar 3673 atmos. Ar = 2.70 %

From monzogranite
(94 E/13) Northwestern Toodoggone map area, north central British Columbia, 57°51′56″N, 127°52′21″W; see Anderson (1984). Sample GAAN-83-42-1, collected and interpreted by R.G. Anderson.

The sample was collected from white weathering, well-foliated, fresh, inclusion- and schlieren-bearing equigranular, medium-grained biotite-hornblende monzogranite (typical of unit uTmd of Anderson (1984)). The sample contains alkali feldspar (30 %), plagioclase (27 %), quartz (25 %), green hornblende (11 %), brown biotite (5 %) and accessory minerals (1 %) apatite, sphene and opaque minerals (see also GSC-220).

For interpretation GSC 87-222.

GSC 87-220 Hornblende, K-Ar age
217 ± 4 Ma

K = 0.67 %, radiogenic
Ar = 60.02 × 10^{-7} cc/gm
K-Ar 3674 atmos. Ar = 9.45 %

(94 E/13) Details as for GSC 87-219

For interpretation see GSC 87-222.

GSC 87-221 Biotite, K-Ar age
215 ± 4 Ma

K = 6.803 %, radiogenic
Ar = 602.98 × 10^{-7} cc/gm
K-Ar 3677 atmos. Ar = 1.72 %

From granite
(94 E/13) Northwestern Toodoggone map area, north central British Columbia, 57°54′58″N, 127°57′13″W; see Anderson (1984). Sample GAAN-83-114-2, collected and interpreted by R.G. Anderson.

The sample was collected from pink-weathering, fresh, massive, homogeneous, siliceous, equigranular, medium-grained, megacrystic hornblende-biotite monzogranite typical of unit uTqm (Anderson, 1984). The sample contains alkali feldspar (37 %), plagioclase (35 %) quartz (22 %), brown biotite (3 %), hornblende (2 %) and accessory minerals less than 1 %) sphene, opaque minerals, zircon and apatite (see also GSC 87-222).

For interpretation see GSC 87-222.

GSC 87-222 Hornblende, K-Ar age
213 ± 4 Ma

K = 0.643 %, radiogenic
Ar = 56.39 × 10^{-7} cc/gm
K-Ar 3678 atmos. Ar = 9.91 %

(94 E/13) Details as for GSC 87-221

INTERPRETATION OF STIKINE BATHOLITH K-AR AGES

Results reported here were determined from 5 samples collected from phases or dykes of the late Triassic suite in the Stikine batholith. The batholith is typical of Late Triassic and Middle Jurassic composite intrusions which underlie part of the Stikine Arch (Anderson, 1984). Heterogeneous, equigranular or seriate, and medium grained, phases of leucocratic, megacrystic monzogranite (unit uTqm; GSC 87-221 and 87-222), mafic monzogranite or quartz monzodiorite (unit uTmd; GSC 87-219, 87-220, 87-224, and 87-225), hornblendite or hornblende clinopyroxenite (unit uTum; GSC 87-223, quartz monzonite or granite and gabbro make up the batholith's Late Triassic phases. The phases were emplaced sequentially into polydeformed Mississippian to Permian chlorite schists, chert and amphibolite (Thorstad, 1980) and massive Upper Triassic porphyritic volcanics (Figure 5). Although the units clearly intrude parts of Middle to Upper Triassic Stuhini Group (Read, 1983), common inclusions of hornblendite and hornblende clinopyroxenite, and steep, easterly-trending (hornblende-) clinopyroxene porphyry feeder dykes in the Late Triassic plutonic phases indicate they are coeval and, in part, comagmatic with the upper part of the volcanics.

A Late Triassic isotopic age, 222 ± 10 Ma (G.S.C. 78-22), determined for hornblende in unit uTqm in earlier studies (C.J. Dodds, *in* Wanless et al. (1979, p. 10), was consistent with observed external and interphase intrusive relations. The present study was undertaken to confirm the date and the suggested consanguinity of phases in the Late Triassic plutonic suite and for comparison with the well- dated and similar Hotailuh batholith, 125 km to the west (R.G. Anderson *in* Stevens et al. (1982, p. 4-11)).

The eight K-Ar age determinations range from 204-217 Ma (average of all pluton ages is 214 ± 6 Ma (1 s.d.); Table 3) Hornblende retains the oldest K-Ar isotopic ages, 213-217 Ma, as expected from its higher blocking temperature compared with biotite. The hornblende isotopic age from the earlier study is concordant with the new ages within experimental error. The relative interphase intrusive chronology described in Anderson (1984) cannot be distinguished isotopically. Except for sample 114-2 (GSC 87-221 and 87-222), K-Ar isotopic ages for biotite are discordant and younger than that for the associated hornblende. The Late Triassic average age for hornblende, 217 ± 3 Ma, is considered the closest estimate of the cooling age for the Stikine batholith and its constituent phases (Figure 6).

The K-Ar isotopic age for the clinopyroxene-hornblende porphyry dyke is slightly younger but concordant with the hornblende K-Ar ages for the plutonic phases. The age supports the contention that these east-trending, steep dykes,

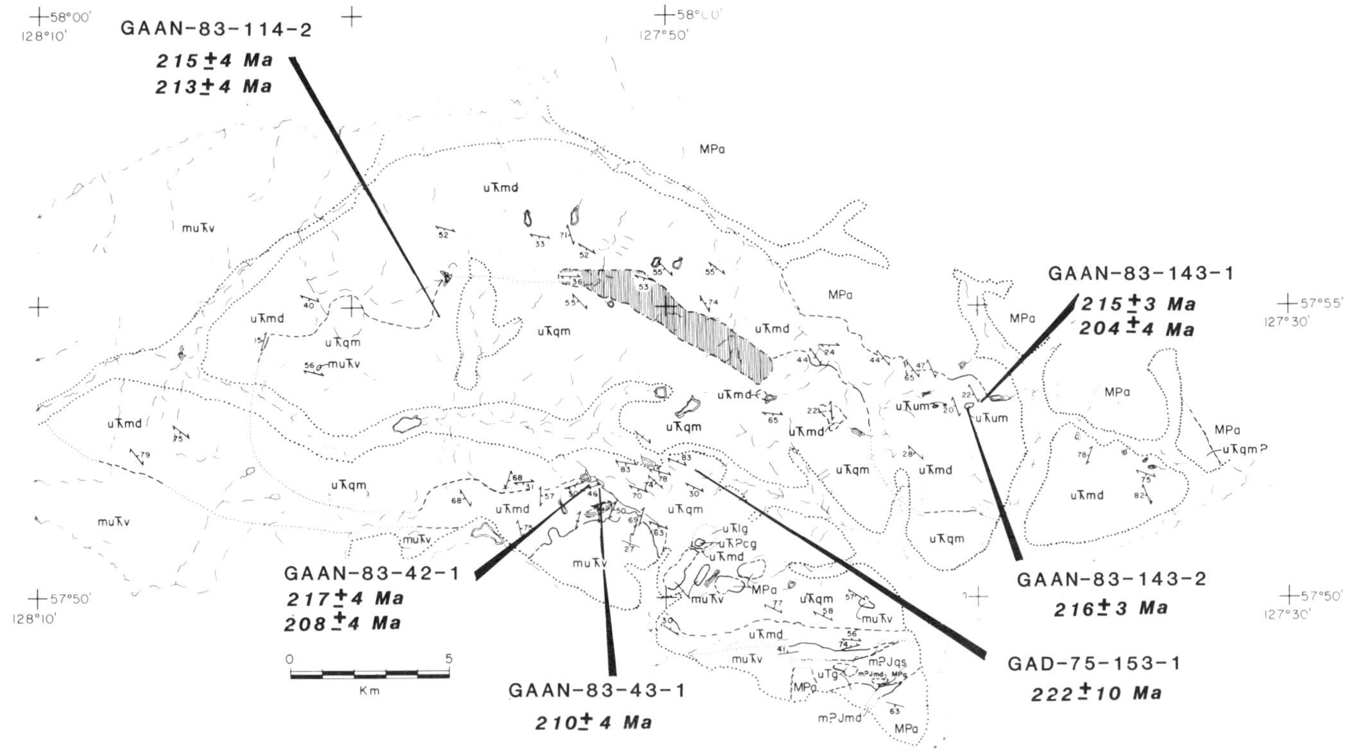

MID(?)-JURASSIC

m?Jqs	fine to medium grained alkali feldspar quartz syenite, quartz syenite
m?Jmd	fine to medium grained quartz monzodiorite, diorite

UPPER TRIASSIC

uTRg	clinopyroxene gabbro with porphyroblastic(?) biotite
uTRqm	megacrystic and minor equigranular biotite-hornblende quartz monzonite and granite
uTRum	heterogeneous hornblendite and hornblende clinopyroxenite
uTRmd	melanocratic or mesocratic, heterogeneous, biotite-hornblende quartz monzodiorite, monzodiorite, diorite
uTR?cg	granitoid-bearing volcanic conglomerate
uTRlg	leucocratic granite

heterogeneous zone of unit uTR md screens and unit uTR qm apophyses

metamorphosed and granitized unit muTRv and apophyses of unit uTR md

MIDDLE TO UPPER TRIASSIC

muTRv	STUHINI GROUP: green or light grey aphanitic or clinopyroxene porphyry basalt; minor greywacke

MISSISSIPPIAN AND PERMIAN

MPa	undivided green chlorite phyllite and schist, rhyolite, amphibolite; minor carbonate

45 bedding or compositional layering

45 mineral or mafic schlieren foliation, inclined, vertical

geological contact: defined, approximate, gradational, inferred

fault, inferred

limits of outcrop

lakes, creeks

Figure 5. Geological map of Stikine batholith and location of dated samples (after Anderson, 1984).

which characterize Late Triassic batholiths along the eastern Stikine Arch, are feeders for the upper part of the compositionally similar, Middle to Upper Stuhini Group volcanism and coeval with Late Triassic plutonism.

The biotite K-Ar ages may have been partly reset by the mid(?)-Jurassic plutonism common to the Stikine Arch and represented in the Stikine batholith by a small, undated mid(?)-Jurassic pluton in the batholith's southeastern lobe (Anderson, 1984).

The range of hornblende K-Ar isotopic ages for the Stikine batholith overlaps the younger part of the 218-230 Ma range for the Hotailuh batholith's less precise hornblende K-Ar isotopic ages (Figure 6 and Stevens et al. (1982, p. 8, Table 3)).

Lithology, structure, associated subvolcanic dykes, intrusive relations, and petrography of Late Triassic phases in the Stikine batholith compare closely with the Triassic plutonic suite of the Hotailuh batholith 125 km to the west along the Stikine Arch. Their isotopic ages are also closely similar.

Table 3. Compilation of GSC K-Ar isotopic ages for Triassic Plutonic Suite in Stikine Batholith

GSC No.	Field Number	Location	Lithology	Mineral Dated	Age (Ma)
GSC 87-218	GAAN-83 43-1	northwestern Toodoggonemap area	clinopyroxene-hornblende porphyry dyke	Hb	210 ± 4
GSC 87-221	GAAN-83 114-2	northwestern Toodoggone map area	megacrystic monzogranite	Bi	215 ± 4
GSC 87-222	"	"	"	Hb	213 ± 4
GSC 78-221	"	"	"	Hb	222 ± 10
GSC 87-223	GAAN-83 143-2	northwestern Toodoggone map area	hornblendite	Hb	216 ± 3
GSC 87-219	GAAN-83 42-1	northwestern Toodoggone map area	monzogranite	Bi	208 ± 4
GSC 87-220	"	"	"	Hb	217 ± 4
GSC 87-224	GAAN-83 143-1	northwestern Toodoggone map area	monzogranite	Bi	204 ± 4
GSC 87-225	"	"	"	Hb	215 ± 3

[1] C. Dodds, *in* Wanless et al. (1979, p. 10)

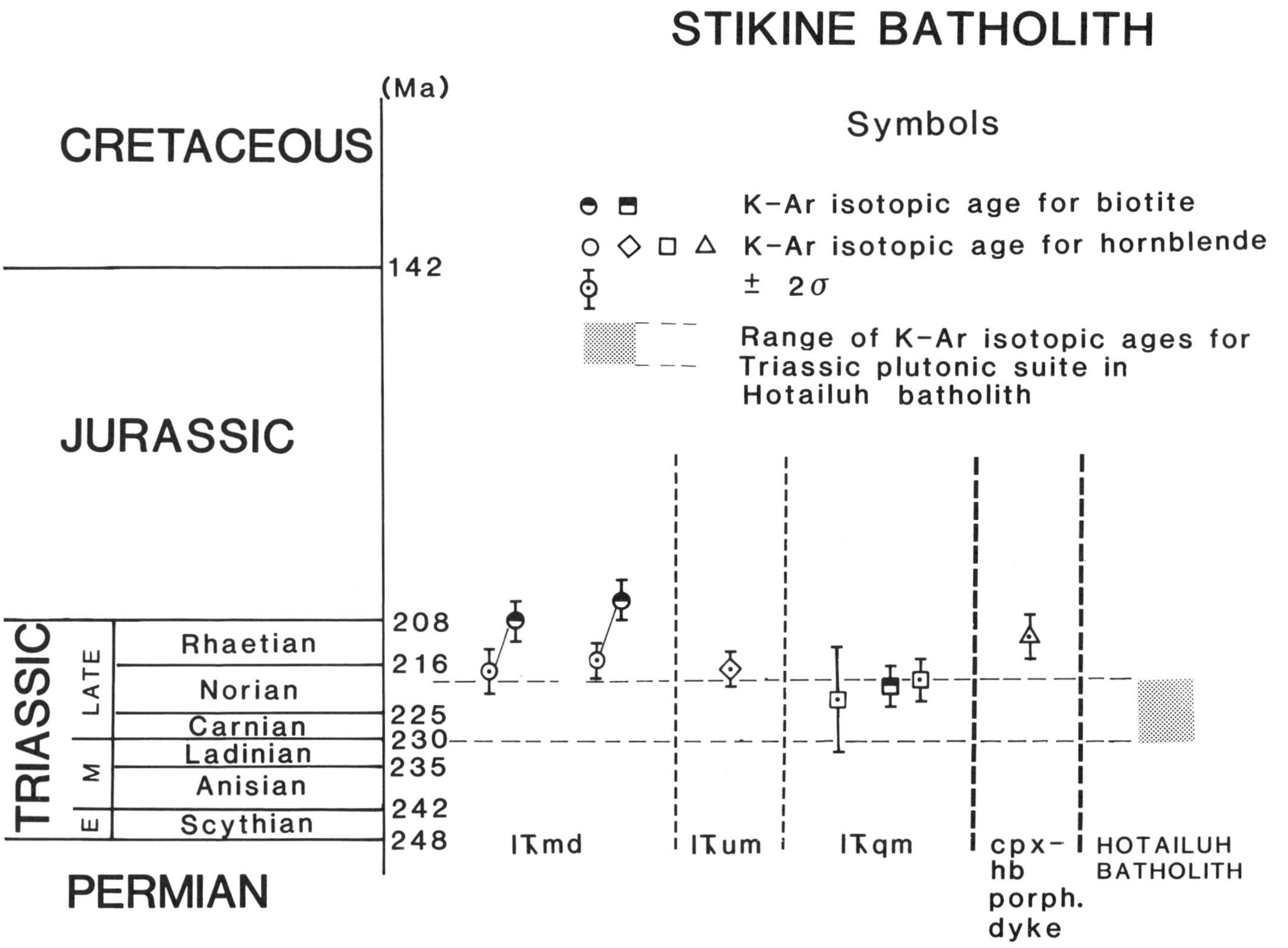

Figure 6. Plot of K-Ar isotopic ages (and 2 s.d. uncertainties) with respect to time scale of Armstrong (1978, 1982) and the range of isotopic ages for the Hotailuh batholith (shaded box).

REFERENCES

Anderson, R.G.
1984: Late Triassic and Jurassic magmatism along the Stikine Arch and the geology of the Stikine Batholith, north-central British Columbia; *in* Current Research, Part A, Geological Survey of Canada, paper 84-1A, p. 67-73.

Armstrong, R.L.
1978: Pre-Cenozoic Phanerozoic time scale computer file of critical dates and consequences of new and in-progress decay-constant revisions; *in* Contributions to the Geologic Time Scale, eds., G.V. Cohee, W.F. Glaessner and H.D. Hedberg, American Association of Petroleum Geologists, Studies in Geology, no. 6, p. 73-91.

1982: Late Triassic-Early Jurassic time scale calibration in British Columbia, Canada; *in* Numerical Dating and Stratigraphy, ed. G.S. Odin, John Wiley and Sons, Ltd., New York p. 509-513.

Stevens, R.D., Delabio, R.N. and Lachance G.R.
1982: Age determinations and geological studies, Geological Survey of Canada, Paper 82-2, p. 4-11.

Thorstad, L.
1980: Upper Paleozoic volcanic and volcaniclastic rocks in northwest Toodoggone map area, British Columbia; in Current Research, Part B, Geological Survey of Canada, Paper 80-1B, p. 207-211.

Wanless, R.K., Stevens, R.D., Lachance, G.R. and Delabio, R.N.
1979: Age determinations and geological studies; Geological Survey of Canada, Paper 79-2, p. 10.

GSC 87-223 Hornblende, K-Ar age
216 ± 3 Ma

K = 1.027 %, radiogenic
Ar = 91.63 × 10^{-7} cc/gm
K-Ar 3672 atmos. Ar = 8.73 %

From hornblendite
(94E/14) Northwestern Toodoggone map area, north central British Columbia, 57°53′17″N, central British Columbia, 57°53′17″N, 127°40′10″W; see Anderson (1984). Sample GAAN-83-143-2, collected and interpreted by R.G. Anderson.

The sample was collected from green weathering, massive, seriate, medium- to coarse-grained hornblendite (unit uTum of Anderson (1984)) which forms small intrusions in the quartz monzodiorite phase (unit uTmd) of the Stikine Batholith. The sample contains light to dark greenish brown hornblende (85 %), biotite (1 %), interstitial plagioclase (1 %) and sphene (less than 1 %).

For interpretation see GSC 87-222.

GSC 87-224 Biotite, K-Ar age
204 ± 4 Ma

K = 7.341 %, radiogenic
Ar = 617.77 × 10^{-7} cc/gm
K-Ar 3675 atmos. Ar = 3.28 %

From monzogranite
(94E/14) Northwestern Toodoggone map area, north central British Columbia, 57°53′23″N, 127°40′00″W; see Anderson (1984). Sample GAAN-83-143-1, collected and interpreted by R.G. Anderson.

The sample was collected from white-weathering, well foliated, fresh, melanocratic, mafic, equigranular, medium-grained biotite-hornblende monzogranite (typical of unit uTmd of Anderson (1984)). The sample contains plagioclase (43 %), alkali feldspar (25 %), quartz (18 %), green hornblende (9 %), brown biotite (3 %) and accessory minerals (less than 1 %) apatite, sphene and opaque minerals (see also GSC 87-225).

For interpretation see GSC 87-222.

GSC 87-225 Hornblende, K-Ar age
215 ± 3 Ma

K = 1.082 %, radiogenic
Ar = 96.04 × 10^{-7} cc/gm
K-Ar 3676 atmos. Ar = 5.14 %

(94 E/14) Details as for GSC 87-224.
For interpretation see GSC 87-222.

GSC 87-226 Hornblende, K-Ar age
200.1 ± 3 Ma

K = .669 %, radiogenic
Ar = 55.04 × 10^{-7} cc/gm
K-Ar 3583 atmos. Ar = 8.1 %

From a granodiorite separate
(94L) Ridge crest south of Pitman River at elevation 6500′, Kechika map-area, Cassiar Mountains, B.C., 58°00′30″N, 127°34′30″W. Sample GAD 81-G7C, collected by C.J. Dodds and interpreted by H. Gabrielse.

The sample is atypical of a suite of Early Jurassic plutons assigned to Quesnellia in north-central British Columbia. This confirms that the lithology is not sufficiently distinctive to discriminate between these rocks and associated mid-Cretaceous granite.

GSC 87-227 Whole rock, K-Ar age
100 ± 2 Ma

K = 5.06 %, radiogenic
Ar = 2.022 × 10^{-5} cc/gm
K-Ar 3355 atmos. Ar = 3.2 %

From a sericitized granodiorite porphyry
(94 M) Boya tungsten-molybdenum deposit, 6 km west of south end of Graveyard Lake, Rabbit River map area, British Columbia, 59°13′55″N, 127°29′05″W. Sample SYA80-43 from Kidd Creek Mines Limited drill hole B14 at a depth of 162.1 to 163.5 m, collected and interpreted by W.D. Sinclair.

The rock consists of approximately 15 % anhedral quartz phenocrysts, 35 % altered feldspar phenocrysts and 10 % altered mafic phenocrysts in a matrix of very fine grained quartz and K-feldspar. The feldspar phenocrysts are partially to completely replaced by muscovite or sericite and carbonate. The mafic phenocrysts are altered to pale chlorite or epidote.

Pyrite (2 %) forms anhedral, irregular grains associated with the altered mafic phenocrysts. Zircon and sphene are prominent accessory minerals.

The Boya deposit consists of scheelite and molybdenite in quartz vein stockworks and, to a lesser extent, in epidote-quartz skarn. It is associated with a sill-like complex of quartz diorite to granodiorite composition that intruded Proterozoic to Ordovician fine clastic rocks and intercalated limestone (Moreton et al., 1983).

The sample dated is from a sericitized granodiorite porphyry dyke that is unmineralized and appears to post-date the tungsten-molybdenum deposit. The 100 ± 2 Ma, age, therefore, is a minimum age for the deposit. This age should be treated with caution as felsic rocks are commonly unsatisfactory for whole rock determination due to argon loss. However, it is consistent with the mid-Cretaceous ages of other plutons in the Selwyn plutonic suite (cf. Anderson, 1982).

REFERENCES

Anderson, R.G.
1982: Geology of the Mactung pluton in Niddery Lake map area and some of the plutons in Nahanni map area, Yukon Territory and District of Mackenzie; *in* Current Research, Part A, Geological Survey of Canada, Paper 82-1A, p. 299-304.

Moreton, E.P., Clark, A.H. and Peatfield, G.R.
1983: The Boya claim group: a tectonically-transported W-Mo stockwork-skarn occurrence, NE British Columbia; *in* Geological Association of Canada, Program with Abstracts, v. 8, p. A48.

GSC 87-228 Biotite, K-Ar age
51.9 ± 1.3 Ma

K = 7.57 %, radiogenic
Ar = 1.549 × 10^{-5} cc/gm
K-Ar 3383 atmos. Ar = 43.9 %

From granodiorite.
(103 I/12) See GSC 87-229 for description, and interpretation.

GSC 87-229 Hornblende, K-Ar age
53.7 ± 3 Ma

K = 1.06 %, radiogenic
Ar = 2.246 × 10^{-6} cc/gm
K-Ar 3428 atmos. Ar = 36.8 %

From granodiorite.
(103 I/12) 1829 m summit of Lluvia Peak, about 6.5 km north-northwest of Redcap Mountain, Prince Rupert map-area, British Columbia, 54°44.3′N, 129°46.4′W. Map-unit F, GSC Map 1472A. Sample WV-ML-103, collected and interpreted by M.C. Hill.

The sample is massive coarse-grained granodiorite, with about 10 % biotite (slightly chloritized), 5 % hornblende, and accessory sphene, apatite, epidote, and zircon.

The sample is from a small granodiorite pluton which caps Lluvia Peak. The pluton cuts folded metasediments of unknown age. Although the contact is discordant and sharp, there is no apparent contact aureole in the metasediments (Hill, 1982).

The biotite age is slightly younger than the hornblende age, but the two are equivalent within stated error limits. These ages are typical of K-Ar ages for plutonic rocks in this area.

These dates constrain the age of metamorphism and deformation in this part of the coast Mountains. As this pluton cuts across major tight folds formed during the peak of metamorphism, this folding event and synchronous metamorphism were at least prior to 53.7 Ma.

REFERENCES

Hill, M.L.
1982: Geology of Redcap Mountain area, British Columbia, *in* Current Research, Part B; Geol. Surv. Can., Paper 82-1B, p. 267-268.

Hutchison, W.W.
1982: Geology of the Prince Rupert — Skeena map area, British Columbia. Geol. Surv. Can., Memoir 394, 116 p.

GSC 87-230 Biotite, K-Ar age
104.6 ± 1.6 Ma

K = 7.65 %, radiogenic
Ar = 320.5 × 10^{-7} cc/gm
K-Ar 3563 atmos. Ar = 6.4 %

From a pegmatite separate
(104 I) On Eagle River where crossed by Kutcho fault zone. B.C., 58°49′N, 129°40′W, Sample GAD 81-45-1. Collected by C.J. Dodds and interpreted by H. Gabrielse.

The sample is a strongly rodded muscovite-biotite pegmatite from the Kutcho fault zone. The mid-Cretaceous age supports determinations made on minerals along other dextral faults in the Cassiar regions and further suggests that dextral strike-slip faulting was active during mid-Cretaceous time about coincident with the emplacement of the widespread granitic plutons.

GSC 87-231 Hornblende, K-Ar age
214.1 ± 3.2 Ma

K = .351 %, radiogenic
Ar = 30.99 × 10^{-7} cc/gm
K-Ar 3564 atmos. Ar = 23.2 %

From a dioritic pegmatite separate
(104 I) About 6 km east of Eagle River on crest of E-W trending ridge. Cry Lake map-area, Cassiar Mountains. B.C., 58°47′N, 129°36′W, Sample GAD 81-44, collected by C.J. Dodds and interpreted by H. Gabrielse.

This sample represents one phase of a complex granitic terrane lying between the Thibert and Kutcho faults in northwestern Cry Lake map-areas. This diorite pegmatite yields the oldest age so far obtained from the complex but is consistent with a latest Triassic-Early Jurasic age for hornblende granodiorite in this part of the Quesnel Terrane.

GSC 87-232 Hornblende, K-Ar age
155.0 ± 2.5 Ma

K = .672 %, radiogenic
Ar = 42.25 × 10^{-7} cc/gm
K-Ar 3568 atmos. Ar = 14.3 %

From a quartz monzonite separate
(104 I) About 7 km east of Eagle River, Cry Lake map-area, Cassiar Mountains. 58°44′N, 129°37′30″W. Sample GAD 81-42, collected by C.J. Dodds and interpreted by H. Gabrielse.

The age obtained for this sample is consistent with those obtained elsewhere for pink-weathering hornblende- bearing rocks between the Thibert and Kutcho faults. Three distinct groupings of ages are apparent in the plutonic complex (Late Triassic-Early Jurassic, Middle Jurassic and mid-Cretaceous). The results are important for a synthesis of plutonism and tectonism along the suture between Quesnellia and ancestral North America.

GSC 87-233 Biotite, K-Ar age
106.1 ± 1.6 Ma

K = 7.41 %, radiogenic
Ar = 314.6 × 10^{-7} cc/gm
K-Ar 3569 atmos. Ar = 3.8 %

From a granodiorite separate
(104 I) 5.5 km east of Eagle River, Cry Lake map-area, Cassiar Mountains, British Columbia, 58°46′30″W, 129°37′30″W. Sample GAB 81-57-1, collected by D. Brown and interpreted by H. Gabrielse. See GSC 87-237 for description and interpretation.

GSC 87-234 Hornblende, K-Ar age
108.8 ± 1.7 Ma

K = .505 %, radiogenic
Ar = 22.01 × 10^{-7} cc/gm
K-Ar 3570 atmos. Ar = 20.3 %

From a granodiorite separate
(104 I) See GSC 87-233 for location and GSC 87-237 for description and interpretation

GSC 87-235 Biotite, K-Ar age
111.2 ± 1.7 Ma

K = 6.3 %, radiogenic
Ar = 280.9 × 10^{-7} cc/gm
K-Ar 3579 atmos. Ar = 3.7 %

From a granodiorite separate
(104 I) About 8 km east of Eagle River, Cry Lake map-area. Cassiar Mountains, B.C., 58°44′30″N, 129°37′W. Sample GAD 81-43-1, collected by C.J. Dodds and interpreted by H. Gabrielse.

The sample of granodiorite is characteristic of a pink-weathering Middle Jurassic suite of granitic rocks along the Kutcho Fault. These rocks are cut by mid-Cretaceous granite which, locally, have reset K-Ar dates.

GSC 87-236 Hornblende, K-Ar age
157.5 ± 2.4 Ma

K = .818 %, radiogenic
Ar = 52.34 × 10^{-7} cc/gm
K-Ar 3580 atmos. Ar = 10.7 %

From a granodiorite separate
(104 I) See GSC 87-235 for description and interpretation.

GSC 87-237 Biotite, K-Ar age
107.4 ± 1.6 Ma

K = 6.91 %, radiogenic
Ar = 296.9 × 10^{-7} cc/gm
K-Ar 3585 atmos. Ar = 3.4 %

From a granite separate
(104 I) 5.5 km east of Eagle River, Cry Lake map-area, Cassiar Mountains, B.C. 58°46′30″N, 129°37′30″W. Sample GAB 81-57, collected by D. Brown and interpreted by H. Gabrielse.

The sample GSC 87-237 was collected from a granite typical of the Cassiar Suite which has intruded granodiorite (GSC 87-233 and 87-234) in a granitic complex along the Kutcho Fault. The ages confirm the mid-Cretaceous age of granite and indicate complete re-setting of the granodiorite ages which elsewhere cluster around 155 Ma (GSC 87-232).

GSC 87-238 Biotite, K-Ar age
157.8 ± 2.4 Ma

K = 7.77 %, radiogenic
Ar = 498 × 10^{-7} cc/gm
K-Ar 3649 atmos. Ar = 5.8 %

From a granodiorite separate
(104 I) About 2.5 km east of N end of Glacial Lake on ridge south of McBride River, elev. 5800′. SW Cry Lake map-area, Cassiar Mountains. Map unit mJgd. Reference O.F. 610. Sample GA-83-22A, collected and interpreted by H. Gabrielse.

This medium grained, biotite-hornblende quartz granodiorite is typical of Middle to Late Jurassic plutons near Tachilta Lakes in the Dease Lake (104 J) map-area and north of the Hotailuh Batholith in the Cry Lake (104 I) map-area that cut structures associated with southerly directed thrusting along the King Salmon Fault. Their ages, ranging from about 173 Ma (GSC 80-69) to about 147 Ma (GSC 80-1) are consistent with stratigraphic data which indicate uplift of the Cache Creek Terrane in Middle Jurassic time.

GSC 87-239 Hornblende, K-Ar age
160.8 ± 2.5 Ma

K = .353 %, radiogenic
Ar = 23.08 × 10^{-7} cc/gm
K-Ar 3650 atmos. Ar = 23.5 %

From a granodiorite separate
(104 I) See GSC 87-238 for location, description and interpretation.

GSC 87-240 Biotite, K-Ar age
266.5 ± 3.9 Ma

K = 6.1 %, radiogenic
Ar = 681 × 10^{-7} cc/gm
K-Ar 3654 atmos. Ar = 2.6 %

From a granite separate
(104 I) On low ridge about 2 km west of Four Mile River, north-central Cry Lake map-area Cassiar Mountains, B.C., 58°59′30″N, 129°13′30″W, Map unit Kgd, GSC, open file 610. GA-83-42. Collected and interpreted by H. Gabrielse.

The sample represents one of the three distinctive lithologies in the Four Mile Batholith. The nature of the contacts between the bodies is unknown. Unlike the other lithologies the granite in the area of this sample is not foliated. The K-Ar determination confirms the presence of Permian plutonic rocks with considerable potassium content in the otherwise oceanic Sylvester Allochthon of the Slide Mountain Terrane. The age is supported by a U/Pb date on zircon of $277.5 \pm ^{20}_{8}$ Ma.

GSC 87-241 Hornblende, K-Ar age
298.2 ± 4.5 Ma

K = .216 %, radiogenic
Ar = 27.22 × 10^{-7} cc/gm
K-Ar 3576 atmos. Ar = 22.4 %

From a diorite separate
(104 J) About 4 km southwest of Northwest Mountain, on a southwest trending spur at elevation 5000 feet. British Columbia. 58°55′N, 130°07′W. Sample GAD 81-10d, collected by C.J. Dodds and interpreted by H. Gabrielse.

The age on this sample is clearly anomalous. A U/Pb age on zircon carried out after this determination was made suggests a date of about 228 Ma consistent with the stratigraphic relationships between the diorite body and overlying Upper Triassic volcanics.

GSC 87-242 Biotite, K-Ar age
108.7 ± 1.7 Ma

K = 6.35 %, radiogenic
Ar = 276.7 × 10^{-7} cc/gm
K-Ar 3577 atmos. Ar = 4.1 %

From a quartz diorite separate
(104 J) On Cassiar-Stewart Highway about 1 km south of Packer Tom Creek, Dease Lake, map area. B.C. 58°53′30″N, 130°01′W. Sample 6AEN 81-12, collected by T. England and interpreted by H. Gabrielse.

The sample is representative of a suite of mesocratic granodiorite to dioritic rocks intimately associated with Upper Triassic volcanics.

The anomalous biotite perhaps reflects metamorphism by the nearby mid-Cretaceous granitic rocks. The ages are in accord with the field data which demonstrate a complex interrelationship between three suites of granitic rocks comprising earliest Jurassic (hornblende granodiorite), Middle Jurassic (hornblende biotite granodiorite) and mid-Cretaceous (granite).

GSC 87-243 Hornblende, K-Ar age
127 ± 2 Ma

K = .503 %, radiogenic
Ar = 25.73 × 10^{-7} cc/gm
K-Ar 3578 atmos. Ar = 17.1 %

From a quartz diorite separate
(104 J) See GSC 87-242 for description and interpretation.

GSC 87-244 Biotite, K-Ar age
156.3 ± 2.4 Ma

K = 6.72 %, radiogenic
Ar = 426.6 × 10^{-7} cc/gm
K-Ar 3586 atmos. Ar = 1.7 %

From a quartz-biotite schist separate
(104 J) Crest of ridge SW of Gouthorn Creek 7.5 km NE of Northwest Mountain, Stikine Plateau, Dease Lake map-area, B.C. 58°57′30″N, 131°01′30″W. Sample GAD-81-01, collected by C.J. Dodds and interpreted by H. Gabrielse.

The schist represented by this sample was collected from a thick sequence of metasedimentary rocks lying between the Kutcho and Klinkit faults in northeastern Dease Lake (104J) map area. The age suggests that regional metamorphism took place well before the emplacement of the widespread mid-Cretaceous plutons. The metasediments are believed to be a distal facies of ancestral North American strata and their age of metamorphism is important relative to the accretion of Quesnellia.

GSC 87-245 Biotite, K-Ar age
69.2 ± 1.6 Ma

K = 7.36 %, radiogenic
Ar = 2.030 × 10^{-5} cc/gm
K-Ar 3360 atmos. Ar = 9.6 %

From a granite porphyry separate
(104 P) Mount Haskin molybdenum-tungsten deposit, 21 km east-northeast of Cassiar, on the northwest side of Mount Haskin, 59°20′46″N, 129°30′10″W. Sample SYA80-101, collected and interpreted by W.D. Sinclair.

The rock is a crowded porphyry consisting of 20 % quartz phenocrysts, 30 % orthoclase phenocrysts, 20 % plagioclase phenocrysts and 2 % biotite phenocrysts in a fine grained matrix of quartz and orthoclase. Accessory minerals are apatite, sphene, zircon pyrite and magnetite. The biotite forms light to dark brown, anhedral grains from 1 to 2 mm in size. The edges are ragged and very slightly chloritized.

The sample is from a small stock that intrudes Lower Cambrian sedimentary rocks of the Atan Group. Stockwork and disseminated molybdenite and scheelite occur in the granite porphyry and in associated skarn and hornfels. The age of the granite porphyry (52.6 ± 1.6 Ma) is probably equivalent to the age of the associated mineralization. Comparable ages of 49.7 ± 1.5 and 50.5 ± 1.5 Ma were obtained by Christopher et al., (1972) from the same granite porphyry stock. The nearby Mount Reed intrusion gave similar ages of 48.7 ± 1.9 and 50.2 ± 1.6 Ma (Christopher et al., 1972).

REFERENCE

Christopher, P.A., White, W.H., and Harakal, J.E.
1972: Age of molybdenum and tungsten mineralization in Northern British Columbia; Canadian Journal of Earth Sciences, v. 9, p. 1727-1734.

APPENDIX

The numbers listed below refer to the individual sample determination numbers, e.g. (GSC) 62-189, published in the Geological Survey of Canada age reports listed below:

	Determinations		Determinations
GSC Paper 60-17, Report 1	59-1 to 59-98	GSC Paper 71-2, Report 10	70-1 to 70-156
GSC Paper 61-17, Report 2	60-1 to 60-152	GSC Paper 73-2, Report 11	72-1 to 72-163
GSC Paper 62-17, Report 3	61-1 to 61-204	GSC Paper 74-2, Report 12	73-1 to 73-198
GSC Paper 63-17, Report 4	62-1 to 62-190	GSC Paper 77-2, Report 13	76-1 to 76-248
GSC Paper 64-17, Report 5	63-1 to 63-184	GSC Paper 79-2, Report 14	78-1 to 78-230
GSC Paper 65-17, Report 6	64-1 to 64-165	GSC Paper 81-2, Report 15	80-1 to 80-208
GSC Paper 66-17, Report 7	65-1 to 65-153	GSC Paper 82-2, Report 16	81-1 to 81-226
GSC Paper 67-2A, Report 8	66-1 to 66-176	GSC Paper 87-2, Report 17	87-1 to 87-245
GSC Paper 69-2A, Report 9	67-1 to 67-146		

GSC Age determinations listed by N.T.S. co-ordinates

1-M	62-189, 190; 63-136, 137; 66-170, 171; 70-145, 146, 147, 152
1-N	65-150; 70-156
2-C	70-155
2-D	59-94, 95, 96, 97, 98; 60-151, 152; 63-182; 65-142, 143; 66-172; 70-153, 154
2-E	62-187, 188; 63-168, 169, 170, 171, 183, 184; 64-159; 65-144, 145, 146, 147, 148, 149; 67-144; 70-151; 78-229, 230
2-F	70-148; 80-206
2-L	72-158, 159
2-M	66-173; 73-192, 193, 194
3-D	63-161
10-N	72-163
11-D	70-122, 123
11-E	66-156, 157, 158; 70-124, 125; 78-209; 87-14
11-F	62-168, 169; 78-211; 80-200; 87-14
11-J	78-212
11-K	66-159, 160, 161; 78-210; 80-199
11-L	65-133, 134, 135; 66-163; 70-128, 129, 130; 72-124, 125, 126; 76-231, 232, 233, 234, 235, 236, 237, 238, 239
11-N	78-206, 207, 208; 81-206, 207
11-O	61-202; 63-162; 65-138, 139, 140, 141; 66-168
11-P	67-143
12-A	67-142; 70-120, 121; 72-160, 161; 73-197, 198; 81-212, 213, 214, 215, 216, 217
12-B	60-147; 61-199; 62-186; 63-166, 167; 81-218, 219
12-E	65-129; 66-153; 70-102, 103, 104, 105; 72-95
12-H	60-148; 61-203, 204; 70-143, 144, 149
12-I	60-149; 61-200, 201; 64-158; 66-169; 70-150; 72-153, 154, 155, 156, 157; 73-195, 196
12-L	60-133, 134, 143
12-M	78-202, 203, 204, 205
12-O	60-135
12-P	73-191
13-C	66-167; 67-138
13-D	60-132
13-E	64-160; 70-133; 80-201
13-F	60-145; 67-136, 137
13-H	60-146; 67-141
13-I	70-138, 142; 72-140, 150
13-J	70-134, 135, 136, 137; 72-139; 78-228; 87-1, 2, 3, 4, 5
13-K	60-144; 61-196; 62-183, 184, 185; 63-178, 179; 72-141, 142, 143; 73-168, 169; 76-241, 242, 244, 247, 248; 81-208, 209, 211
13-L	61-197; 62-177; 63-148, 163, 177; 64-157; 65-151; 73-163, 164, 167; 76-240, 245
13-M	63-174; 64-162; 70-131, 132; 73-174; 76-243
13-M/3	87-19
13-M/6	87-6
13-M/11	87-7
13-M/14	87-8
13-N	62-178; 63-172; 73-176, 177, 178, 179, 183, 184; 76-246; 81-210
13-N/6	87-9
13-N/9	87-10
13-O	62-179, 180, 181, 182; 67-133, 134, 135; 70-140, 141; 72-144, 145, 146, 147, 148, 149, 151, 152; 73-180, 181, 185, 186, 187, 188, 189, 190
13-0/13	87-11
14-C	72-138; 73-182
14-D	60-143; 63-175; 65-122, 152; 73-166
14-D/3	87-12
14-E	61-195; 62-172; 63-181; 64-164; 65-153; 66-166; 72-134; 73-165, 172
14-F	62-171; 63-180; 64-163; 72-135, 136, 137
14-L	63-173, 176; 64-165; 67-130, 131, 132, 140; 73-171, 175; 78-213, 214, 215, 216, 217, 218, 219, 220, 221, 222, 223, 224, 225, 226, 227
14-M	67-129; 72-133; 73-170, 173

OTHER PUBLICATIONS CONTAINING GEOCHRONOLOGICAL DATA

Brandon, M.T., Orchard, M.J., Sutherland Brown, A., Yorath, C.J. and Parrish, R.R.

1986: Fossil ages and isotopic dates from the Paleozoic Sicker Group and associated rocks, Vancouver Island, British Columbia; *in* Current Research, Part A, Geological Survey of Canada, Paper 86-1A, p. 683-696.

Carr, S.D., Parrish, R. and Brown, R.L.

1987: Eocene structural development of the Valhalla Complex, southeastern British Columbia; Tectonics, v. 6, p. 175-196.

Cassidy, R.M., Miller, F.C., Knight, C.H., Roddick, J.C. and Sullivan, R.W.

1986: Evaluation of dynamic ion exchange for the isolation of metal ions for characterization by mass and α spectrometry; Analytical, Chemistry, v. 58, p. 1389-1394.

Chandler, F.W., Sullivan, R.W. and Currie, K.L.

1987: The age of the Springdale Group, western Newfoundland, and correlative rocks — evidence for a Llandovery overlap assemblage in the Canadian Appalachians; Royal Society of Edinburgh, Transactions: Earth Sciences, v. 78, p.41-49.

Currie, K.L. and Loveridge, W.D.

1985: Geochronology of retrogressed granulites from Wilson Lake, Labrador; *in* Current Research, Part B, Geological Survey of Canada, Paper 85-1B, p. 191-197.

Emslie, R.F., Loveridge, W.D. and Stevens, R.D.

1983: The Mealy dykes, Labrador: petrology, age, and tectonic significance; Canadian Journal of Earth Sciences, v. 21, p. 437-446.

Evenchick, C. and Gabrielse, H.

1984: Precambrian gneiss and late Proterozoic sedimentation in north-central British Columbia; Geology, v. 2, p. 233-237.

Frith, R.A. and Loveridge, W.D.

1982: Ages of Yellowknife Supergroup volcanic rocks, granitoid intrusive rocks and regional metamorphism in the northeastern Slave Structural Province; *in* Current Research, Part A, Geological Survey of Canada, Paper 82-1A, p. 225-237.

Frith, R.A., Loveridge, W.D. and van Breemen, O.

1986: U-Pb ages on zircon from basement granitoids of the western Slave Structural Province, northwestern Canadian Shield; *in* Current Research, Part A, Geological Survey of Canada, Paper 86-1A, p. 113-119.

Jamieson, R.A., van Breemen, O., Sullivan, R.W. and Currie, K.L.

1986: The age of igneous and metamorphic events in the western Cape Breton Highlands, Nova Scotia; Canadian Journal of Earth Sciences, v. 23, p. 1891-1901.

LeCheminant, A.N., Roddick, J.C., Tessier, A.C. and Bethune, K.M.

1987: Geology and U-Pb ages of early Proterozoic calcalkaline plutons northwest of Wager Bay, District of Keewatin; *in* Current Research, Part A, Geological Survey of Canada, Paper 87-1A, p. 773-782.

Loveridge, W.D.

1986: Measurement of biases in the electron multiplier ion detection system of a Finnigan-MAT model 261 mass spectrometer; International Journal of Mass Spectrometry and Ion Processes, v. 74, p. 197-206.

1986: U-Pb ages on zircon from rocks of the Lac de Morhiban map area, Quebec; *in* Current Research, Part A, Geological Survey of Canada, Paper 86-1A, p. 523-530.

Loveridge, W.D. and Chorlton, L.B.

1985: Rb-Sr age measurement on volcanic rocks from the Georges Brook Formation, La Poile Bay area, southwest Newfoundland; *in* Current Research, Part B, Geological Survey of Canada, Paper 85-1B, p. 89-93.

Mortensen, J.K.

1986: U-Pb ages for granitic orthogneiss from western Yukon Territory: Selwyn Gneiss and Fiftymile Batholith revisited; *in* Current Research, Part B, Geological Survey of Canada, Paper 86-1B, p. 141-146.

1987: Preliminary U-Pb zircon ages for volcanic and plutonic rocks of the Noranda-Lac Abitibi area, Abitibi Subprovince, Quebec; *in* Current Research, Part A, Geological Survey of Canada, Paper 87-1A. p. 581-590.

Roddick, J.C. and Loveridge, W.D.

1986: Low level analysis of radiogenic lead using the advantage of simultaneous Faraday/Multiplier ion collection. Finnigan-MAT Appl. Note 66, 11 p.

Parrish, R.R.

1984: Slocan Lake Fault: a low angle fault line bounding the Valhalla gneiss complex, Nelson W(1L/2) map area, southern British Columbia; *in* Current Research, Part A, Geological Survey of Canada Paper 84-1A, p. 323-330.

Parrish, R.R. and Roddick, J.C.

1985: Geochronology and isotope geology for the geologist and explorationist; Geological Association of Canada, Short Course Number 4, 71 p.

Sullivan, R.W., Sage, R.P. and Card, K.D.

1985: U-Pb zircon age of the Jubilee Stock in the Michipicoten Greenstone Belt near Wawa, Ontario; *in* Current Research, Part B, Geological Survey of Canada, Paper 85-1B, p. 361-365.

Tella, S., Heywood, W.W. and Loveridge, W.D.
1985a: A U-Pb age on zircon from a quartz syenite intrusion, Amer Lake map area, District of Keewatin, NWT; *in* Current Research, Part B, Geological Survey of Canada, Paper 85-1B, p. 367-370.

1985b: A U-Pb age on zircon from a dacite porphyry, Amer Lake map area, District of Keewatin, NWT; *in* Current Research, Part B, Geological Survey of Canada, Paper 85-1B, p. 371-374.

Thorpe, R.I., Guha, J., Franklin, J.M. and Loveridge, W.D.
1984: Use of the Superior Province lead isotope framework in interpreting mineralization stages in the Chibougamau District; in Chibougamau-Stratigrapy and Mineralization, Canadian Institute of Mining and Metallurgy, Special Volume 34, p. 496-516.

Trettin, H.P. and Parrish, R.R.
1987: Late Cretaceous bimodal magmatism, northern Ellesmere Island: isotopic age and origin; Canadian Journal of Earth Sciences, v. 24, p. 257-265.

Trettin, H.P., Parrish, R.R. and Loveridge, W.D.
1987: U-Pb age determinations on Proterozoic to Devonian rocks from northern Ellesmere Island, Arctic Canada; Canadian Journal of Earth Sciences, v. 24, p. 246-256.

van Breemen, O. and Parrish, R.R.
1986: Zircons record ancient geological processes; GEOS, v. 15, p. 18-21.

van Breemen, O. and Hanmer, S.
1986: Zircon morphology and U-Pb geochronology in active shear zones: studies on syntectonic intrusions along the northwest boundary of the Central Metasedimentary Belt, Grenville Province, Ontario; *in* Current Research, Part B, Geological Survey of Canada, Paper 86-1B, p. 775-784.

van Breemen, O., Davidson., A., Loveridge, W.D. and Sullivan, R.W.
1986: U-Pb zircon geochronology of Grenville tectonites, granulites and igneous precursors, Parry Sound, Ontario; *in* The Grenville Province, ed. J.M. Moore, A. Davidson, and A.J. Baer; Geological Association of Canada, Special Paper 31, p. 191-207.

van Breemen, O., Henderson, J.B., Loveridge, W.D. and Thompson, P.H.
1987: U-Pb zircon and monazite geochronology and zircon morphology of granulites and granite from the Thelon Tectonic Zone, Healey Lake and Artillery Lake map areas, N.W.T.; *in* Current Research, Part A, Geological Survey of Canada, Paper 87-1A, p. 783-801.

Whalen, J.B. and Roddick, J.C.
1987: K-Ar geochronology of the McGerrigle plutonic complex, Gaspésie Peninsula, Quebec; *in* Current Research, Part A, Geological Survey of Canada, Paper 87-1A, p. 375-380.

Whalen, J.B., Currie, K.L. and van Breemen, O.
1987: Episodic Ordovician-Silurian plutonism in the Topsails igneous terrane, western Newfoundland; Royal Society of Edinburgh, Transactions: Earth Sciences, v. 78 p. 17-28.

Williams, H., Gillespie, R.T. and van Breemen, O.
1985: A late Precambrian rift-related igneous suite in western Newfoundland; Canadian Journal of Earth Sciences, v. 22, p. 1727-1735.

Woodsworth, G.J., Loveridge, W.D., Parrish, R.R. and Sullivan, R.W.
1983: Uranium-lead dates from the Central Gneiss Complex and Ecstall pluton, Prince Rupert map area, British Columbia; Canadian Journal of Earth Sciences, v. 20, p. 1475-1483.